Energy Consumption in a Smart City

Energy Consumption in a Smart City

Editors

Benedetto Nastasi
Andrea Mauri

MDPI • Basel • Beijing • Wuhan • Barcelona • Belgrade • Manchester • Tokyo • Cluj • Tianjin

Editors
Benedetto Nastasi
Department of Planning,
Design & Technology of
Architecture
Sapienza University of Rome
Rome
Italy

Andrea Mauri
Laboratoire d'InfoRmatique
en Image et Systèmes
d'information (LIRIS)
Université Claude Bernard
Lyon 1
Lyon
France

Editorial Office
MDPI
St. Alban-Anlage 66
4052 Basel, Switzerland

This is a reprint of articles from the Special Issue published online in the open access journal *Energies* (ISSN 1996-1073) (available at: www.mdpi.com/journal/energies/special_issues/energy_consumption_in_a_smart_city).

For citation purposes, cite each article independently as indicated on the article page online and as indicated below:

LastName, A.A.; LastName, B.B.; LastName, C.C. Article Title. *Journal Name* **Year**, *Volume Number*, Page Range.

ISBN 978-3-0365-5964-3 (Hbk)
ISBN 978-3-0365-5963-6 (PDF)

Contents

About the Editors

Benedetto Nastasi

Assistant Professor in Smart Energy Systems for the Built Environment at the Department of Planning, Design and Technology of Architecture at Sapienza University of Rome, Italy. He is an alumnus of Sapienza University of Rome, from which he graduated Summa Cum Laude in Architectural and Building Engineering in 2011 and obtained his Ph.D. with Honors in Energy Saving and Distributed Microgeneration in 2015. From 2015 to 2019, he worked as a Researcher at TU/e Eindhoven University of Technology, Guglielmo Marconi University and TU Delft University of Technology. He has also worked as an Energy and Sustainability Consultant at ISES International Solar Energy Society, Italy, and ANEV, the Italian Wind Energy Association. He has won several awards for his research, including the Best Poster Award at SEE SDEWES, 2016, Best Invited Session Chairman at SEB, 2017, and Best Senior Researcher at the Smart Energy Systems Conference, 2018, and Best Young Investigator in Building Science, 2022. His research interests are in innovative energy systems from the building to urban scale, transition pathways to zero-carbon built environments, pioneering hydrogen applications, open data, and energy analytics. He has been ranked in the top 2% of researchers worldwide in both "Building & Construction" and "Energy" fields by the Stanford University study in 2019, in 2020 and in 2021.

Andrea Mauri

Junior Professor (Chaire de Professor Junior) at Université Claude Bernard Lyon 1, affiliated with the Liris Research Lab. Previously, Postdoctoral Researcher within the Knowledge and Intelligence Design Group at the Sustainable Design Engineering Department at the Delft University of Technology and a Research fellow at the Amsterdam Institute for Advanced Metropolitan Solutions. His research lies at the intersection of Human-Computer Interaction and Data Management. Interested in investigating how to integrate human perspective and computational methods in the design, development and deployment of data-intensive applications, with a special focus on the health domain, with the goal of making them efficient, effective, and aware of people's characteristics, needs, and values.

energies

Editorial

Energy Consumption in a Smart City

Benedetto Nastasi [1,*] and Andrea Mauri [2]

[1] Department of Planning, Design and Technology of Architecture, Sapienza University of Rome, Via Flaminia 72, 00196 Rome, Italy
[2] Laboratoire d'InfoRmatique en Image et Systèmes d'Iinformation (LIRIS), Université Claude Bernard Lyon 1, 43 Bd du 11 Novembre 1918, 69100 Villeurbanne, France
* Correspondence: benedetto.nastasi@outlook.com

Citation: Nastasi, B.; Mauri, A. Energy Consumption in a Smart City. *Energies* **2022**, *15*, 7555. https://doi.org/10.3390/en15207555

Received: 6 October 2022
Accepted: 11 October 2022
Published: 13 October 2022

Publisher's Note: MDPI stays neutral with regard to jurisdictional claims in published maps and institutional affiliations.

1. Overview of the Articles in This Special Issue

Increasing and inexorable urbanization calls for the involvement of all the stakeholders. This includes energy providers, policymakers in the municipality, facility managers, and the citizens themselves, as energy consumption is influenced by this interconnected network of actors. The Smartness of the City enhances the interactions between those actors because its architecture already integrates elements to collect data and connect to the citizens. Furthermore, the proliferation of Web platforms (e.g., social media, Web fora) and the increased affordability of sensors and IoT devices (e.g., smart meters) make data related to a large and diverse set of users accessible, as their activities in the digital world reflect their real-life actions. These new technologies can be of great use for the stakeholders as, on one hand, it provides them with semantically rich inputs and frequent updates at a relatively cheap cost and, on the other, it allows them to have a direct channel of communication with the citizens.

This Special Issue aims to provide insights on original multidisciplinary research works about AI, data science methods, and their integration with existing design/decision-making processes in the domain of energy consumption in a Smart City. For this purpose, the Special Issue "Energy Consumption in a Smart City" has been designed and launched, intended for researcher, planners, users of the broad domain of the Smart City. Among a very high number of submissions, 11 articles have been accepted and published.

The first paper by Dharani et al. [1] presents the strategies to manage energy flows in a limited portion of the urban environment, i.e., the University campus, by an intelligent use of time-varying electrical load via developing efficient energy utilization patterns using demand-side management (DSM).

The second paper by Esfandiari et al. [2] investigates and assess the quality of the indoor environment of Platinum-certified office buildings in a tropical climate through surveys and questionnaire highlighting the human-centric perception of comfort that, whatever technologies are going to be included, is the requirement of the built environment.

The third paper of this Special Issue by Mousavi Motlagh et al. [3] propose a way for acquiring the foremost window allocation scheme to have the best trade-off among energy, environmental, and comfort criteria in a building by an advanced decision-making tool.

The fourth paper by Bazazzadeh et al. [4] focuses on the impact of climate change on the heating and cooling energy demands of buildings as influential variables in building energy consumption by the statistical downscaling method accounting for the future forecasted weather data for 2050 and 2080 and derived increased cooling demand calling for serious measures to control it.

The fifth paper by Csáky [5] focuses on the future cooling demand increase in buildings explaining how it greatly depends on the building structure, window coverage, and orientation. The detailed findings in the paper will be useful in the future for building renters and operators that have to manage the cooling systems during the—future more often—torrid days.

The sixth paper of this Special Issue, by Cumo et al. [6] dealt with the different limitations in the transition of historical buildings to near zero energy (nZEB), including invasive interventions, historical and architectural structure constraints and impact on heritage value. The integration of renewable energy technology is seen as the way to reduce the energy footprint of such buildings.

In the seventh paper, Agostinelli et al. [7] discusses an infrastructure digitization policy to manage and optimize the energy transition process to transform ports area into zero energy districts. The strategies are not only aimed at the energy transition but have implications on the environmental, economic and social spheres, setting the port area as the epicenter and extending to the city.

In the eighth paper, Sureshkumar et al. [8] focus on induction heating for melting applications assisted by electronic power control as appealing technology since it provides higher efficiency, zero pollutants, non-contamination of material, etc. in comparison with conventional heating.

The ninth paper by Casini [9] is a review study on virtual reality (VR), augmented reality (AR), and mixed reality (MR) technologies and applications for smart building operation and maintenance. The application of XR in building and city management is showing promising results in enhancing human performance in technical O&M tasks.

In the tenth paper, Bruck et al. [10] propose a new method to compare European neighborhoods to plan the energy infrastructures needed for the energy transition. They identified a set of parameters to build a QGIS-based visualization that allows them to compare different areas easily and directly.

Finally, the last article by Liu et al. [11] analyzed the Smart City Policy (SCP) set in China as significant tool to reduce the carbon emission intensity of enterprises in urban contexts. The mechanism analysis finds that digital transformation, innovation by enterprises, and urban green innovation all strengthen the impact of SCP on the carbon emission intensity of enterprises via a smart way.

Author Contributions: Conceptualization, B.N. and A.M.; writing—original draft preparation, B.N. and A.M.; writing—review and editing, B.N. All authors have read and agreed to the published version of the manuscript.

Funding: This research received no external funding.

Conflicts of Interest: The authors declare no conflict of interest.

References

1. Dharani, R.; Balasubramonian, M.; Babu, T.S.; Nastasi, B. Load Shifting and Peak Clipping for Reducing Energy Consumption in an Indian University Campus. *Energies* **2021**, *14*, 558. [CrossRef]
2. Esfandiari, M.; Mohamed Zaid, S.; Ismail, M.A.; Reza Hafezi, M.; Asadi, I.; Mohammadi, S.; Vaisi, S.; Aflaki, A. Occupants' Satisfaction toward Indoor Environment Quality of Platinum Green-Certified Office Buildings in Tropical Climate. *Energies* **2021**, *14*, 2264. [CrossRef]
3. Mousavi Motlagh, S.F.; Sohani, A.; Djavad Saghafi, M.; Sayyaadi, H.; Nastasi, B. Acquiring the Foremost Window Allocation Strategy to Achieve the Best Trade-Off among Energy, Environmental, and Comfort Criteria in a Building. *Energies* **2021**, *14*, 3962. [CrossRef]
4. Bazazzadeh, H.; Pilechiha, P.; Nadolny, A.; Mahdavinejad, M.; Hashemi safaei, S.s. The Impact Assessment of Climate Change on Building Energy Consumption in Poland. *Energies* **2021**, *14*, 4084. [CrossRef]
5. Csáky, I. Analysis of Daily Energy Demand for Cooling in Buildings with Different Comfort Categories—Case Study. *Energies* **2021**, *14*, 4694. [CrossRef]
6. Cumo, F.; Nardecchia, F.; Agostinelli, S.; Rosa, F. Transforming a Historic Public Office Building in the Centre of Rome into nZEB: Limits and Potentials. *Energies* **2022**, *15*, 697. [CrossRef]
7. Agostinelli, S.; Cumo, F.; Nezhad, M.M.; Orsini, G.; Piras, G. Renewable Energy System Controlled by Open-Source Tools and Digital Twin Model: Zero Energy Port Area in Italy. *Energies* **2022**, *15*, 1817. [CrossRef]
8. Sureshkumar, A.; Gunabalan, R.; Vishnuram, P.; Ramsamy, S.; Nastasi, B. Investigation on Performance of Various Power Control Strategies with Bifilar Coil for Induction Surface Melting Application. *Energies* **2022**, *15*, 3301. [CrossRef]
9. Casini, M. Extended Reality for Smart Building Operation and Maintenance: A Review. *Energies* **2022**, *15*, 3785. [CrossRef]

10. Bruck, A.; Casamassima, L.; Akhatova, A.; Kranzl, L.; Galanakis, K. Creating Comparability among European Neighbourhoods to Enable the Transition of District Energy Infrastructures towards Positive Energy Districts. *Energies* **2022**, *15*, 4720. [CrossRef]
11. Liu, Y.; Li, Q.; Zhang, Z. Do Smart Cities Restrict the Carbon Emission Intensity of Enterprises? Evidence from a Quasi-Natural Experiment in China. *Energies* **2022**, *15*, 5527. [CrossRef]

energies

MDPI

Article

Load Shifting and Peak Clipping for Reducing Energy Consumption in an Indian University Campus

Rajavelu Dharani [1], Madasamy Balasubramonian [1], Thanikanti Sudhakar Babu [2] and Benedetto Nastasi [3,*]

1 Department of Electrical and Electronics Engineering, Alagappa Chettiar Government College of Engineering and Technology, Karaikudi, India; dharaniacgcet@gmail.com (R.D.); balu941rec@gmail.com (M.B.)
2 Department of Electrical Power Engineering, Institute of Power Engineering, Universiti Tenaga National (UNITEN), Kajang, Malaysia; sudhakarbabu66@gmail.com
3 Department of Planning, Design, and Technology of Architecture, Sapienza University of Rome, Via Flaminia 72, 00196 Rome, Italy
* Correspondence: benedetto.nastasi@outlook.com

Abstract: This paper analyzes the intelligent use of time-varying electrical load via developing efficient energy utilization patterns using demand-side management (DSM) strategies. This approach helps distribution utilities decrease maximum demand and electrical energy billing costs. A case study of DSM implementation of electric energy utility for an educational building Alagappa Chettiar Government College of Engineering and Technology (ACGCET) campus was simulated. The new optimum energy load model was established for peak and off-peak periods from the system's existing load profile using peak clipping and load shifting DSM techniques. The result reflects a significant reduction in maximum demand from 189 kW to 170 kW and a reduction in annual electricity billing cost from $11,340 to $10,200 (approximately 10%) in the upgraded system. This work highlights the importance of time of day (TOD) tariff structure consumers that aid reduction in their distribution system's maximum demand and demand charges.

Keywords: building energy flexibility; HOMER software; peak clipping; load shifting; energy saving

Citation: Dharani, R.; Balasubramonian, M.; Babu, T.S.; Nastasi, B. Load Shifting and Peak Clipping for Reducing Energy Consumption in an Indian University Campus. *Energies* **2021**, *14*, 558. https://doi.org/10.3390/en14030558

Received: 8 December 2020
Accepted: 19 January 2021
Published: 22 January 2021

Publisher's Note: MDPI stays neutral with regard to jurisdictional claims in published maps and institutional affiliations.

1. Introduction

The rising energy demands of further population and industrial growth, shortage of fuel for power plants, and increasing costs of electrical energy require a more advanced, secure power system structure all over the world for economic growth [1] and satisfaction of energy consumers as well as conceiving innovative technologies [2]. This leads to a higher value of energy required to supply the needs of industrial, commercial, and all other categories of end-user [3]. Like in most of the countries, India also encourages the initiate energy efficiency activities (Energy conservation Act 2001) based on Japan's Energy Conservation (EC) Guidelines through Bureau of Energy Efficiency (BEE), which is an agency of the Indian Minister of Energy. Activities carried out to prepare EC guidelines are (i) performance audits and sectoral study reports, (ii) original equipment manufacturers, (iii) industries, (iv) sectoral experts, (v) stakeholder consultations with industries and industry associations, and (vi) secondary sources such as relevant websites. It provides direction for the preparation of its own "Energy Management planning Handbook" for the efficient operation of various types of energy consumer such as commercial, industrial, and educational institute buildings. The main components of the guidelines are management and control; it plays a vital role in the energy conservation process in India [4]. As result of the follow-up to these guidelines, the actual power deficit was reduced from 0.6% to 0.5% for the financial year 2019–2020, and the peak power deficit was reduced from 0.8% to 0.7%, as reported by the Central Electricity Authority (CEA) [5]. Figure 1 shows the present Indian Energy scenario: power generation and peak demand deficit. In this scenario, to further reduce the deficit like other developing countries, a simple and cheap alternative is essential to provide reliable electricity service and to attain the objective of becoming a nation of surplus electricity for the growth of a country's economy [6].

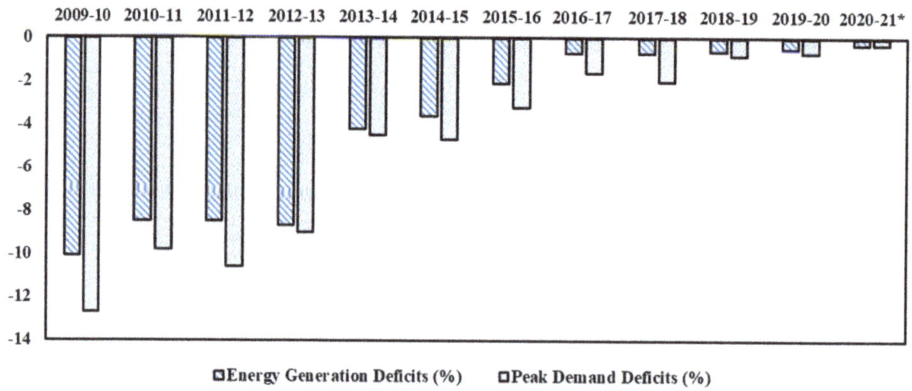

Figure 1. Present Indian Energy scenario: power generation and peak demand deficit from 2009–2021 as per the Central Electricity Authority (CEA).

The continued growth and development of new power plants in particular of renewable energy plant units can minimize the gap between supply and demand. However, the integration of renewable electricity generation systems in most countries is impossible because of resource restrictions like wind, solar radiation, and storage capacity or local geographical limitation for energy harvesting [7,8]. To control this status, a well-developed form of a power management system is essential to provide for the continued supply and trade of electricity in a country. The power management system is classified into two main categories such as supply-side management and demand-side management (DSM) [9]. The first one focuses on enhancing the operating efficiency of electricity production, transmission and distribution. A mainly efficient generation of energy at a minimum cost, by meeting the energy demand without adding a new plant, decreases the impact on the environment by operating the electricity system efficiently [10]. On the other hand, DSM deal with consumers/utilities load levels and usage models and it is insensitive to external factors. Therefore, for the past few decades, many countries decided to implement energy conservation activities mainly through the DSM approach in addition with other simple approaches such as using energy-efficient equipment/appliances, smart power strips, and setting up a programmable or intelligent thermostat in various types of consumers by giving contributions in electricity bill tariffs and offer energy motivation programs that help make energy efficiency a more affordable and easy model [11]. Among these, a lot of research focused on the DSM approach in terms of load control techniques, economical benefits, effects on various kinds of consumers, role in the electricity market, impact on the power system reliability, etc. [12–15]. It is implemented with exciting outcomes in different countries. DSM is has been demonstrated as an essential strategy to achieve energy conservation and to reduce emissions in many studies in China [16]. It also plays a leading role in real-time electricity markets to balance power, maintain security, and keep electricity grids safe and reliable as shown in a study in North America [17] and others in Gulf countries [18,19]. The demand side response approach is further classified into two main categories: time- and incentive-based programs [20]. Time-based programs are intended for monitoring power consumption during peak hours. At the same time, incentive programs are managed by utilities and the provision of incentives for consumers.

The DSM time-based strategies are highly involved in the electricity market for energy efficiency [21,22]. These DSM programs are more common: load shifting, peak clipping, valley filing, load building, and energy conservation programs [23]. These algorithms differ for each utility because these are dependent on the number of customers, the nature of load (commercial building, industrial, or residential), the benefits of this program, and the level of customers' reaction or satisfaction with the program applied. The few numbers of research suggest that energy conservation in the housing/commercial sector as a result of behavioral change strategies is more rapid than any other intervention. This includes direct feedback for smart meters and indirect feedback to improve billing [24,25]. Furthermore, during the peak period, energy saving through behavioral change ranged from 15.7% to 21.7% [26]. The DSM flexibly manages the changing load profile on the consumer side concerning time and use [27]. As a result, the load profile and supply conditions flattening at a location over the day are easily achieved by maintaining a satisfactory end-user level [28–31]. The DSM load modeling objective includes load-shaping measures such as peak clipping, valley filling, load shifting, strategic conservation, strategic load growth, and flexible load shape that may change the time model and level of the utility's profile [32]. A load shifting approach is a form of DSM program that increases the power supply's reliability in an electricity system [33].

Although it is factual that the DSM is a key strategy for minimizing peak demand, developing countries have taken much less consideration of the development of a suitable strategy for the DSM, because the deployment of DSM presents some challenges and barriers. They are a lack of smart metering, communications infrastructure, control techniques, information technologies, understanding the advantages of the DSM, competitiveness issues in relation to traditional approaches, an increase in the complication of the system's operation, and inadequate market incentives among these capital costs and lead times for implementation [34–36]. Hence, it is preferable to choose the best DSM approach which overcomes the issues. Load sharing and peak clipping energy efficiency programs have the advantages of covering the needs of utilities to maintain a reliable, stable, and efficient electrical network and customers' needs to maintain their comfort, safety, and unique lifestyle [37]; and where the energy resources are not sufficient to satisfy the load requirements, load shifting or peak clipping may be an appropriate and better method for DSM [38,39]. In this study, load shifting and peak clipping DSM approaches are applied; the first one is the best solution from the point of view of utility companies where customers are encouraged for this by the cheaper tariff in off-peak hours. The second is quite significant in developing countries and a problem with the need to invest in new installations and generation capacities.

In most Indian educational institutes, energy resources are inadequate to meet the needs; in this situation, the impact of DSM on the distribution system is important to ensuring a balance between demand and supply in addition to a reduced electricity bill. With this motivation, the authors analyzed the educational institute of Alagappa Chettiar Government College of Engineering and Technology (ACGCET) campus on how to upgrade the existing energy system through the peak clipping and load shifting DSM methodology. It is analyzed using a simple solver MS-Excel and HOMER (Hybrid Optimization of Multiple Energy Resources) (2.8, HOMER Energy, Boulder, CO 80301 USA) optimization tool [40]. Initially, the daily load profile is measured, and thereafter, the monthly or seasonal load curves are obtained by HOMER to identify and analyze of peak load and off-peak load period, average load, peak load demand in an hour/month/year, then executed by the direct load control. The utility can change the load model by turning off the power supply in an intelligent manner to a specific category of loads (shiftable load) at chosen time intervals. This study offers the benefits of reducing implementation complexity, operational costs, etc., compared to other DSM approaches. The result demonstrates the enhancement of the load factor and the load profile curve of the campus.

The detailed expected effects of implementing the DSM are:

1. The reduction in energy bills can be achieved compared to the existing billing cost and similar contract and energy price conditions by reducing unnecessary consumption and paying attention to energy consumption behavior on campus;
2. The maximum demand charges can be minimized by reducing the sanctioned peak demand of the electricity supplier;

3. The possibility to avoid the enlargement of the electricity meter and a subsequent increase in the electricity price per kWh can also be optimized;
4. The efficient energy load pattern can derive.

The paper is organized as follows: Section 2 provides the details of the case study and load profile details. Section 3 deals with the methodology used to reduce peak load and load factor for a day/month. Simulation results and discussion are presented in Section 4, and the outcome and conclusions of the work carried out are discussed in Section 5.

2. Case Study Area: ACGCET Campus

The case study was carried out on ACGCET Campus, Karaikudi, where the latitude and longitude are 10.091° N and 78.797° E, respectively. Its location is in the Indian state of Tamilnadu. The power distribution networks of ACGCET Campus consist of a grid line, solar photovoltaic and diesel generator, distribution panels, automatic power factor correction panels, circuit breakers, cables, and loads. The maximum permitted demand on campus is 210 KVA from the Tamilnadu Electricity Board (TNEB). By observing Tamilnadu Generation and Distribution Corporation Limited's (TANGEDCO-Electricity provider, No 825, Link Rd, Anna Salai, Triplicane, Chennai, Tamil Nadu 600002, India) bill (November 2019) for the ACGCET campus [41], the cost of energy (COE) is $2857 per month, the penalty fee, which includes the maximum demand charge, is $943 per month, and compensation fee for the low power factor per month of approximately $100. The total amount payable is approximately $48,000 per year [42,43]. From this data, it can be seen that the minimization of the energy cost of existing systems is mandatory in this institution. As a result, this work's focus is to reduce the demand load by reducing energy consumption during peak hours. Thus, the flattening of the load profile and the state of energy saving behavior of the supply of a site during a day is carried out.

2.1. Load Assessment

The ACGCET campus brief infrastructure details and number of courses offered can be found in Table 1. It is well known that the institute's working days are the only possibility to observe peak loads in relation to public holidays and examination days. As a result, the average working day's energy meter readings are observed in the month of December 2019 and the typical load consumption pattern of the campus is shown in Figure 2. The average load profile of the campus is approximately 44.46 kW.

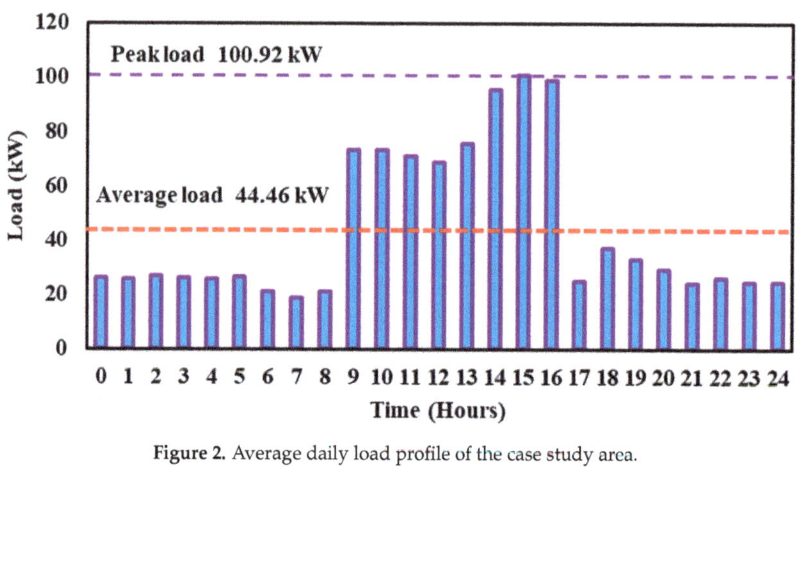

Figure 2. Average daily load profile of the case study area.

Table 1. Courses and Infrastructure details of the campus.

UG Programs	PG Programs	Blocks	Classrooms	Staff Rooms	Laboratories
5	7	21	42	45	40

2.2. Problem Formulation

Usually peak demand requirement is applied by the end-user to the supply provider according to their load in the campus. In this campus, the requested maximum demand is 210 KVA from TNEB (Tamilnadu Electricity Board-provider) to meet our maximum energy demand of the campus. The rules and regulation of TNEB (supply provider) state that a constant supply equal to 90% of recorded demand affects the tariff of a particular end-user—in this case the campus. Even having a lower amount of recorded demand in the month (as per analysis during the month, it is only 100.92 KVA), TNEB considered only 90% of sanction demand as a demand charge. To address this constraint by reducing the peak and related costs, the most viable solution is to implement the DSM approach to the electrical network. DSM entails the supervision of loads on the utility side concerning the occasion and the amount of use to reduce costs. The system's total peak greatly reduces and supports the load pattern itself much closer to the regular load consumption pattern. The successful execution of DSM employment provides significant assistance in managing customer demand at the lowest cost of electricity. In the valley filling DSM approach, some controllable devices are required to operate in different times of the day, allowing the effective integration of various renewable sources [44,45] and fossil-based ones [46], including improving the management of other sectors, such as heat [47,48]. The load shifting and peak clipping approach are the best solutions for utility companies. End-users are encouraged to do so through the lower-cost tariff during peak and subsequent risk of congestion [49].

3. Methodology

The DSM analysis method is described in the following subsections. Firstly, DSM is performed using the peak clipping and load shifting technique. In the peak clipping approach, the maximum demand reduction is obtained by the clip of the unwanted load during peak hours. Second, the HOMER software tool, built by NREL (National Renewable Energy Laboratory) for cost effective and reliable microgrids, shows the maximum load reduction and load factor for a day/month.

3.1. Demand-Side Management

To manage energy demand for the existing system in the campus DSM approach, two techniques, such as peak clipping and load shifting, are utilized as tools in this case study. Based on hourly and average demand, the peak period was clipped (peal clipping approach), and the end-user determined the load or load that could be shifted (in the load shifting approach).

In this case study, loads are categorized as primary, permanent, and suitable loads and are presented in Figure 3. The power consumptions rating of each load is given in Table 1. The primary load consists of light and fan loads; the permanent load consists of computers. The primary load can be turn-on or turn-off based on the end-user requirement. However, the permanent load requires continuous power with constant power consumption, including computers and server rooms. Shiftable loads are used on time. In the present study, three main devices are considered to be shiftable loads. They are all department laboratories, water well pumps, and air-conditioners. At any period 't', the hourly and average load demand can be computed by the sum of the primary loads (PPL), permanent loads (PPR), and deferrable or shiftable loads PSL. It appears in Equations (1)–(4).

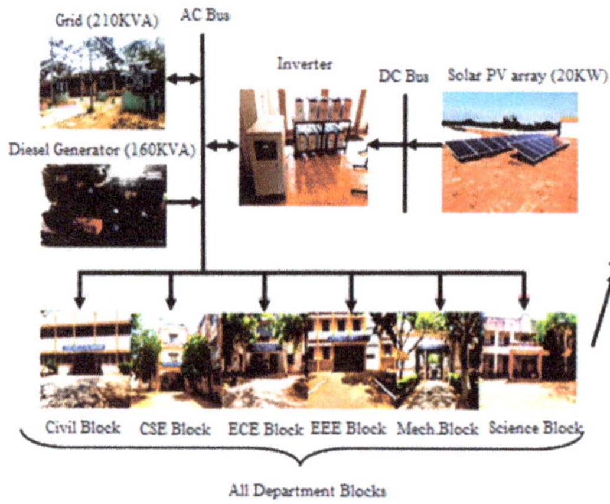

Figure 3. Load assumption for demand-side management (DSM).

$$P_{hr\_dem}(t) = P_{PL}(t) + P_{PR}(t) + \sum_{n=1}^{N} [P_{SL}^{n}(t) * X_{SL}^{n}(t)], \quad (1)$$

$$X_{SL}^{n}(t) = BV_{SL}^{n}(t) * BP_{SL}^{n}(t), \quad (2)$$

$$\sum_{n=1}^{N} \sum_{t=1}^{24} [BV_{SL}^{n}(t) * BP_{SL}^{n}(t)] = 1, \quad (3)$$

$$P_{avg\_dem}(t) = \frac{\sum_{t=1}^{24} P_{(demand)(t)}}{24}, \quad (4)$$

where:

- $P_{PL}(t)$ is the Primary Load;
- P_{PR} (t) is the Permanent Load;
- $P_{SL}(t)$ is the Shiftable Load;
- $BV_{SL}(t)$, $BP_{SL}(t)$ are the Binary Variable;
- $P_{hr\_dem}(t)$ is the Hourly Load;
- $P_{avg\_dem}(t)$ is the average energy demand;
- N is the No. of Shiftable Loads.

Using data from the observed hourly/daily load profile, peak hours occurring during the period are recorded. If the hourly energy demand exceeds the average demand, staggered hourly loads are shifted towards off-peak hours based on the end-user's priorities. At this time, transfer, load to off-peak periods by ON and OFF switching, the channel was denoted by '1' and '0', respectively, in the load shifting DSM approach [50,51]. The system's shiftable loads depend on the priorities assigned by the end-user based on their requirement characteristics on the campus load as in Table 2. The order of preference is given in Tables 3 and 4; shifting loads are assigned based on preference by the binary variable shiftable load (BVSL) and binary parameter (BPSL) at a specific time period. In the peak load clipping method, the shiftable load is clipped as per the end-user's design. The time period of the peak clipping is given in Table 5 for this case study work.

Table 2. Power consumption of the primary, permanent, and shiftable load.

Load Types	Power in kW
Primary	10.032
Permanent	31.194
Shiftable load	147.274

Table 3. End-user priority for shiftable loads.

Shiftable Loads	Priority
Laboratories	I
Air-conditioner	II
Water pump	III

Table 4. End-user priority for Laboratories.

Departments	Laboratories	Priority
All departments	Computer Lab	I
CSE	Graphics and Multimedia Lab	II
Mechanical	Mechatronics Lab	III
	CAD lab	IV
EEE	Electrical Machines and Drives lab	V

Table 5. Twenty-five kilowatt load—peak clipped in the afternoon session.

Time (Hours)		Load (kW)		
From	To	Before DSM	After DSM	Difference
14	15	95.7	90.7	5
15	16	100.922	90.922	10
16	17	99.212	89.212	10
		Total		25

Using this prioritization of loads and subsequent process, the energy demand of the campus with DSM strategy was computed through HOMER tool. The outcomes are evaluated according to the peak load demands. By the peak clipping DSM technique, the maximum value of peak load reduced from 100.92 kW to 90.92 kW, and the average load reduced from 44.46 kW to 43.42 kW. Figure 4 shows the load profile of the system after the peak clipping DSM approach. In the load shifting approach, the deferrable load shifted during the off-peak period such that the maximum peak load was reduced by keeping the same average load. Figure 5 shows the load profile of the existing system after load shifting and is presented in Table 6. It demonstrates that the DSM strategy provides a minimum peak load in comparison to the system without DSM.

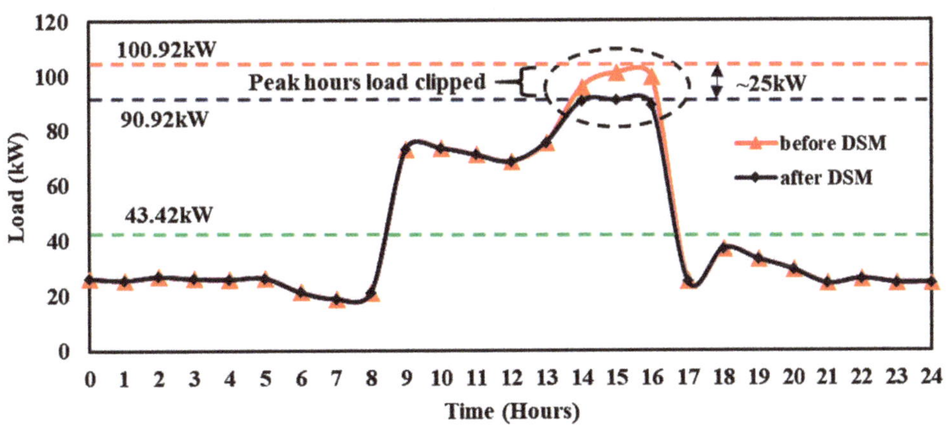

Figure 4. Peak clipping in daily load profile.

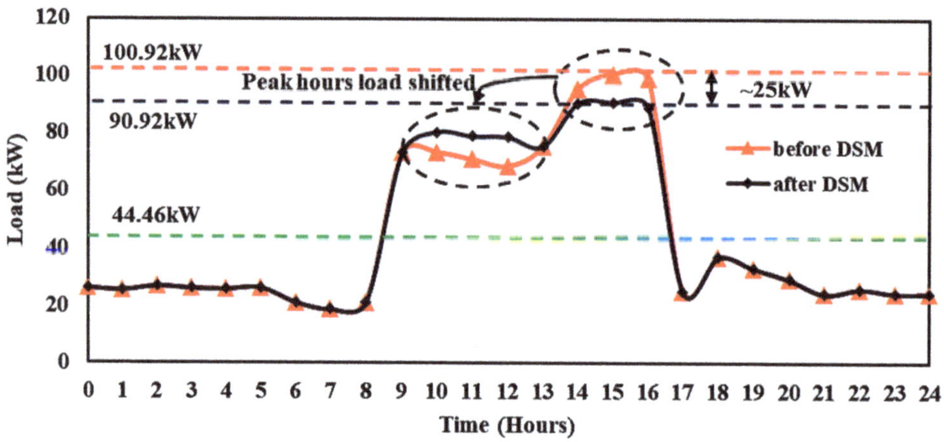

Figure 5. Load shifting in daily load profile.

Table 6. Peak clipped 25 kW load shifted from afternoon to morning.

Time (Hours)		Load (kW)		
From	**To**	**Before DSM**	**After DSM**	**Difference**
10	11	73.607	80.607	−7
11	12	71.2	79.2	−8
12	13	68.849	78.849	−10
		Total		−25

3.2. Determination of Monthly Peak Load—HOMER Tool

The HOMER tools can perform a renewable hybrid system analysis based on various sensitive parameters such as technical, economic, and environmental and hybrid system costs. It determines the peak load for a month based on the scaled annual average method. HOMER computes the peak load, saving energy using Equations (5) and (6)—National Renewable Energy Laboratory (NREL) [52]. It calculates the load factor through average load and peak load value.

$$\Delta kWgross = units \times RLF \times \left[\left(\frac{kW}{unit}\right)_{base} - \left(\frac{kW}{unit}\right)_{ee}\right] \times DF \times CF \times (1 + HVACd), \quad (5)$$

$$\Delta kWhgross = units \times RLF \times \left[\left(\frac{kW}{unit}\right)_{base} - \left(\frac{kW}{unit}\right)_{ee}\right] \times FLH \times (1 + HVACd). \quad (6)$$

- $\Delta kWgross$ is the Gross demand savings;
- *Units* indicate units of measure installed in the program;
- *RLF* refers to the rated load factor;
- *kW/unit* is the unit demand of measure;
- *DF* represents the diversity factor;
- *CF* is the coincidence factor;
- *FLH* is the full load hours;
- *HVACd* is for HVAC system interaction factor for demand.

In HOMER, to add randomness to the load data to make it more realistic, the random variability data are day to day and time step. First, it compiles the annual load data based on daily load profiles [53,54]. Then, it goes through that time series, and in each time step, it multiplies the value of that time step by a perturbation factor α, given in Equation (7):

$$\alpha = 1 + \delta_d + \delta_{ts}, \tag{7}$$

where,

- δ_d indicates the daily perturbation value;
- δ_{ts} represents time stepperturbation value;

4. Results and Discussion

This section discusses the possibilities of inefficient load clipping/shifting in the real-time system and the importance of the load factor. Using pre-analysis of the load profile data per day, the peak hours occurring time period is identified. During this period, shift load to off-peak periods by ON and OFF switching in a smart manner following the prioritization, the sequence was denoted by '1' and '0', respectively, in the load shifting DSM technique. The system's shiftable loads depend on the end-user's priorities and are clearly explained in Section 3. Similarly, in the peak load clipping method, the shiftable load is clipped as per the end-user's design. Mostly, the peak load source in the afternoon session of this institute can be clipped or shifted to the morning session by the below three possibilities (uses shiftable load) to control the maximum demand in peak hours:

1. Altering the laboratory schedule between morning and afternoon sessions;
2. Timer setting of an air-conditioner;
3. Timer setting of water pumps for buildings.

Currently, most of the laboratory classes at this educational institute are conducted in the afternoon session, and there is no timer setting for the air-conditioner and water pumps. Therefore, these three possibilities have a strong influence on reducing the peak load during peak hours. Other options to reduce the peak load are

1. Setting fixed maximum demand in the circuit breaker;
2. Restructuring the entire electrical distribution network and circuit breaker programming;
3. Increasing the renewable energy generation on the campus.

The first two are firmly not recommended from the above three possibilities, and the next one is increasing in renewable energy generation and is partially recommended. However, the main problem of installing a renewable energy plant requires very high capital cost. Therefore, it is concluded that the DSM approach significantly influenced the existing system and reduced peak load, resulting in a reduction in monthly demand charges of about \$93. It is represented in Table 6 or before and after the DSM approach, and a reduction in maximum demand is noted [55,56]. The annual cost reduction of the existing energy system after the implementation of the DSM approach is approximately \$1140, around 10%, per year in the electricity bill as per the Tamilnadu (a state of Indian Government) Generation, and Distribution Corporation (TANGEDCO) Limited revised tariff rates from the TNERC (Tamilnadu Electricity Regulatory Commission) Tariff Order.

The results also confirmed that the DSM strategy reduces peak load and reduction in the electric bill and prevents the installation of new power plants compared to the system without DSM.

The DSM data response is given to the HOMER tool to calculate the monthly peak load. Figure 6a,b shows the simulation result of the daily load profile improvement in peak clipping and load shifting DSM approaches. The following observations are taken as follows: first, the maximum demand is reduced during peak hours, so that demand charges are definitely reduced, and also the load factor of the system is significantly increased; second, the utility will reschedule its investments (installing the new power source) to meet the load requirements. On top of that, it stimulates the user to consume less power during peak hours. The obtained monthly peak load for both methods is reduced from 189 kW to 170 kW (approximately 10%), around 19 kW for the same amount of daily load utilization (about 1060 kWh/d), which reduces the electricity charge approximately by $95 per month (Figure 7). Seasonal profiles for monthly peak loads for a year, both before and after DSM (peak clipping and load shifting), are obtained using the HOMER tool, which clearly shows that the measured demands of the reduction in peak demand for each month of the year did not exceed the mean values. Thus, a DSM approach is justified, which will lead to a significant reduction in the cost of electricity energy bills per year.

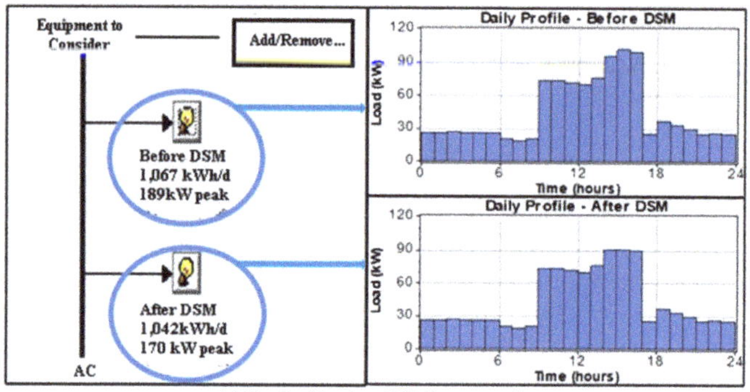

(**a**) Peak clipping DSM approach

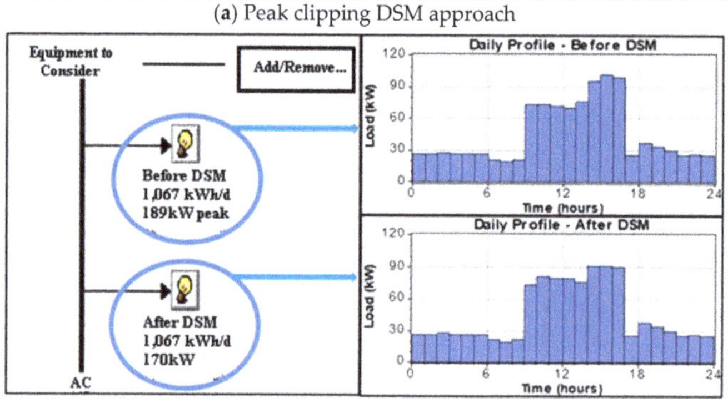

(**b**) Load shifting DSM approach

Figure 6. Daily Load Profile before and after DSM using HOMER.

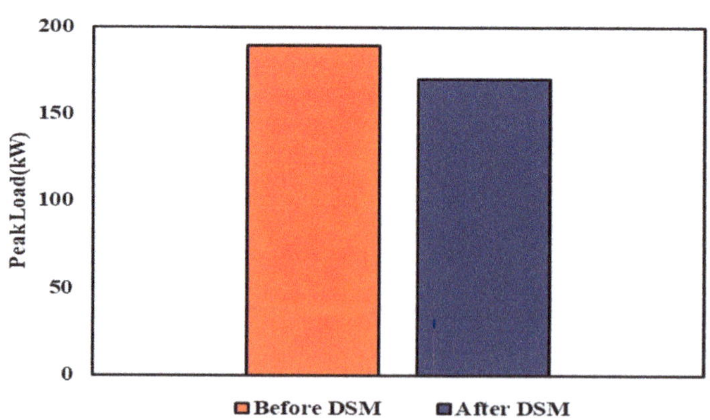

Figure 7. Monthly maximum demand reduction using Hybrid Optimization of Multiple Energy Resources (HOMER).

The flattening of the load profiles not only reduced the maximum demand, but also improved the load factor. The improvement of the load factor of the profile enables better use of the energy of the system. The load factor (*LF*) is obtained through average load and the maximum load in a given time period and is reported in Equation (7). The higher value of *LF* indicates that the load uses the electric system more efficiently.

$$LF = \frac{Average\ Load}{Peak\ Load}. \tag{8}$$

The average load can be calculated using Equation (4). The load factor is obtained through daily and monthly load profile curves of the existing energy system, before and after the DSM approach, and is compared in Figures 8 and 9. Table 7 shows that the average load demand is 44, and the peak demand is 90.92 kW for the daily load profile. Consequently, the load factor is increased by about 1.5%, taking into account the daily load profile and by 3.8% for the monthly load profile [57–61]. It is confirmed that increasing the load factor enhances the energy efficiency measures on campus.

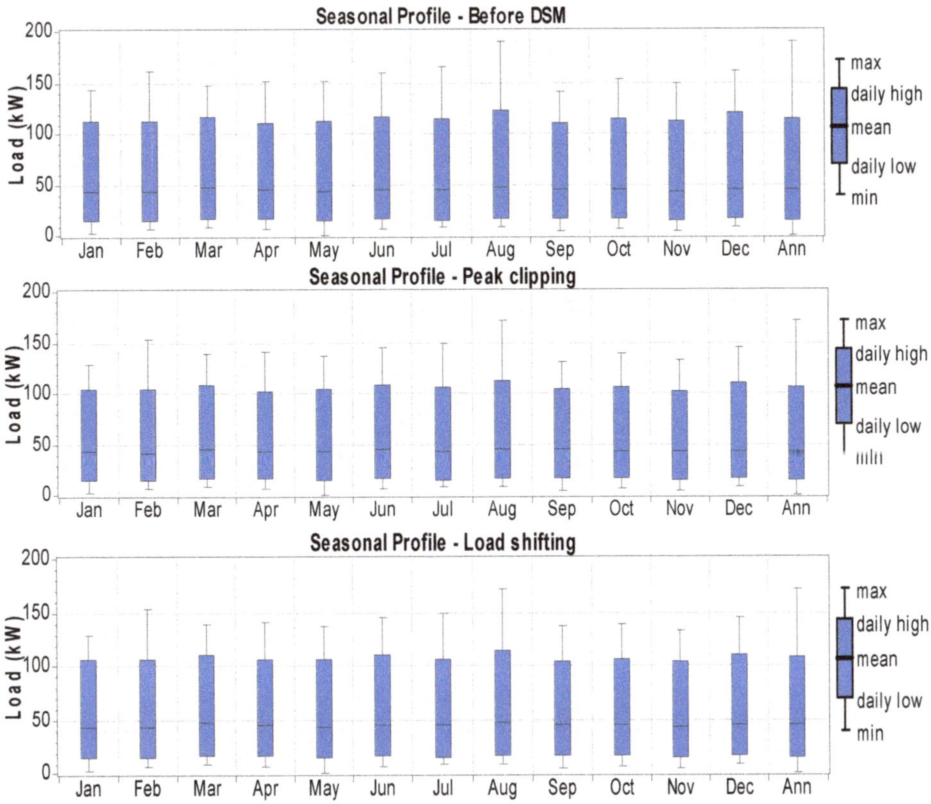

Figure 8. Seasonal profile before and after DSM using HOMER.

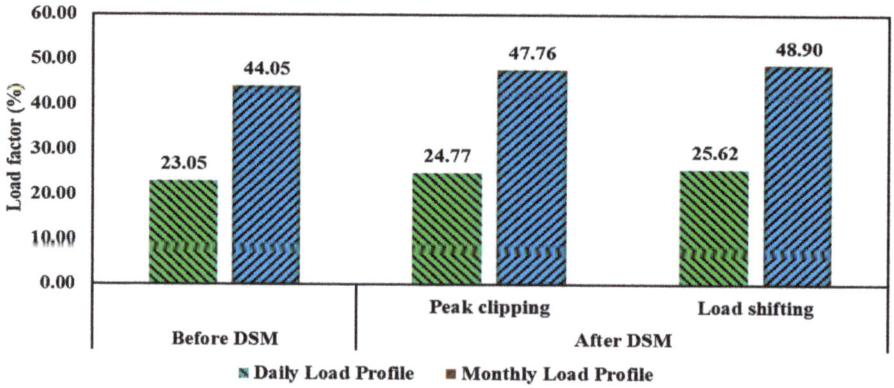

Figure 9. Comparison of load factor—before and after DSM from daily and monthly load profile curve.

Table 7. Comparison of average load, peak load, and electricity bill (EB) reduction–before and after DSM.

Description		Average Load (kW)		Peak Load (kW)		EB Bill (Dollar)	
		Working Day	Monthly	Working Day	Monthly	Monthly	Annual
Before DSM		44.46	32,405	100.92	189	945	11,340
After DSM	Peak clipping	43.42	31,330	90.92	170	850	10,200
	Load shifting	44.46	32,405	90.92	170	850	10,200

Furthermore, Figure 10 shows a comparison of the DSM approaches in the campus' daily load curve. It is observed that the peak load is 90.92 kW, and the average loads of peak clipping and load shifting approaches are 43.42 kW and 44.46 kW, respectively. It clearly shows that the load shifting DSM approach gives a slightly better result as compared to the peak clipping technique while maintaining a satisfactory level of reliability. This change can be monitored by measurement and verification techniques in buildings [62].

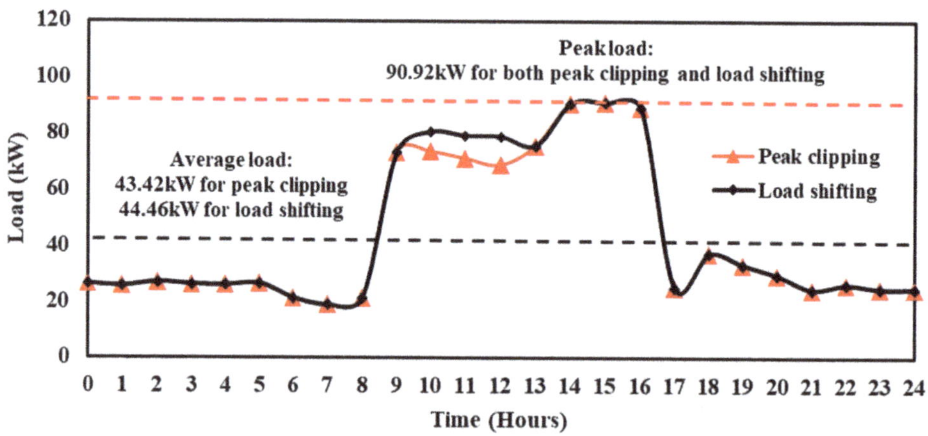

Figure 10. Comparison of DSM approaches in daily load profile curve.

It should be noted that the maximum demand period occurs roughly in 8–12% of the load profile and consumes approximately 8% more than the normal value of the total power of the system. Hence, installing a new power plant to satisfy this requirement is too expensive. The DSM approach was an effective solution to this issue. In the annual cost comparison of the existing energy system before and after implementing the DSM approach, approximately $1140 was saved per year on the electricity bill.

The overall benefits provided by DSM in the institute included an economic and technical point of view—avoiding the capital cost for installing a new power plant to meet out the maximum demand, efficient utilization of devices during peak hours, low-cost energy during energy consumption in off-peak periods, reduced electricity bill cost, up-gradation of the existing system, as well as improved system flexibility and balance.

5. Conclusions

In India, energy extraction from resources is insufficient to meet the current load demand; in this situation, the impact of DSM on the distribution network is analyzed. In this work, our DSM analysis was performed using peak clipping and load shifting technique in the case study location (ACGCET Campus, India) power utilization network. Using this approach, the load pattern was modified by clipping or changing shiftable load from peak time to non-peak time with a simple solver. Further, these DSM approaches' benefits are no need for smart metering, information, and communication infrastructure, the less complexity of system operation, and minimum cost of implementation. This case study strategy has reduced the peak load demand and demand charges, which are performed using the HOMER simulation tool. At this campus, peak demand was reduced from 210 KVA to 190 KVA, or approximately 20 KVA, after implementing the DSM technique. By the energy supplier's standards, the maximum demand charge is $5 per KVA, resulting in a reduction in the demand fee per month of approximately $95. In this investigation, several benefits can be obtained. Moreover, the objectives achieved with the proposed work are listed below:

(i) Reduced the institute's annual electricity consumption by 10% by sorting out the peaks shaved into the annual energy consumption;
(ii) The approximate cost saved from the maximum demand is $95 per month (for annual cost saved is the cost of the demand charge in $/kW * Energy saved * 12 = $1140);
(iii) The load factor of the energy network of the institute is increased up to 1.5% per day and 3.5% per month;
(iv) Indirectly, greenhouse gas emissions (CO_2) are reduced due to the shaving of peak demand energy. It is around 207.48 kg per annum (0.91 kg/kWh × 228 kWh) in India considering its Power Grid emission factor.

DSM analysis in the institutional building contributes to the sustainability of the active power distribution network rather than new plant facilities. The satisfactory simulation results confirmed that it might be executed in real-time on the campus. In general, the DSM approach improved the reliability and financial performance of the electrical power system network of the similar campus/network; the most significant points that were studied in this paper are highlighted as follows:

(i) Encourage better energy management practices for the demand scheduling procedures that target the peak demand reduction in a public educational institution;
(ii) This work is not only related to the maximum load, but also the average load and energy. These results are more sensible and a comprehensive reflection of DSM impacts on-campus that strengthen positive energy use behaviors;
(iii) This analysis provides a direct way to learn the implications of the load factor;
(iv) Save the demand charge bill amount, resources, and environmental damage by avoiding unnecessary over-investment in the peak load generating capacity power plant by the electricity sector.

Author Contributions: Conceptualization, R.D.; data curation, R.D. and M.B.; formal analysis, R.D. and T.S.B.; funding acquisition, B.N. and T.S.B.; project administration, M.B. and B.N.; investigation, R.D., M.B., and T.S.B.; methodology, R.D. and M.B.; resources, R.D., T.S.B., and B.N.; software, R.D.; supervision, B.N. and T.S.B.; writing—original draft, R.D. and M.B.; writing—review & editing, T.S.B. and B.N. All authors have read and agreed to the published version of the manuscript.

Funding: This research received no external funding.

Institutional Review Board Statement: Not applicable.

Informed Consent Statement: Not applicable.

Acknowledgments: The authors are also thankful to the World Bank-sponsored Technical Education Quality Improvement Program (TEQIP)—Government of India, Institute of ACGCET, Tamilnadu, India, have all supported to complete this research work.

Conflicts of Interest: The authors declare no conflict of interest.The funders had no role in the design of the study; in the collection, analyses, or interpretation of data; in the writing of the manuscript, or in the decision to publish the results.

References

1. Babu, T.S.; Vasudevan, K.R.; Ramachandaramurthy, V.K.; Sani, S.B.; CheMud, S.; Lajim, R.M. A Comprehensive Review of Hybrid Energy Storage Systems: Converter Topologies, Control Strategies and Future Prospects. *IEEE Access* **2020**, *8*, 148702–148721. [CrossRef]
2. Barbarelli, S.; Florio, G.; Zupone, G.L.; Scornaienchi, N.M. First techno-economic evaluation of array configuration of self-balancing tidal kinetic turbines. *Renew. Energy* **2018**, *129*, 183–200. [CrossRef]
3. Kinhekar, N.; Padhy, N.P.; Gupta, H.O. Demand side management for residential consumers. In Proceedings of the 2013 IEEE Power & Energy Society General Meeting, Vancouver, BC, Canada, 21–25 July 2013; pp. 1–15.
4. Energy Conservation Act. *Ministry of Power, New Delhi, India*. 2017. Available online: http://www.powermin.nic.in/acts_notofication/energy_conservation_act/index.html (accessed on 2 September 2020).
5. Load Generation Balance Report. Central Electricity Authority, Ministry of Power, Government of India. 2019–2020. Available online: http://www.cea.nic.in/report.html (accessed on 20 August 2020).
6. Harish, V.; Kumar, A. Demand side management in India: Action plan, policies and regulations. *Renew. Sustain. Energy Rev.* **2014**, *33*, 613–624. [CrossRef]
7. Barbarelli, S.; Florio, G.; Scornaienchi, N. Developing of a small power turbine recovering energy from low enthalpy steams or waste gases: Design, building and experimental measurements. *Therm. Sci. Eng. Prog.* **2018**, *6*, 346–354. [CrossRef]
8. Piras, G.; Pini, F.; Garcia, D.A. Correlations of PM10 concentrations in urban areas with vehicle fleet development, rain precipitation and diesel fuel sales. *Atmos. Pollut. Res.* **2019**, *10*, 1165–1179. [CrossRef]
9. Groppi, D.; Astiaso Garcia, D.; Lo Basso, G.; De Santoli, L. Synergy between smart energy systems simulation tools for greening small mediterranean islands. *Renew. Energy* **2019**, *135*, 515–524. [CrossRef]
10. United Nations Industrial Development Organization. Supply Side Management—Module 13. Sustainable Energy Regulation and Policy Making for Africa. Available online: http://africa-toolkit.reeep.org/modules/Module13.pdf (accessed on 23 July 2020).
11. Mancini, F.; Nardecchia, F.; Groppi, D.; Ruperto, F.; Romeo, C. Indoor Environmental Quality Analysis for Optimizing Energy Consumptions Varying Air Ventilation Rates. *Sustainability* **2020**, *12*, 482. [CrossRef]
12. Kirschen, D.S. Demand-side view of electricity markets. *IEEE Trans. Power Syst.* **2003**, *18*, 520–527. [CrossRef]
13. Esther, B.P.; Kumar, K.S. A survey on residential Demand Side Management architecture, approaches, optimization models and methods. *Renew. Sustain. Energy Rev.* **2016**, *59*, 342–351. [CrossRef]
14. Behrangrad, M. A review of demand side management business models in the electricity market. *Renew. Sustain. Energy Rev.* **2015**, *47*, 270–283. [CrossRef]
15. Vardakas, J.S.; Zorba, N.; Verikoukis, C.V. A survey on demand response programs in smart grids: Pricing methods and optimization algorithms. *IEEE Commun. Surv. Tutor.* **2015**, *17*, 152–178. [CrossRef]
16. Ming, Z.; Song, X.; Mingjuan, M.; Lingyun, L.; Min, C.; Yuejin, W. Historical review of demand side management in China: Management content, operation mode, results assessment and relative incentives. *Renew. Sustain. Energy Rev.* **2013**, *25*, 470–482. [CrossRef]
17. Wang, Q.; Zhang, C.; Ding, Y.; Xydis, G.; Wang, J.; Østergaard, J. Review of real-time electricity markets for integrating distributed energy resources and demand response. *Appl. Energy* **2015**, *138*, 695–706. [CrossRef]
18. Alasseri, R.; Tripathi, A.; Rao, T.J.; Sreekanth, K. A review on implementation strategies for demand side management (DSM) in Kuwait through incentive-based demand response programs. *Renew. Sustain. Energy Rev.* **2017**, *77*, 617–635. [CrossRef]
19. Zehir, M.A.; Batman, A.; Bagriyanik, M. Review and comparison of demand response options for more effective use of renewable energy at consumer level. *Renew. Sustain. Energy Rev.* **2016**, *56*, 631–642. [CrossRef]
20. Jabir, H.J.; Teh, J.; Ishak, D.; Abunima, H. Impacts of Demand-Side Management on Electrical Power Systems: A Review. *Energies* **2018**, *11*, 190. [CrossRef]
21. Torriti, J. *Peak Energy Demand and Demand Side Response*, 1st ed.; Routledge: New York, NY, USA, 2016. [CrossRef]
22. Newsham, G.R.; Bowker, B.G. The effect of utility time-varying pricing and load control strategies on residential summer peak electricity use: A review. *Energy Policy* **2010**, *38*, 3289–3296. [CrossRef]
23. Khan, I. Energy-saving behaviour as a demand-side management strategy in the developing world: The case of Bangladesh. *Int. J. Energy Environ. Eng.* **2019**, *10*, 493–510. [CrossRef]

24. Gilbert, B.; Graff Zivin, J. Dynamic Salience with Intermittent Billing: Evidence from Smart Electricity Meters. 2013. Available online: http://www.nber.org/papers/w19510 (accessed on 22 October 2018).
25. Allcott, H.; Rogers, T. The Short-Run and Long-Run Effects of Behavioral Interventions: Experimental Evidence from Energy Conservation. *Am. Econ. Rev.* **2014**, *104*, 3003–3037. [CrossRef]
26. Asensio, O.I.; Delmas, M.A. The dynamics of behavior change: Evidence from energy conservation. *J. Econ. Behav. Organ.* **2016**, *126*, 196–212. [CrossRef]
27. Tronchin, L.; Manfren, M.; Tagliabue, L.C. Optimization of building energy performance by means of multi-scale analysis—Lessons learned from case studies. *Sustain. Cities Soc.* **2016**, *27*, 296–306. [CrossRef]
28. Khripko, D.; Morioka, S.; Evans, S.; Hesselbach, J.; De Carvalho, M. Demand Side Management within Industry: A Case Study for Sustainable Business Models. *Procedia Manuf.* **2017**, *8*, 270–277. [CrossRef]
29. Apajalahti, E.-L.; Lovio, R.; Heiskanen, E. From demand side management (DSM) to energy efficiency services: A Finnish case study. *Energy Policy* **2015**, *81*, 76–85. [CrossRef]
30. Mancini, F.; Romano, S.; Lo Basso, G.; Cimaglia, J.; de Santoli, L. How the Italian Residential Sector Could Contribute to Load Flexibility in Demand Response Activities: A Methodology for Residential Clustering and Developing a Flexibility Strategy. *Energies* **2020**, *13*, 3359. [CrossRef]
31. Manfren, M.; Nastasi, B.; Tronchin, L. Linking Design and Operation Phase Energy Performance Analysis through Regression-Based Approaches. *Front. Energy Res.* **2020**, *8*, 288. [CrossRef]
32. Li, D.; Chiu, W.-Y.; Sun, H. Demand Side Management in Microgrid Control Systems. In *Microgrid*; Butterworth-Heinemann: Oxford, UK, 2017; pp. 203–230. [CrossRef]
33. Jabir, H.J.; Teh, J.; Ishak, D.; Abunima, H. Impact of Demand-Side Management on the Reliability of Generation Systems. *Energies* **2018**, *11*, 2155. [CrossRef]
34. Nolan, S.; O'Malley, M. Challenges and barriers to demand response deployment and evaluation. *Appl. Energy* **2015**, *152*, 1–10. [CrossRef]
35. Lu, R.; Hong, S.H.; Zhang, X. A dynamic pricing demand response algorithm for smart grid: Reinforcement learning approach. *Appl. Energy* **2018**, *220*, 220–230. [CrossRef]
36. Sharifi, R.; Fathi, S.H.; Vahidinasab, V. A review on Demand side tools in electricity market. Renew. *Sustain. Energy Rev.* **2017**, *72*, 565–572. [CrossRef]
37. Guzman, C.; Cardenas, A.; Agbossou, K. Local Estimation of Critical and Off-Peak Periods for Grid-Friendly Flexible Load Management. *IEEE Syst. J.* **2020**, *14*, 4262–4271. [CrossRef]
38. Syafrudin, M.; Abd, H.A.R. Demand control & monitoring system as the potential of energy saving. In Proceedings of the 2014 IEEE Student Conference on Research and Development, Batu Ferringhi, Malaysia, 16–17 December 2014; pp. 1–6.
39. Tenaga Nasional Berhad, Electricity Pricing, and Tariff. 2014. Available online: https://www.tnb.com.my/commercial-industrial/maximum-demand (accessed on 20 August 2020).
40. Homer Energy, NREL. Available online: http://www.homerenergy.com/ (accessed on 7 September 2020).
41. Tamil Nadu Generation and Distribution Corporation Limited(TANGEDCO). Electricity Service Number: 059094600008. Available online: https://www.tnebnet.org/awp/login (accessed on 7 September 2020).
42. Tamil Nadu Generation and Distribution Corporation Limited Revised Tariff rates from 12.12.2014 Ordered in TNERC Tariff Order Dated 11.12.2014 Subsidized Rate (Payable by the Consumer). Available online: https://www.tangedco.gov.in/linkpdf/Tariff%20payable%20by%20consumer(12.12.2014).pdf (accessed on 1 September 2020).
43. Tamil Nadu Electricity Regulatory Commission, Determination of Tariff for Generation and Distribution Manual, T.P.No.1 of 2013 Order Dated: 20-06-2013. Available online: https://www.tangedco.gov.in/linkpdf/Tariff1.pdf (accessed on 1 September 2020).
44. Salama, M.M.; Saied, E.M.; Mahmoud, H.M.; Abdelhadi, H.A. *Residential Loads and Application of Demand Side Management (DSM) Techniques*; Benha University: Cairo, Egypt, 2013.
45. Barbarelli, S.; Amelio, M.; Florio, G.; Scornaienchi, N. Procedure Selecting Pumps Running as Turbines in Micro Hydro Plants. *Energy Procedia* **2017**, *126*, 549–556. [CrossRef]
46. Mazzoni, S.; Rajoo, S.; Romagnoli, A. A boil-off gas utilization for improved performance of heavy duty gas turbines in combined cycle. *Proc. Inst. Mech. Eng. Part A J. Power Energy* **2019**, *233*, 96–110. [CrossRef]
47. Groppi, D.; Garcia, D.A.; Basso, G.L.; Cumo, F.; De Santoli, L. Analysing economic and environmental sustainability related to the use of battery and hydrogen energy storages for increasing the energy independence of small islands. *Energy Convers. Manag.* **2018**, *177*, 64–76. [CrossRef]
48. Cerri, G.; Chennaoui, L.; Giovannelli, A.; Mazzoni, S. Expander models for a generic 300 MW F class gas turbine for IGCC. In Proceedings of the ASME Turbo Expo 2014: Turbine Technical Conference and Exposition, Dusseldorf, Germany, 16–20 June 2014.
49. Javor, D.; Janjic, A. Application of demand side management techniques in successive optimization procedures. *Commun. Dependabil. Qual. Manag.* **2016**, *19*, 40–51.
50. Vishnupriyan, J.; Manoharan, P.S. Demand side management approach to rural electrification of different climate zones in Indian state of Tamil Nadu. *Energy* **2017**, *138*, 799–815. [CrossRef]
51. Chauhan, A.; Saini, R.P. Techno-economic optimization based approach for energy management of a stand-alone integrated renewable energy system for remote areas of India. *Energy* **2016**, *94*, 138–156. [CrossRef]

52. Kurnik, C.W.; Stern, F.; Spencer, J. *Peak Demand and Time-Differentiated Energy Savings Cross-Cutting Protocol. The Uniform Methods Project: Methods for Determining Energy Efficiency Savings for Specific Measures (No. NREL/SR-7A40-68566)*; National Renewable Energy Lab. (NREL): Golden, CO, USA, 2017.
53. HOMER Energy. *Getting Started Guide for Homer Legacy (Version 2.68)*; Homer Energy and National Renewable Energy Laboratory: Golden, CO, USA, 2011; Available online: https://www.gn-sec.net/content/getting-started-guide-homer-legacy-version-268 (accessed on 2 September 2020).
54. HOMER 2.28 Help Manual. 2015. Available online: https://www.homerenergy.com/pdf/HOMER2_2.8_HelpManual.pdf (accessed on 20 August 2020).
55. Montani, P.D.B.; da Luz, L.T.O.; Thomé, B.A.; Bento, R.G.; Nepomuceno, L.E.R.; Bernardon, D.P.; Canha, L.N. Intelligent Energy Management in Public Institutions. In Proceedings of the 2019 IEEE PES Innovative Smart Grid Technologies Conference Latin America (ISGT Latin America), Gramado, Brazil, 15–18 September 2019; pp. 1–5.
56. Abdul, C.; Mohammed, E.L.; Kiptoo, M.K.; Tomonobu, S.; Paras, M.; Shantanu, C. An Economic Analysis of Demand Side ManagementConsidering Interruptible Load and RenewableEnergy Integration: A Case Study of Freetown Sierra Leone. *Sustainability* **2019**, *11*, 2828.
57. Mancini, F.; Lo Basso, G.; De Santoli, L. Energy Use in Residential Buildings: Impact of Building Automation Control Systems on Energy Performance and Flexibility. *Energies* **2019**, *12*, 2896. [CrossRef]
58. Loganthurai, P.; Rajasekaran, V.; Gnanambal, K. Evolutionary algorithm based optimum scheduling of processing unitsin rice industry to reduce peak demand. *Energy* **2016**, *107*, 419–430. [CrossRef]
59. Kathiresan, A.C.; PandiaRajan, J.; Sivaprakash, A.; Babu, T.S.; Islam, M.R. An Adaptive Feed-Forward Phase Locked Loop for Grid Synchronization of Renewable Energy Systems under Wide Frequency Deviations. *Sustainability* **2020**, *17*, 7048. [CrossRef]
60. Kovács, A. On the Computational Complexity of Tariff Optimization for Demand Response Management. *IEEE Trans. Power Syst.* **2018**, *33*, 3204–3206. [CrossRef]
61. Nawaz, A.; Khadim, G.; Khan, I.; Jan, K.U.; Li, H.; Khan, S.A.; Zahid, W. An Intelligent Integrated Approach for Efficient Demand Side Management with Forecaster and Advanced Metering Infrastructure Frameworks in Smart Grid. *IEEE Access* **2020**, *8*, 132551–132581. [CrossRef]
62. Manfren, M.; Nastasi, B. Parametric Performance Analysis and Energy Model Calibration Workflow Integration—A Scalable Approach for Buildings. *Energies* **2020**, *13*, 621. [CrossRef]

energies

MDPI

Article

Occupants' Satisfaction toward Indoor Environment Quality of Platinum Green-Certified Office Buildings in Tropical Climate

Masoud Esfandiari [1,2,*], Suzaini Mohamed Zaid [1], Muhammad Azzam Ismail [1], Mohammad Reza Hafezi [2], Iman Asadi [3], Saleh Mohammadi [4,5,*], Salah Vaisi [5] and Ardalan Aflaki [6]

1 Faculty of Built Environment, University of Malaya, Kuala Lumpur 50603, Malaysia; suzaini_zaid@um.edu.my (S.M.Z.); ma.ismail@um.edu.my (M.A.I.)
2 Faculty of Architecture and Urban Planning, Shahid Beheshti University, Tehran 69411, Iran; mr-hafezi@sbu.ac.ir
3 School of Mechanical Engineering, College of Engineering, University of Tehran, Tehran 66191, Iran; iman.asadi@ut.ac.ir
4 Department of Architectural Engineering & Technology, Faculty of Architecture and the Built Environment, Delft University of Technology, Julianalaan 134, 2628 BL Delft, The Netherlands
5 Department of Architecture, Faculty of Art and Architecture, University of Kurdistan, Sanandaj 66177, Iran; svaisi@uok.ac.ir
6 Faculty of Architecture and Art, University of Guilan, 41996 Rasht, Iran; ar.aflaki@guilan.ac.ir
* Correspondence: m.esfandiari@um.edu.my (M.E.); saleh.mohammadi@tudelft.nl (S.M.)

Abstract: The quality of the indoor environment has become a vital component for buildings due to the time spent indoors. To this extent, the performance of the indoor environment is considered as part of the greenery criteria by green rating schemes such as the Green Building Index in Malaysia. This study aims to investigate and assess the quality of the indoor environment of Platinum-certified office buildings in a tropical climate. This research applied a case study approach over two Platinum-certified office buildings. Post-occupancy evaluation is employed integrating full-scale measurement with an occupants' survey. The measurement was carried out from May to August, and 112 questionnaires were retrieved to evaluate occupants' satisfaction with aspects of the indoor environment. Thermal comfort, indoor air quality, acoustic, lighting, furniture, and cleanliness are considered as the main study variables. The findings of full-scale measurement indicated high relative humidity, and low air velocity and illuminance. While occupants reported overall indoor environment quality (IEQ) comfort, a significant correlation of variables was observed. The main sources of dissatisfaction were identified as overcooling around 24 °C, high relative humidity (RH), around 70% RH, glare, and background noise around 51.9 dB. Statistically, a significant difference between occupants' responses to IEQ of two cases was identified, although both buildings are labelled with a Platinum certificate.

Keywords: building performance assessment; indoor environment quality; occupants' satisfaction; post-occupancy evaluation; Green Building Index; tropical climate

Citation: Esfandiari, M.; Mohamed Zaid, S.; Ismail, M.A.; Reza Hafezi, M.; Asadi, I.; Mohammadi, S.; Vaisi, S.; Aflaki, A. Occupants' Satisfaction toward Indoor Environment Quality of Platinum Green-Certified Office Buildings in Tropical Climate. *Energies* **2021**, *14*, 2264. https://doi.org/10.3390/en14082264

Academic Editor: Angelo Zarrella

Received: 23 March 2021
Accepted: 15 April 2021
Published: 17 April 2021

1. Introduction

The quality of the environment inside any building is called indoor environment quality (IEQ) [1]. IEQ has been proven to have a significant effect on occupants' wellbeing, health, comfort, productivity, and behaviour; thus, it has been a vital research area for decades [2,3]. The focus of the research in the field of IEQ has been adapting continuously according to human demands. For instance, poor indoor air quality (IAQ) as one of the IEQ parameters, could lead to sick building syndrome (SBS), which has been acknowledged as a health issue since the 1970s [4]. Later, the focus has become wider, from health issues to occupants' comfort. Various studies have investigated parameters of IEQ such as thermal quality, IAQ, acoustic quality, lighting quality, etc., concerning occupants' comfort and satisfaction [5–8]. Also, researchers evaluated IEQ, implying that the acceptable ranges of IEQ factors can be varied case by case. Sadick et al. [9] evaluated the importance of IEQ in offices of Ghana with a tropical climate. This study concluded that improving IEQ can positively impact the occupants' productivity. Besides, indoor air temperature is found to be an effective factor in thermal comfort and the perception of IEQ variables of offices located in tropical climates [10]. Another study in the tropical climate of Singapore showed that the most acceptable temperature is 26 °C in offices using ceiling fans [11]. Nematchoua et al. [12] reported that workers' productivity in Cameroon offices dramatically reduced in temperatures above 28 °C.

After the introduction of green buildings, studies attempted to compare IEQ of green buildings with conventional ones [13,14]. A recent comparative study reported that the SBS prevalence was found to be 38.1% and 53.1% in green and conventional buildings, respectively [3]. Also, Lee et al. [15], in a comparative study, investigated satisfaction and health symptoms, reporting higher IEQ acceptance for employees of green offices.

General beliefs are that green buildings are preferred regarding the performance competition against conventional buildings. Some studies claimed that the occupants of green buildings reported higher satisfaction and healthier in comparison to conventional buildings [14]. On the other hand, there are studies that indicated that there is no significant difference in occupants' satisfaction and wellbeing between IEQ of green and conventional buildings [16,17]. MacNaughton et al. [17] used post-occupancy evaluation (POE) to compare occupants' satisfaction between green and conventional buildings. In this study, occupants moved from a conventional to green building, reporting higher IEQ satisfaction from the green building. The result of this experiment indicated that the green building has higher performance for occupants, however, it is not possible to generalise this finding. For instance, Paul et al. [16] reported that there is no significant difference in occupants' satisfaction from IEQ of green and conventional buildings.

A high-performance green building brings higher occupants' productivity, which could be considered as the ultimate economic outcome [18], considering that studies indicated occupants' higher productivity, healthier, less absenteeism, and lower compensation for health insurance and claim returns, resulted from green buildings [19–22]. Moreover, green buildings are known for the energy-saving benefits, which ultimately reduce energy consumption costs.

However, the energy-efficient design approach might have side effects on green building performance. Providing a thermally comfortable environment usually has a direct influence on the energy consumption of office buildings. The majority of office buildings implemented heating, ventilation, and air-conditioning (HVAC) systems to maintain indoor air temperature, typically between 21.5 and 24 °C [23]. Additionally, HVAC systems attributed to about 43% of total energy consumption in US office buildings [24]. Thus, reducing energy consumption by increasing HVAC set points could lead to the uncomfortable indoor thermal condition [25]. Particularly, green office buildings in a tropical climate, where the HVAC system is often the main source of cooling or ventilation in the buildings, could become a challenge [26].

In addition, conflicts within IEQ parameters might occur with an attempt to provide a high-performance building. For instance, a high ventilation rate from a natural ventilation or mechanical system to improve IAQ could contribute to increasing background noise, resulting in acoustic discomfort [27]. This conflict could also happen between thermal comfort and daylight, to provide visual comfort, particularly for green buildings, as presented by Liang et al. [14]. Occupants in this study reported high satisfaction with lighting, but thermal comfort was less satisfactory. This could be due to the application of daylight resulting in higher heat exchange with the indoor environment and the increase in indoor temperature [28]. Application of daylight also associates with an increase in glare, which negatively affects visual comfort [29]. On the other hand, daylight has the potential to reduce energy consumption and provide health benefits [30].

Concerning IEQ influence on health and productivity, the extent of the effect depends on the IEQ variables. As mentioned in the literature, each variable alone impacts occupants in a particular way. Additionally, the quality of the variables can be varied according to building typology. Besides, green or conventional construction methods have shown diverse performances as there is no consensus on which method performs better. Achieving a high-performance IEQ requires periodical assessment to identify and improve sources of IEQ dissatisfaction [31]. Many studies have investigated IEQ, particularly in the European and American countries, to suggest improvements for future constructions. However, according to the literature, a few studies have focused on IEQ, specifically green buildings, in countries located in tropical regions such as Malaysia [32–34]. Particularly in the tropical climate of Malaysia, energy efficiency is one of the key factors to label buildings as green by the Green Building Index (GBI). Indeed, energy efficiency methods applied to green buildings—as mentioned earlier—could negatively impact occupant satisfaction. Thus, IEQ assessment of green-certified buildings is vital to have an insight into the constructions made to meet GBI criteria.

This study, therefore, aims to investigate and assess green-certified building performance regarding IEQ and occupants' satisfaction in the tropical climate of Kuala Lumpur. Additionally, the investigation is narrowed to office buildings with a Platinum Certificate. Another objective of this study is to make a comparison between the IEQ of Platinum GBI-certified office buildings. To achieve the objectives, the main variables are defined as indoor air quality, thermal quality, lighting quality, and acoustic quality. To perform the assessment, a post-occupancy evaluation (POE) was employed with the integration of full-scale measurements and an occupants' survey, leading to a comprehensive IEQ evaluation. Full-scale measurement was performed using environmental loggers monitoring IEQ variables. The occupants' IEQ satisfaction are assessed using a standardised questionnaire survey. Collected data from measurement and occupant surveys are utilised to address IEQ dissatisfaction sources and recommendations to further improve IEQ. To address this, this paper is structured as follows: after the introduction given in Section 1, green building tools in Malaysia will be briefly explained in Section 2. The methods applied to investigate IEQ, including full-scale measurement and occupants' survey, and cases selected for this study will be elaborated in Section 3. This is followed by the presentation of results in Section 4, and these results are discussed in Section 5. Finally, a conclusion and summary will be provided in Section 6.

2. Green Building Certification in Malaysia

The green building introduction required new construction methods that were different in various aspects from the old techniques. By this introduction, the green rating tools were established to evaluate and certify alleged green constructions. The United Kingdom pioneered and established the first green building rating system, known as the "Building Research Establishment Environmental Assessment Method" (BREEAM) in 1990. The United States then established a rating system eight years later in 1998, known as "Leadership in Energy and Environmental Design" (LEED). More countries have conducted the rating system in the following years, such as Australia (Green Star) in 2003 and Singapore (Green Mark) in 2005 [13,35]. Additionally, with a global approach, the WELL Building Standard was introduced in 2013 to exclusively focus on the impact of the building on human wellbeing and health.

The Malaysia Green Building Confederation (MGBC) was introduced in early 2007. It started when a group of consultants, academics, and representatives from the building industry met to initiate a non-profit organisation promoting sustainable buildings in Malaysia [36]. Following this movement, the GBI was established in 2009, which was formulated by the Malaysian Institute of Architecture, or Persatuan Arkitek Malaysia (PAM), and the Association of Consulting Engineers of Malaysia (ACEM) [36]. The green building assessment system was developed particularly to ensure the welfare of the occupants and the entire environment [37]. Although there are various building assessment tools in Malaysia, GBI is mostly applied to assess green buildings.

GBI assesses buildings through six (6) key criteria, such as Energy Efficiency (EE), Indoor Environment Quality (EQ), Sustainable Site Planning and Management (SM), Materials and Resources (MR), Water Efficiency (WE), and Innovation (IN) [35]. The buildings are assessed and scored between zero (0) to a maximum of 100, and as a result, they are awarded through one of the four categories: (1) Platinum: above 86 points, (2) Gold: between 76 to 85 points, (3) Silver: with 66 to 75 points, and (4) Certified: minimum of 50 points up to 65 points [35]. Hence, according to GBI, any building that achieves a minimum of 50 points in the evaluation process can be called "green".

GBI has organised buildings in seven (7) categories, such as Non-Residential New Construction (NRNC), Residential New Construction (RNC), Industrial New Construction (INC), Non-Residential Existing Building (NREB), Industrial Existing Building (IEB), Interior (ID), and Township (T). According to the study by Zian et al. [38] in 2019, the energy efficiency criteria is the most widely used in green building assessment systems, followed by water efficiency and innovations. Moreover, about 1/5 of the total points were dedicated to the IEQ in the NRNC, where it was 21 points out of 100. The IEQ of NRNC is evaluated through air quality, thermal comfort, lighting, visual and acoustic comfort, and verification (post-occupancy evaluation) during occupancy [35].

3. Methodology

This study utilised a post-occupancy evaluation (POE) method to investigate occupants' satisfaction toward the existing implementation of green design concepts for the office buildings in Malaysia. POE is a widely acceptable systematic method to investigate building performance through occupants' feedback [31,39]. Many believe that this method is the most effective and the best systematic technique of investigation to explore and examine the mutual interaction between the building and the occupants' needs [40–42]. Particularly for office buildings, many studies emphasised the occupants' assessment role in measuring the performance of the workplace [43–45]. Findings from POE can bring suggestions to form an alternative basis method for any construction or renovation in the future; therefore, it could amend and influence design codes, standards, and decisions [40,46]. Although some studies have only used survey to assess and investigate IEQ [14,47], POE in this study utilised occupants' survey integrated with full-scale measurement [31]. This is to comprehensively evaluate the IEQ of the Platinum-certified office buildings. IEQ is evaluated according to the following criteria: indoor thermal environment (thermal comfort, temperature, humidity), IAQ (ventilation, carbon dioxide (CO_2), volatile organic compounds (VOCs)), lighting (daylight, artificial light, glare), acoustic (sound insulation), furniture, and maintenance (cleaning).

3.1. Case Study

As mentioned earlier, the purpose of this study is to investigate the IEQ of Platinum-certified office buildings in Kuala Lumpur. GBI categorizes buildings into separate groups, and the NRNC category includes buildings with functions such as governmental agencies, office buildings or towers, malls, galleries, factories, etc. Besides, buildings to be certified as Platinum must obtain the maximum points from various rigorous criteria. Since green construction is a new growing concept in Malaysia, only limited buildings have been able to integrate sustainable technologies and methods that successfully obtained the Platinum certificate. According to the latest GBI report in December 2020, the buildings awarded a Platinum certificate represent less than 4% of all the certificates issued in NRNC [48]. For this study, however, considering office functionality within the NRNC category, two cases are identified with a Platinum certificate located in Klang Valley [48]. Also, to achieve the study objective, the building must be in the operational phase for more than two years. These Platinum certificate office buildings are named buildings A and B. Building A is a government agency and building B is the headquarters of a private company.

The unique construction of the buildings incorporates various sustainable applications, such as utilisation of daylight, roof light, daylight reflectors, artificial light sensors, passive design strategies, green roof, sustainable materials, photovoltaic panels, rain harvest system, recyclable materials, etc. Table 1 represents the key features of these buildings.

Table 1. Key data for case study buildings A and B.

	Building A	Building B
No. stories	8 + 2	9
Green landscape area	3600 m^2	-
Structure	Steel-reinforced concrete (SRC)	Steel-reinforced concrete (SRC)
Facade finishes	Double glass curtain wall	Double glass curtain wall
Access to public transport	Bus	Bus
Air conditioning system	Air Handling Unit (AHU), Fans	Underfloor Air Distribution (UAFD)
Roof vents	Yes	-
Lighting type	Fluorescent	Fluorescent, Personal LED
Window to wall ratio (WWR)	About 60%	About 55%
Window blind, roller shade, fixed shading system	Yes	Yes
Interior separators	Medium-density fiber-board (MDF) and glass	Medium-density fiber-board (MDF) and glass
Exterior shades	No	No
Atrium	Yes	No
Courtyard	No	Yes
Employees/Occupants	80–100	60–100
Year of GBI certificate	2014 (validated until 2019)	2014 (validated until 2020)

Case study building A is located in Putrajaya, whereas case study building B is located in Shah Alam, both within Kuala Lumpur. Buildings A and B have been in operation since 2010 and 2014, respectively. Each floor of the buildings is dedicated to a few departments, except for parking and the ground floor. Within each department, employees are involved in sedentary activities, such as typical office tasks like writing, reading, computer typing, and occasional walking within the workstations and rooms. There is no significant heat emission from body activities. To carry out office tasks, the majority of employees work within open-plan offices, and the minority in private rooms, making the open-plan offices the main work areas in buildings. Also, the seminars and meetings are held in conference rooms. Generally, the occupants' appearance was observed to be trousers, long sleeve or short sleeve shirts, socks, and shoes. The clothing value of the occupants was 0.5 to 0.7 clo during the investigation.

Building A has an 11,473 m^2 net floor area and 14,230 m^2 gross floor area, with 75% floor efficiency (Figure 1a). Building B includes a 14,087 m^2 net floor area and 33,798 m^2 gross floor area (Figure 1b). Buildings A and B accommodate up to 400 and 250 occupants, respectively. For building A, a combination of a radiant cooling system with pipes embedded in the concrete slabs and an air handling unit (AHU) with fans has been employed to provide thermal comfort. Building B maintains thermal comfort with an underfloor air distribution (UFAD) system. Raising the floor to provide ventilation and distributing conditioned air through vents or diffusers into the indoor environment is known as UFAD. Both buildings have utilised a centralised ventilation system, with no occupant control over temperature or air velocity. In a tropical climate, thermal comfort significantly impacts occupants' satisfaction, thus, due to the centralised system, buildings should be investigated as one unit [14,49,50]. The facade of both buildings is mainly constructed by a glass curtain without any operable window for occupants.

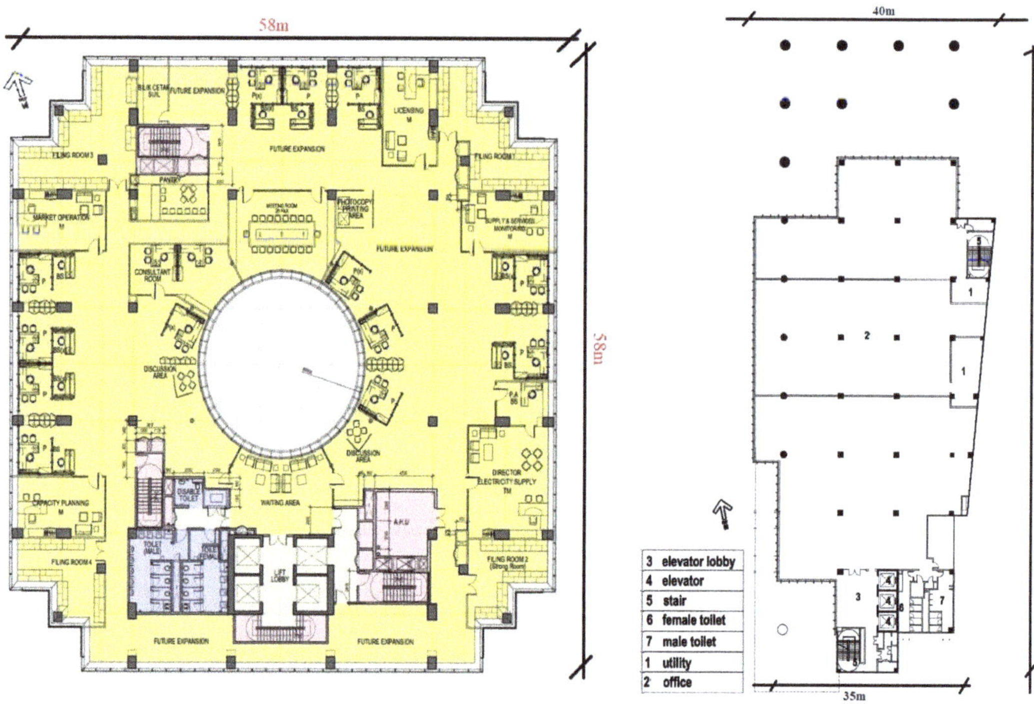

(**a**) Building A (**b**) Building B

Figure 1. Plan layout of buildings.

3.2. *Occupants' Satisfaction Survey*

This study implemented a questionnaire survey to subjectively investigate IEQ satisfaction of green-certified office buildings. This would allow quantifying the subjective opinions. In this regard, the latest version of the Building Use Studies (BUS) questionnaire, specialised for non-domestic buildings in a tropical climate, was employed. BUS is a structured questionnaire with several versions depending on the building's typology, study design, and microclimate. The original questionnaire survey and database was established in 1985 because of the working group which tried to cover the surveying of 4300 office workers in 50 buildings in the United Kingdom in response to the SBS. The questionnaire has been constantly updated and modified to reduce the flaws, hence different sections of the questionnaire have been changed [51]. This standard questionnaire has been used to evaluate over 850 buildings worldwide up to 2021 [52]. This questionnaire uses a seven (7)-point satisfaction scale to evaluate various aspects of the building performance, but the "perceived productivity" was measured with a nine-point scale.

As is the norm for short-term POE, the survey was conducted one time on-site [3,31,43,47], where occupants were informed of the survey 24 h in advance. The questionnaires were distributed to occupants by the start of working hours and collected by the end of working hours on the same day. Table 2 presents a summary of the collected data.

Table 2. Questionnaire parameters.

Category	Question
Background information	Age, Sex, Years of working in the building, Type of the working office
Satisfaction with building design	Cleaning, Furniture
Satisfaction with IEQ parameters	Overall comfort, Perceived health, Perceived productivity Overall thermal comfort, Temperature, Temperature stability Overall air quality, Air humidity, Air freshness, Odour Overall noise, Noise from colleagues, Noise from other people, Other noise inside, Noise from outside Overall lighting, Natural light, Glare from sun and sky, Artificial light, Glare from light

3.3. Full-Scale Measurement of the Indoor Environment

Full-scale measurement or objective measurement was performed to monitor IEQ in the study. This aided to provide a holistic perspective of the circumstances occupants responded to in the questionnaire survey. Additionally, full-scale measurement integrated with a survey would create a platform to identify and improve any potential source of IEQ dissatisfaction [31,53]. The parameters for monitoring were air indoor temperature (T_a), relative humidity (RH), air velocity (m/s), CO_2 concentration, TVOC concentration, illumination level, and indoor sound pressure level (SPL). Also, regarding mean radiant temperature, the difference between the indoor air temperature and mean radiant temperature is negligible under moderate outdoor conditions. Thus, the mean radiant temperature can be assumed equal to the indoor air temperature [54–56].

Kuala Lumpur is located between latitude 3.12 °N and longitude 101.55 °E, with a 13:11 solar noon. The daily climate is hot and humid, which is fairly consistent throughout the entire year, with minor temperature fluctuation [57]. Considering the stability of the outdoor microclimate throughout the year, the full-scale measurement was carried out from May to August 2019, which could be extended to the yearly data. Monitoring IEQ was performed with five-minute intervals of data logging, limited to operational days, and working hours 8:00 to 18:00 of the buildings.

Table 3 illustrates the specification of equipment used for full scale measurement. Ten sets of equipment were used to collect environmental data. Each floor, starting from the first floor, was divided into zones—rooms without occupancy were excluded—and loggers were impended to monitor the zones for seven working days. This helped to monitor occupied spaces in a sequence covering floors and buildings.

The environmental loggers attached to portable poles (Figure 2a,b) in the height of 1.1 m above the finished floor (level of sedentary activities), as suggested by ISO 16000-1 [58], close to occupants, 0.3 m. For the horizontal distribution of measurements, the devices were positioned approximately 0.8 m away from the devices and walls (Figure 2d) to avoid radiative temperature. This permitted simulation exposure of occupants to indoor air near the breathing zone when seated. Furthermore, portable light meters were placed on the working desk of the occupants to measure the lighting level (Figure 2c). This was to record a realistic approximation of the illumination received by occupants. A daily average of each parameter (121 data points) was generated for the period of measurement to present buildings' indoor environment conditions.

(**a**): Air temperature, air velocity, RH measurement

(**b**): Portables poles with loggers

(**c**): Illuminance, CO2, TVOC measurement

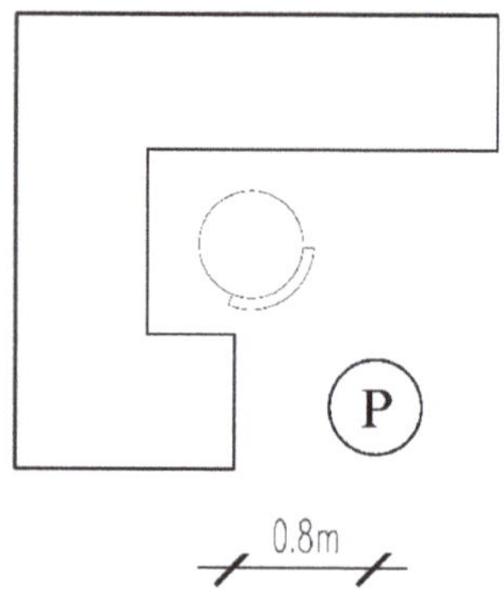

(**d**): Sample of poles' location close to workstation

Figure 2. Demonstration of equipment installed to monitor indoor environment quality (IEQ).

Table 3. Specifications of equipment for physical measurement of indoor environment parameters.

IEQ Parameter	Environmental Sensor	Range and Accuracy
Air temperature	HOBO-U12-012 [1], Onest	−20 to 70 °C, accuracy ±0.35 °C, from 0 to 50 °C,
Relative humidity		±2.5%, 3.5% maximum, from 10% to 90% RH
Air velocity	T-DCI-F900-S-O [1], Onest	Accuracy greater than 10% of the reading or ±0.05 m/s or 1% full-scale
Illuminance	TM-203 Datalogging [2], TENMARS	20 to 200 K Lux, ±3%
Sound pressure level	Solo 1092 01dB-METRAVIB [2], ACOEM	20 to 137 dB (A) class 1, or 30 to 137 dB (A) class 2
Carbon dioxide	98,123 J [2], MIC	0 to 50 PPM with a resolution of 1 PPM, ±30 PPM + 5% of reading
TVOCs	98, 519 [2], MIC	0–50 PPM with 0.01 PPM resolution, ±30 PPM + 5% of reading,

[1] Sensor was tested for accuracy in a pilot study. [2] Sensor was calibrated before the monitoring in the laboratory of the Faculty of Built Environment, University of Malaysia.

3.4. Questionnaire Respondents

Completed questionnaires were considered for the final analysis to have more accurate results. A total of 174 questionnaires were distributed, of which 112 valid questionnaires were retrieved, with a response rate of 64%. According to Cochran [59], the sample size and response rate comes with a 94% confidence level. Among the responses, 57 (50.9%) were completed by males and 55 (49.1%) were completed by females, 47 (42%) of the respondents reported age under 30 years and 65 (58%) reported age 30 or over. Further details on the distribution of the respondents, including respondents from each building, work experience in the building, and type of the work office are summarised in Table 4.

Table 4. Background and demographic information of valid respondents.

		Building A	Building B	Total
Sex	Female	30	25	55
	Male	29	28	57
Age	Under 30	21	26	47
	30 and above	38	27	65
Work experience in building	Less than a year	20	20	40
	A year or more	39	33	72
Type of the work office	Normally occupied by 1	16	8	24
	Shared with 1 other	1	0	1
	Shared with 2–4 others	6	22	28
	Shared with 5–8 others	16	15	31
	Shared with more than 8 others	20	8	28

4. Results

As explained in Section 1, the POE approach in this study integrated a survey with full-scale measurement to investigate IEQ performance of study buildings. Collected data during full-scale measurement are utilised to report a daily average of the variables and questionnaire survey to quantify occupants' responses.

4.1. Full-Scale Measurement of the Indoor Environment

The results from the full-scale measurement of buildings have been presented to make a comparison between the study outcomes, recommendations, and standards. Figure 3 presents the mean and variance of the parameters such as air temperature (T_a), relative humidity (RH), air velocity, CO_2 concentration, illuminance level, and sound pressure level (SPL) for buildings A and B. T_a was observed to be in a similar range for both buildings, as for building A, T_a ranged from 23.3 to 24.6 °C with a mean of 24.2 °C, and T_a for building B ranged from 23.6 to 24 °C, with mean of 23.8 °C (Figure 3a). T_a for building B has a lower variance in comparison to building A. Lower mean T_a and low variance of T_a contributed to higher unconformable and neutral sensation mean vote from occupants for building B. It seems that both buildings have tried to maintain operation temperature (T_o) around 24 °C, an appropriate range according to MS 1525 [60]. Malaysian standard MS 1525 [60] recommend dry bulb temperature to be in a range of 24 to 26 °C, with a minimum of 23 °C for office buildings. The level of monitored RH is significantly different for buildings A and B, ranging from 58% to 64.5% with a mean of 60%, and 72% to 78.5% with a mean of 72.4%, respectively (Figure 3b). RH level for building B is above the recommended range by MS 1525 [60], which is recommended to be in the range of 50% to 70%. Particularly, the RH level is recommended not to exceed 70% [60]. The main reason for the high RH level in building B in comparison to building A was due to the quality of the HVAC system to maintain the indoor RH level, as natural ventilation or operable windows are not included in both buildings. This in compliance with Aziz, Sumiyoshi [61], who implied that the implementation of normal air handling units is unable to adequately dehumidify the indoor environment in the tropical climate of Malaysia.

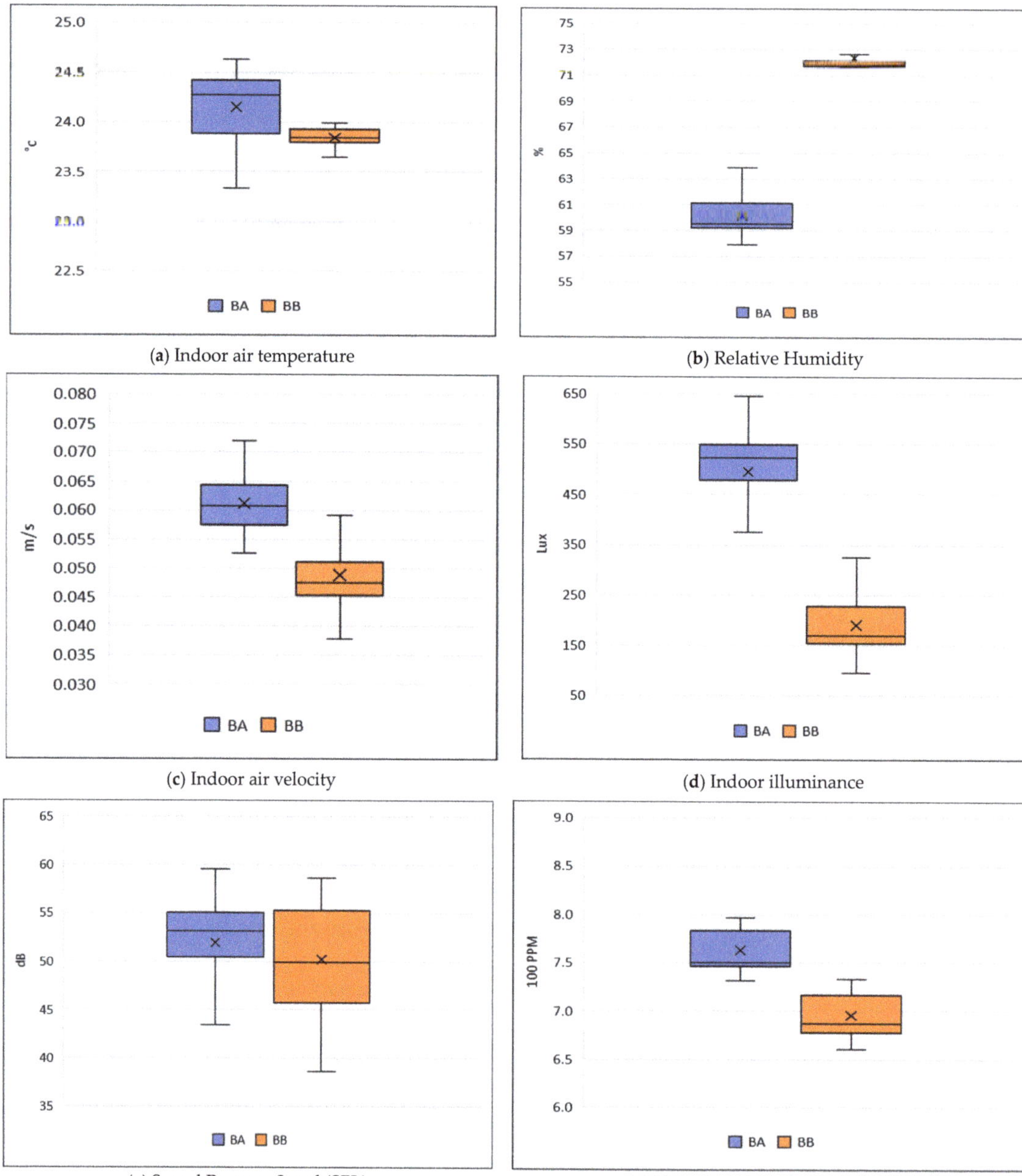

(**a**) Indoor air temperature (**b**) Relative Humidity

(**c**) Indoor air velocity (**d**) Indoor illuminance

(**e**) Sound Pressure Level (SPL) (**f**) Indoor CO_2 concentration

Figure 3. Distribution of IEQ parameters for building A (BA) and building B (BB).

Air velocity for both buildings is observed to be similar and considered below the recommended range (0.15–0.50 m/s) by MS 1525 [60]. Air velocity for building A ranged from 0.05 to 0.07 m/s, with a mean of 0.06 m/s, and ranged between 0.04 and 0.07 m/s with a mean of 0.05 m/s for building B (Figure 3c). The main reason for low air velocity within building A was observed to be due to the location of vents in the top of the rooms and over the corridors, which resulted in low air velocity around the breathing area of occupants.

Building B is equipped with adjustable vents to distribute air from the underfloor; however, it was observed that most of the vents adjacent to workstations were closed by occupants, resulting in low air velocity near the breathing area of a seated person. The main reason for occupants to close the vents was found to be the cool air temperature, as higher air velocity can extend the level of coolness feeling for the occupants.

The illuminance level for building B is significantly lower than building A. It was in the range of 95 to 325 lux with a mean of 191 lux for building B. In parallel, building A provided illuminance in the range of 375 to 646 lux, with a mean of 494 lux (Figure 3d). In this regard, various standards and studies suggested that acceptable illuminance for a normal office work should be between 500 and 1000 lux [62,63]. In addition, the average minimum daylight available in Kuala Lumpur is above 10,000 lux, from 10 a.m. to 6 p.m. [57]. Thus, both buildings have tried to utilise daylight as the main source of lighting. It was observed that occupants sitting near the window received a high illuminance level from daylight. To control the level of illuminance, blinds were used by occupants sitting near the window to reduce the intensity of daylight. However, this action resulted in dim spaces within a range from the window, in which occupants suffered from low-level daylight and luminance level. To diminish this issue, occupants were provided with task lighting in building B and an atrium is applied in the design of building A. Particularly, task lighting significantly enhances the individual's visual quality, although the indoor environment is dim. This is supported by the illuminance level being very low, although the occupants reported overall comfort from light quality in building B. In addition to the distance of the window within the inner part of the building, the daylight was trapped by workspaces with high-level partitions before reaching the inner workspaces, resulting in dim spaces for inner parts.

Regarding background noise from outdoors, buildings were located near the street. Additionally, the neighbourhood buildings and transportation density are observed to be lower for building B. However, the study buildings are fully air-conditioned with no operable windows, which significantly reduces noise penetration. SPL is similar for both buildings and ranges from 43.45 to 59.55 dB, with a mean of 51.9 dB, and 38.65 to 58.65, with a mean of 50.2 dB, for buildings A and B, respectively (Figure 3e). Mui and Wong [64] reported that the neutral SPL for aural comfort in typical air-conditioned offices should be between 45 and 70 dB, with a mean of 57.5 dB. Mean CO_2 was detected to be 763 and 695 ppm for buildings A and B, respectively (Figure 3f). CO_2 lower than 1000 ppm is considered harmless [65]. No vestige of TVOC was detected in both buildings, with 0 ppm measured.

A similar AC management scheme is applied to both buildings A and B. This scheme activates AC only during operation hours of the office building (8 a.m. to 6 p.m.). Besides, the buildings mainly included open-plan office workstations as well as a few private rooms, conference rooms, and storage spaces.

Given the typical office work of the occupants, such as sedentary activities, there are no significant indoor heat gains from occupants' activities. Personal fans were observed to be used by occupants in both buildings, particularly for private rooms, which could be a sign of low indoor air velocity.

Another issue that should be considered is the tropical climate of Malaysia, which directly influences the indoor thermal condition. Malaysia is located in the equatorial region of South East Asia between 1° N and 7° N, and longitude 0° and 119° E [57]. Kuala Lumpur as a base location for this study is located between the latitude of 3.12° N and longitude of 101.55° E, with a 13:11 solar noon [57]. The daily climate in Malaysia is described as hot and humid, which is consistent throughout the entire year, and the heavy rain comes during monsoon seasons. The average dry bulb temperature is around 26.92 °C for the entire year. The highest dry bulb temperature fluctuation happens around 14:00 to 18:00 [57]. The effect of this fluctuation was observed in the samples collected during the monitoring period, in which around 14:00 to 16:00, the mean air temperature of the indoor environment was at the crest. Moreover, it was observed that relative humidity started to peak by the start of the operation hour of the buildings, then it declined to a lower percentage after the start of the HVAC system to maintain relative humidity; however, building B presented a very high RH level.

4.2. Occupants' Satisfaction with Indoor Environment Quality

Despite the high RH and low illuminance for building B, findings from measuring indoor environment parameters show general compliance with standards such as MS 1525 [60] for both buildings. However, the findings cannot speculate occupants' comfort from IEQ. Thus, the occupants' IEQ feedback would give a better perspective.

The data from the questionnaire survey were analysed individually to make a comparison between IEQ of buildings and to correlate the results of the measurements with the findings of the survey. To perform the statistical analysis, responses to "5, 6, 7, or +1, +2, +3" transformed into a "comfortable" category and "1, 2, 3, or −3, −2, −1" into the "uncomfortable" category. Figure 4 shows the frequency of the occupants' responses to the overall IEQ comfort. As indicated in the results, about 91.5% of occupants were comfortable with the overall IEQ of building A, and only 3.8% of occupants were uncomfortable with the IEQ of building B (Figure 4). Despite the high neutral percentage (43.40%) of occupants for building B, generally, both green buildings were able to deliver an indoor environment condition to meet occupants' needs and fulfil their satisfaction. A comparison between buildings A and B indicates a high ratio of occupants in the "comfortable" category for building A (91.5% to 52.8%). The mean scores for buildings A and B are 1.66 and 0.58, which is another index to indicate higher satisfaction from building A's performance.

Besides occupants' satisfaction from overall IEQ, IEQ variables were assessed and analysed. As shown in Figure 5, occupants' satisfaction was analysed using frequency distribution over 'overall thermal comfort', 'overall indoor air quality', 'overall noise quality', 'overall lighting quality', 'furniture quality', and 'cleaning quality'.

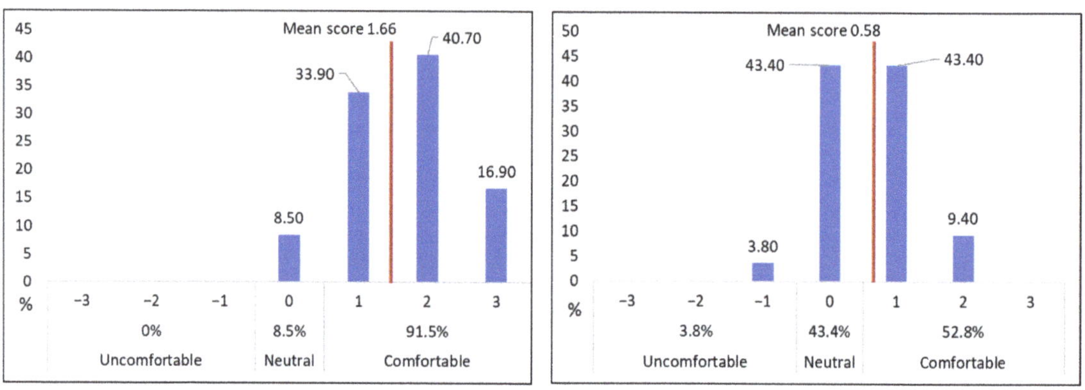

(**a**) Building A (**b**) Building B

Figure 4. Frequency distribution of occupants' vote for overall comfort.

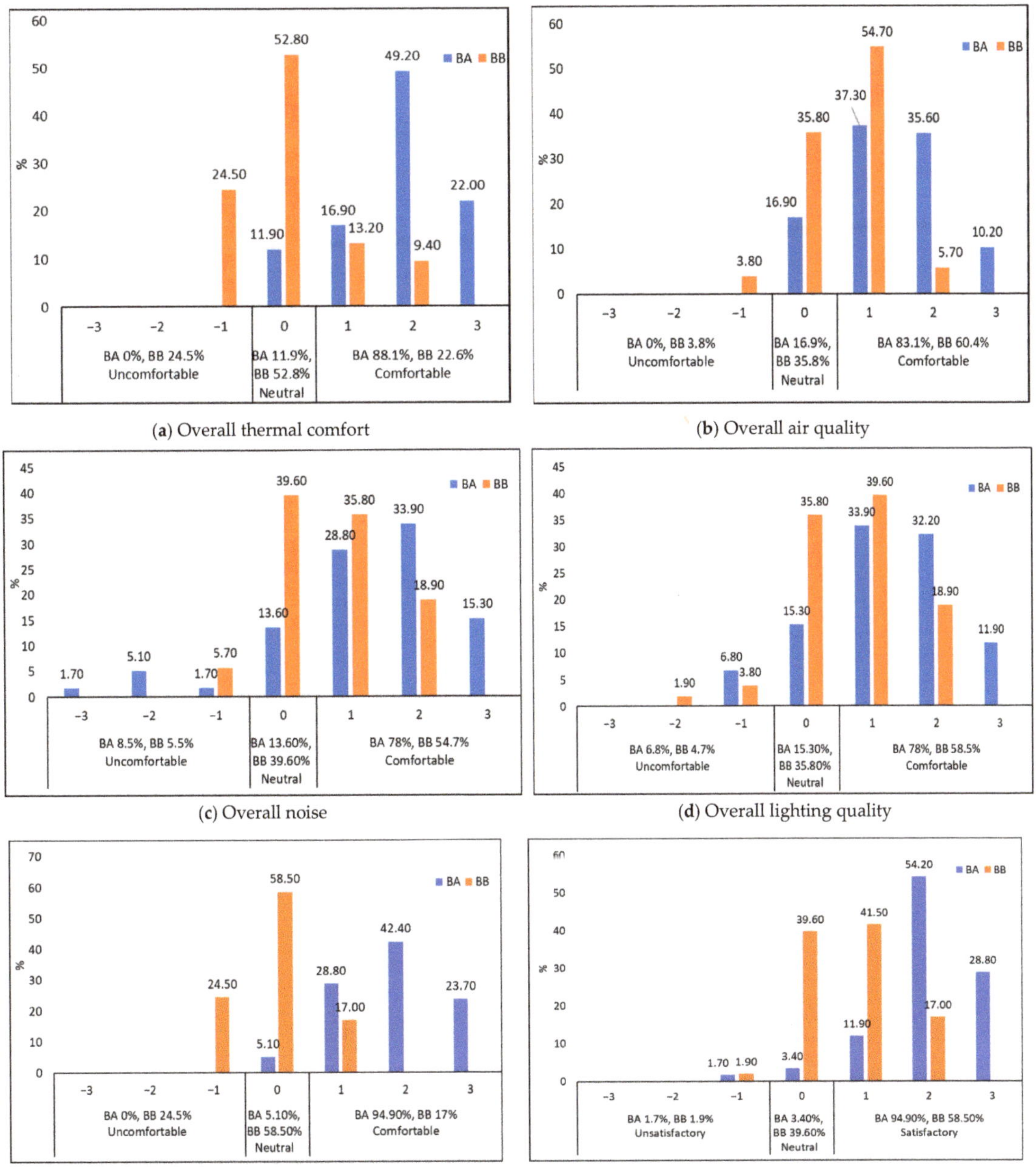

(**a**) Overall thermal comfort (**b**) Overall air quality

(**c**) Overall noise (**d**) Overall lighting quality

(**e**) Furniture quality (**f**) Overall cleaning quality

Figure 5. Frequency distribution of occupants' votes over IEQ parameters.

Satisfaction from overall thermal comfort is measured using a seven-point scale from uncomfortable to comfortable. This gives an overview of the portion of the occupants satisfied with thermal comfort. The finding indicated that 88.1% of respondents reported overall thermal comfort for building A, but 22.6% of respondents reported overall thermal comfort for building B. There is no thermal comfort dissatisfaction report of building A, whereas 24.5% of thermal comfort dissatisfaction was observed for building B (Figure 5a).

ASHRAE Standard 55 [66] suggests a minimum of 80% satisfaction for acceptable thermal comfort, in which building B fails to meet the condition.

Responses to the overall air quality could be triggered by factors such as air velocity, humidity, temperature, and indoor pollutants [67,68]. Regarding the air quality, 83.1% and 60.4% of occupants reported comfort for buildings A and B, respectively. Only a minority of occupants (3.8%) of building B reported dissatisfaction from IAQ (Figure 5b). The high relative humidity level (mean of 72.4% RH) of building B is observed to be the major reason for less satisfied occupants of building B in comparison to building A. The overall noise satisfaction for buildings A and B is observed at 78% and 54.7% and minority dissatisfaction at 8.5% and 5.5%, respectively (Figure 5c). For overall lighting quality, 78% and 58.5% of occupants reported satisfaction for buildings A and B, respectively. A minority of occupants, 6.8% (building A) and 4.7% (building B), reported dissatisfaction with overall lighting quality (Figure 5d). Although the mean indoor environment illuminance level in building B was measured to be low (190.88 lux), the application of task lighting increased the occupants' satisfaction for overall lighting. Another significant factor of indoor environment quality is furniture, which contributes to the occupants' overall IEQ comfort. Furniture for buildings A and B were reported to be 94.90% and 17% comfortable, respectively. However, 24.5% reported uncomfortable furniture for building B (Figure 5e). The cleanliness satisfaction of buildings A and B were reported as 94.90% and 58.50%, respectively. Dissatisfaction from cleanliness was reported as being minor, as 1.7% and 1.9% for buildings A and B, respectively (Figure 5f). Generally, most occupants reported satisfaction for the overall quality of parameters, which is aligned with the overall comfort. However, it should be noted that a significant group of occupants reported neutral from "overall IEQ" as well as parameters of IEQ, where any changes in the indoor environmental quality could push them to comfort, or in a worse-case scenario, to the uncomfortable range.

A comparison between the frequency of IEQ parameters contributed to 'uncomfortable' and 'neutral' for buildings A and B is demonstrated in Figure 6. 'Uncomfortable' is determined based on the ratio of the votes '-3 to -1' to the total responses. In building A, 'stuffy air' is the top dissatisfaction source underlying 'uncomfortable' and 'neutral' that is indicated by around 61% of occupants. This agrees with 'humidity' as the top dissatisfaction source for building B, with more than 81% of responses. This is followed by 'glare from sun and sky' with 59.3% of responses and 'cold' with 77.4% of responses for buildings A and B, respectively. 'Cold', 'other noise inside', 'noise from colleagues', 'smelly air', 'humid', 'too much natural light', 'temperature varies during the day', and 'noise from other people' are the third to eighth parameters in sequence for building A, with considerable percentages. For building B, 'temperature varies during the day', 'smelly air', 'stuffy air', 'too much natural light', and 'glare from sun and sky' are the third to fifth parameters, with more than 40% for 'uncomfortable' and 'neutral' responses. For building A, 'too little artificial light', 'dry', and 'warm' were addressed by less than 20% of occupants. 'Noise from other people', 'other noise inside', 'noise from outside', 'too little artificial light', 'dry', and 'warm' were addressed by less than 20% for building B.

High percentages of occupants voted 'uncomfortable' and 'neutral' for 'overall thermal comfort' (total 77.4%, Figure 5a). It can be explained by issues such as 'high humidity' and 'cold temperature' (Figure 6) as the main sources of uncomfortable thermal quality. 'Too much glare from sun and sky' and 'too much natural light' (Figure 6) are the main sources for 22.1% of occupants who did not vote 'comfortable' for 'overall lighting' (Figure 5d). 'Stuffy air' is also the main reason for 16.9% of 'neutral' votes to the overall air quality of building A, whereas 'smelly air' and 'stuffy air' are the main causes for 39.6% of votes in building B. Finally, 'noise from colleagues' is the main reason for occupants 'uncomfortable' or 'neutral votes' for both buildings (Figures 5 and 6).

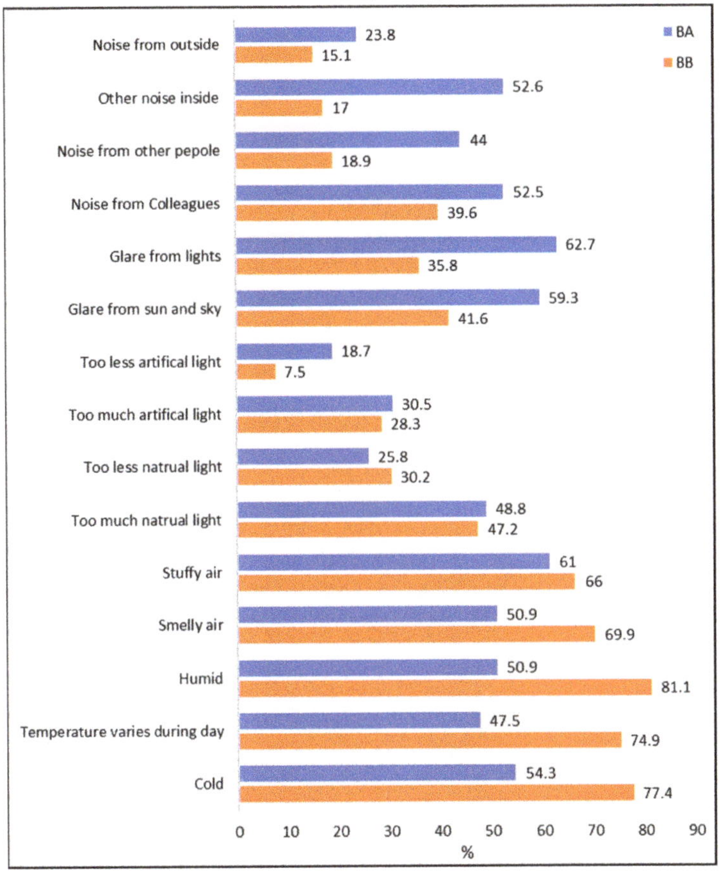

Figure 6. Comparison between the frequency of votes to the IEQ parameters contributing to uncomfortable and neural sensation.

Productivity is defined as " ... the ability of people to enhance their work output through increases in the quantity and/or quality of the product or service they deliver", by Leaman et al. [69]. Since it is difficult to achieve a meaningful evaluation of the productivity or work output, a self-report of productivity evaluation is suggested, known as "perceived productivity' [69]. Similarly, feeling healthy in the buildings has been self-reported and evaluated in the POE studies [70–72]. To measure perceived productivity and health, therefore, occupants were asked in the questionnaire survey to self-report IEQ impact on perceived productivity and heath. For productivity, a nine-point scale measurement was used, as productivity decreased by 1: −40% or more, 2: −30%, 3: −20%, 4: −10%, 5: 0%, and productivity increased by 6: + 10%, 7: +20%, 8: +30%, 0: +40% or more for IEQ. Also, occupants reported feeling healthier or less healthy in the questionnaire (seven-point scale, with 1 as less healthy to 7 as healthier) in the buildings to measure perceived health. The measuring scale was transformed into −4 to +4 from 1 to 9 for productivity and −3 to 3 from 1 to 7 for health. Also, votes over −4 to −1 were considered as less productive and +1 to +4 as more productive, as well as −3 to −1 as less healthy and 1 to 3 as healthier. Figure 7 demonstrates the responses' frequency distribution for perceived productivity and health: 86.40% and 45.30% of occupants reported being more productive in buildings A and B respectively, whereas only 18.9% of respondents reported decreased productivity in building B (Figure 7a). Similar to the perceived productivity,

67.80% and 62.30% of respondents reported feeling healthier, with only a minority reporting less healthy, 10.2% and 11.30% for buildings A and B, respectively.

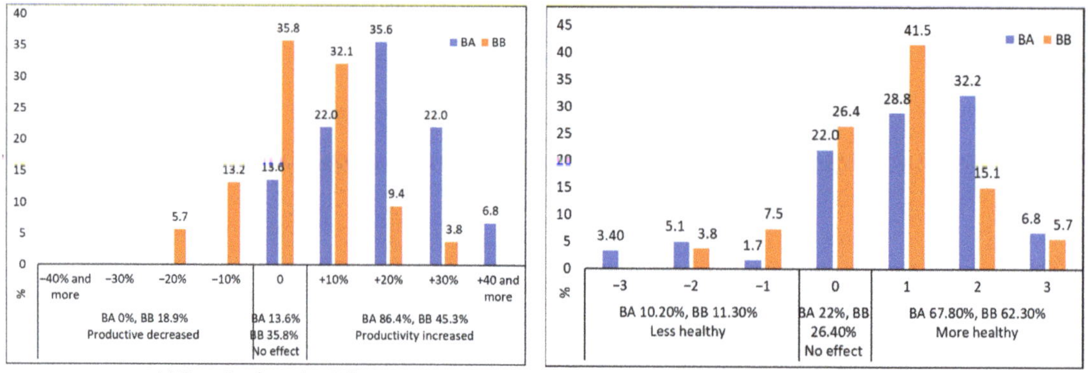

(**a**) Perceived productivity (**b**) Perceived health

Figure 7. Frequency distribution of perceived health and productivity.

4.3. Variation in Satisfaction between Buildings A and B

To statistically make a comparison between occupants' mean vote of buildings A and B for variables such as comfort, thermal, IAQ, noise, lighting, furniture, cleanliness, productivity, and health, an independent sample *t*-test was used. This test statistically compares the significant difference between the means of two groups [73]. This test has also been used to make a comparison between IEQ mean votes of green and non-green buildings [14]. For this study, to conduct the independent *t*-test and compare the mean between two Platinum-certified office buildings, IBM SPSS statistical package version 24 was employed. As shown in Table 5, the mean vote for all the concerned parameters, overall comfort, perceived productivity, and health is higher for building A. *t*-test results also indicate that statistically, there is a significant difference ($p < 0.05$) for the mean vote of the parameters, except for perceived health. As mentioned above, both buildings are evaluated by GBI and certified as Platinum. However, occupants' mean votes over IEQ parameters indicate a significantly better IEQ for building A in comparison to building B. The mean score of 'cleanliness' (2.05) for building A is the highest mean among all the parameters for both buildings, whereas the mean of −0.7 for 'furniture' of building B is the lowest mean vote. Furniture is the sole parameter with a negative mean value, indicating unsatisfactory furniture quality of building B. Besides, all the mean values for building A are around 1 to 2, highlighting the perception of slightly satisfied or satisfied, whereas mean values for building B are mostly between around 0.5 to 0.75, which stays in the neutral range, except the 'furniture', with a negative value.

Table 5. Comparison of IEQ parameters between buildings A and B.

IEQ Parameter	Building A	Building B	*p*-Value [a]
Overall comfort	1.66	0.58	0.000 [b]
Overall thermal comfort	1.81	0.07	0.000 [b]
Overall air quality	1.39	0.62	0.000 [b]
Overall Noise	1.25	0.68	0.009 [b]
Overall lighting	1.27	0.70	0.003 [b]
Furniture	1.85	−0.75	0.000 [b]
Overall cleaning quality	2.05	0.74	0.000 [b]
Perceived productivity	1.86	0.38	0.000 [b]
Perceived health	0.92	0.74	0.449

[a] Two-tailed *t*-test was employed to compare the mean score between buildings A to B over IEQ parameters. [b] Highlights the statistical significance of test result at a level of $p < 0.05$.

5. Discussion

There are many studies focused on assessing the quality of the indoor environment [74]. Findings of the assessment can be used to improve buildings' performance [53]. This study focused on the IEQ assessment of the Platinum-certified office buildings in Kuala Lumpur. As mentioned in Section 1, to obtain a GBI certificate, office buildings should be assessed by GBI through six categories, including IEQ. According to assessment results, buildings could be awarded one of four different certificates, with Platinum being the best award. This indicates that the IEQ of the Platinum-certified office buildings should be of a high quality compared to non-certified buildings. Concerning all six GBI assessment criteria, however, IEQ could be negatively influenced. This makes it essential to assess IEQ of Platinum-certified office buildings.

To perform the assessment, the IEQ variables were investigated through full-scale measurement and an occupants' survey. Figure 8 shows the distribution of occupants' votes over thermal sensation, dry–humid air, stuffy–fresh air, and smelly–odourless air. These variables are measured with a seven-point scale. ASHRAE [66] suggests a seven-point scale with −3 = Cold to +3 = Hot. The mean vote between −0.5 to +0.5 is considered as a neutral thermal feeling [66]. Mean votes out of the suggested neutral range are considered as uncomfortable thermal sensation, which leads to a high rate of dissatisfied occupants from thermal comfort. The TSV means for occupants of buildings A and B are 0.64 and −1.25 (equivalent to slightly cool), respectively (Figure 8a). TSV for both buildings is out of the suggested neutral range [66], indicating an uncomfortable cool thermal sensation. This indicates the overcooled indoor environment. Cheung, Schiavon [47] also suggested an overcooled indoor environment as one of the major contributors to occupants' dissatisfaction in green-certified office buildings in Singapore, with a tropical microclimate. As summarised in Table 6, the air temperature of 24.2 and 23.8 °C are the sources for the thermal sensation mean vote. Association of thermal sensation mean vote with the air temperature of the buildings suggests that to have a neutral thermal sensation mean vote, a higher air temperature is required. This complies with Malaysian standard MS 1525 [60], that suggested a minimum of 24 °C for dry bulb temperature of the indoor environment of office buildings. Buonocore et al. [75] suggested an optimised air temperature of 26 °C to have thermal comfort and energy efficiency for air-conditioned faculties in a hot and humid climate. Besides, office air temperature was suggested to be set at 28 °C and with low clothing value for occupants, which resulted in energy-saving and thermal satisfaction [76]. Aghniaey et al. [77] suggested that office air temperature should be increased to 26 °C to have less than 10% PPD. In compliance with the literature, this study also found that air temperature around 24 °C is one of the major sources of thermal dissatisfaction.

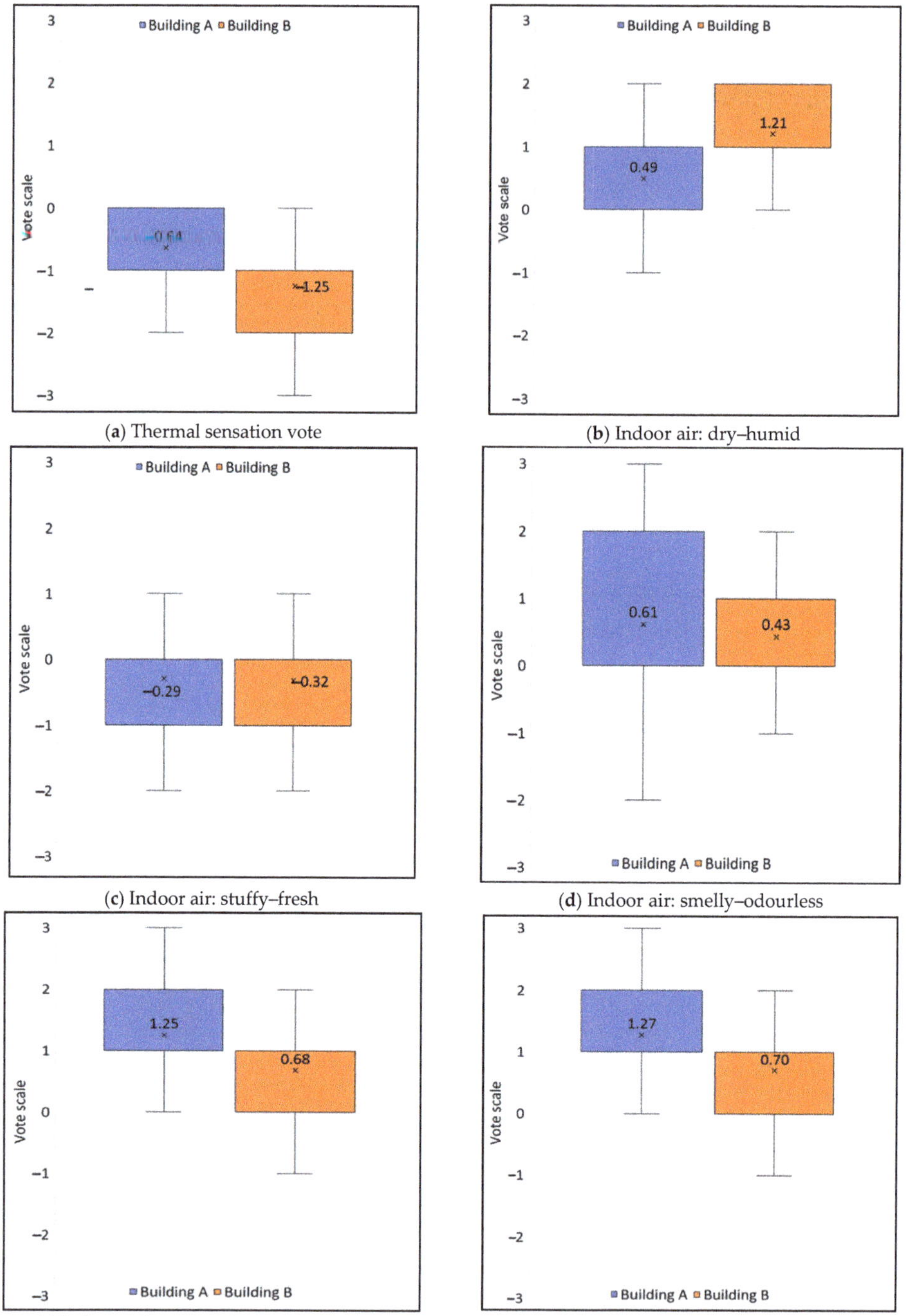

(**a**) Thermal sensation vote (**b**) Indoor air: dry–humid (**c**) Indoor air: stuffy–fresh (**d**) Indoor air: smelly–odourless (**e**) Indoor noise unsatisfactory–satisfactory (**f**) Indoor lighting unsatisfactory–satisfactory

Figure 8. Distribution of occupants' votes.

The humidity levels of 60% (building A) and 72.4% RH (building B) (Table 6) have resulted in the occupants' mean votes of 0.49 (equivalent to neutral) and 1.21 (equivalent to slightly humid) for buildings A and B, respectively (Figure 8b). Malaysian standard MS 1525 [60] suggests a maximum RH lower than 70%. However, according to the occupants' mean vote, RH around 60% also seems to be at the edge of occupants' tolerance, in which lower RH for higher satisfaction should be considered. The circumstance of building B (72.4% RH) led temperature and humidity to be the main concerns and sources of dissatisfaction. Aziz, Sumiyoshi [61] suggest the usage of a Dual Air Handling Unit system to control indoor humidity around 50% and achieve energy efficiency goals. In alignment with this study, Ravindu et al. [78] found that high humidity is the main IEQ dissatisfaction source for a LEED Platinum-certified factory in a tropical microclimate. It should also be considered that thermal parameters such as air temperature and relative humidity have a relationship with each other, in which higher air temperature results in lower relative humidity and vice versa [57]. This is because warm air can contain more water vapour than cool air [57]. The low difference of air temperature (around 0.4 °C) for buildings A and B is not a significant factor and contributes to around a 12% difference in RH level.

Table 6. Summary of results from measurement and survey of IEQ parameters.

<table>
<tr><th colspan="2" rowspan="2">IEQ Parameter</th><th colspan="2">Occupants' Vote Mean Score</th><th colspan="2">Measurement Results</th><th colspan="2">First Three Main Sources of Dissatisfaction</th></tr>
<tr><th>BA</th><th>BB</th><th>BA</th><th>BB</th><th>BA</th><th>BB</th></tr>
<tr><td rowspan="3">Thermal comfort quality</td><td>Temperature</td><td>−0.64</td><td>−1.24</td><td>24.2 °C</td><td>23.8 °C</td><td>☑</td><td>☑</td></tr>
<tr><td>Air: dry–humidity</td><td>0.49</td><td>1.20</td><td>60%</td><td>72.4%</td><td>-</td><td>☑</td></tr>
<tr><td>Air velocity</td><td>-</td><td>-</td><td rowspan="3">0.06 m/s</td><td rowspan="3">0.05 m/s</td><td>-</td><td>-</td></tr>
<tr><td rowspan="2">Indoor air quality</td><td>Air: stuffy–fresh</td><td>−0.29</td><td>−0.32</td><td>☑</td><td></td></tr>
<tr><td>Air: smelly–odourless</td><td>0.61</td><td>0.43</td><td>-</td><td>-</td></tr>
<tr><td colspan="2">Noise quality</td><td>1.25</td><td>0.68</td><td>51.9 dB</td><td>50.2 dB</td><td>-</td><td>-</td></tr>
<tr><td colspan="2">Lighting</td><td>1.27</td><td>0.70</td><td>494 lux</td><td>191 lux</td><td>☑</td><td>-</td></tr>
</table>

Low rate of air velocity (0.06 and 0.05 m/s for buildings A and B respectively, see Table 6) in both buildings resulted in occupants reporting stuffy air with a mean of −0.29 and −0.32 for buildings A and B, respectively (Figure 8c). Particularly, stuffy air was found to be one of the main sources of dissatisfaction for building A. Occupants' mean votes are positive for smelly–odourless air (0.61 and 0.43), indicating a neutral feeling for air freshness. Based on the findings for air quality variables, higher air velocity is required to have fresher air and be more odourless. In line with this study finding, low air velocity is also found to be a major contributor to occupants of the office buildings in Singapore [47]. To improve the air quality and thermal satisfaction, a higher air velocity rate is suggested. Melikov et al. [68] implied that an increase in the air velocity can widen occupants' acceptance level for humidity up to 60% and temperature up to 26 °C, however, an increase in the air velocity did not reduce the SBS symptoms, but the replacement of a clean and fresh air did. This study also concluded that improved air velocity is more critical for hot and humid conditions [68].

Regarding acoustic quality, both buildings have a similar result to a typical office building within a range of 45 to 70 dB. The SPL for building A was determined to be around 51.9 dB (Table 6), which resulted in occupants' mean vote of 1.25 (equivalent to slightly satisfactory) (Figure 8e). SPL of 50.2 dB for building B comes with occupants' mean vote of 0.7 (equivalent to slightly satisfactory). Under the noise circumstances of both buildings, lower SPL is suggested for higher satisfaction. Particularly, the reduction of noises from colleagues is critical, as it was one of the main dissatisfaction sources in this study, as well as findings from Banbury et al. [79]. Menadue et al. [80] indicated that noise satisfaction is higher in the non-Green Star office buildings compared to Green

Star-certified office buildings. Also, Lee [81] indicated that acoustic quality in open-plan offices with high cubicle stations is lower than private rooms or shared rooms in LEED-certified office buildings. Passero et al. [82] suggested the use of divider panels between the cubicles/workstations and high sound-absorber materials for the roof finishing to improve the acoustic quality of the open-plan offices. Virjonen et al. [83] also indicated that SPL could be as low as 39 dB in open-plan office buildings with a robust acoustic design plan, whereas Hongisto et al. [84] implied that typical open-plan offices show an SPL range between 50 and 60 dB.

Regarding the lighting, the desk surface illuminance level of 493.84 lux (Table 6) in building A comes with occupants' mean vote of 1.27 (equivalent to slightly satisfactory). Building B presented with 190.88 lux as a desk surface mean illuminance level, which resulted in an occupants' mean vote of 0.70 (equivalent to slightly satisfactory) (Figure 8f). Cheung, Schiavon [47] reported that electrical and natural light quality is among the IEQ factors with high satisfaction in commercial buildings in Singapore. Although the mean surface illuminance level is low in building B, task lighting has improved overall light satisfaction. This is inconsistent with Lim et al. [85] findings indicating that task lightning and daylight combined can improve visual comfort for paper works as well as saving energy. However, to have higher light satisfaction, a range of 500 to 1000 lux is suggested for the indoor visual comfort, for which both buildings failed to meet the minimum range, particularly building B. Also, ceilings with indirect lens types are suggested to increase visual comfort [86]. The same study also recommends a surface illuminance level of 406 to have the highest satisfaction in office buildings. It should be mentioned that due to the green constructions, both buildings have tried to maximise daylight application. For building A, it was observed that parts of the building in the middle distance from window to atrium seemed dim. For building B, it was observed that windows were mostly covered with blinds by occupants sitting near the window to avoid glare from the sun and sky, and this action blocked the daylight from penetrating to spaces with distance from the window. Also, glare is reported to be the main source of dissatisfaction for both buildings. Similar to this, Kwong [87] reported dissatisfaction from sky glare as well as usage of blinds to control the daylight near the windows. Discomfort due to glare is found to be a typical source of dissatisfaction for green buildings, as was supported by Hirning et al. [29]. To control the discomfort from glare, Luo et al. [88] suggested the application of automated shading control for the blinds.

Relationship between Overall IEQ Comfort and IEQ Variables

The relationship of full-scale measurement with a survey is discussed, and results indicated temperature and relative humidity to be among the primary sources of dissatisfaction (Figure 6). To statistically investigate the relationship between overall comfort and IEQ variables, thus, Pearson's correlation coefficient was employed. As shown in Table 7, all the variables were correlated with the overall comfort at a statistically significant level ($p < 0.001$). The lowest Pearson correlation coefficient value is the noise quality, whereas the thermal comfort shows the highest value. The importance of the thermal comfort impact on overall comfort was also reported by Liang et al. [14], who investigated IEQ of green office buildings in the tropical climate of Taiwan.

Table 7. Pearson correlation between overall IEQ comfort and IEQ variables.

Overall Comfort	Thermal Comfort	Indoor Air Quality	Noise	Light	Furniture	Cleaning
Pearson correlation coefficient	0.625 **	0.467 **	0.329 **	0.455 **	0.552 **	0.486 **
p-value	0.000	0.000	0.000	0.000	0.000	0.000

** Correlation is significant at the 0.01 level (2-tailed).

6. Conclusions

The green building movement has altered the construction industry to promote sustainable construction, requiring unique methods of building design. This movement led to the establishment of organisations around the world to systematically evaluate and certify potential buildings as green. However, energy-saving has become the main focus of majority research on green building, with minority studies focusing on occupants' satisfaction with the quality of the indoor environment. Thus, this study has focused on the limited research from IEQ of the GBI Platinum-certified office buildings, to investigate IEQ performance and occupants' satisfaction. In this regard, the IEQ of the selected case studies was evaluated by the utilisation of post-occupancy evaluation as well as full-scale measurement. The primary findings of this study include:

- The onsite monitoring of indoor environment parameters indicates general compliance of parameters with international standards as well as Malaysian standards. However, RH and illuminance level, particularly for building B, were found to be out of the suggested range of the standards.
- The results of the questionnaire survey indicated a general comfort (a mean score of 1.66 on a 7-point scale, −3 to +3) from overall IEQ for building A and a neutral feeling for occupants of building B (mean score of 0.58).
- Regarding the parameters of IEQ, the majority of occupants reported a comfortable or neutral feeling for overall thermal comfort, overall IAQ, overall acoustic, overall lighting, furniture comfort, and cleanliness.
- Most occupants reported being more productive and healthier in the buildings.
- Sources of dissatisfaction were identified to be cool air temperature and high relative humidity, especially with an extreme condition for building B. Besides, low air velocity resulted in stuffy or smelly air, which was reported to be another main source of dissatisfaction. Glare and noise from colleagues were also found to be the main sources of dissatisfaction.
- Finally, although both buildings have been awarded Platinum certificates, the results of the *t*-test show a statistically significant difference between the mean vote of occupants' for IEQ variables, with higher satisfaction from building A.

Findings from this study provide a holistic view of the IEQ performance of the GBI Platinum-certified office buildings in a tropical climate. Limited office buildings obtained GBI Platinum certificates due to the robust and strict GBI criteria. Thus, the findings of this study can be generalised to the GBI-certified office buildings, particularly results related to the IEQ dissatisfaction sources. This is because NRNC buildings should follow the criteria posed by GBI for obtaining the certificate. Given that the cases of this study have been evaluated with the highest points and certified as Platinum, the IEQ of the future constructions looking to obtain a GBI certificate or already certified office buildings should have the same dissatisfaction sources as this study highlighted. Thus, the findings of this study can be applied to further improve national and international guidelines in green construction, particularly for the tropical climate of south-east Asia. This would help to move to a more sustainable and satisfactory indoor environment. This could be achieved by further studies on how to dimmish sources of occupants' dissatisfaction. Also, for additional investigation, a comparison study between IEQ of GBI office buildings with other green rating tools such as LEED or Green mark should be performed. Besides, the findings of this study indicated that more investigations on thermal quality are needed to design a thermally comfortable indoor environment.

Author Contributions: Conceptualisation, M.E.; methodology, M.E., S.M.Z. and M.A.I.; software, M.E., A.A. and I.A.; formal analysis, S.M.Z., M.R.H., I.A., S.M. and S.V.; investigation, M.E., S.M.Z. and M.A.I.; resources, S.M.Z., M.R.H. and S.M.; data curation, M.E., I.A. and A.A. writing—original draft preparation, M.E., S.M.Z. and M.A.I.; writing—review and editing, M.R.H., I.A., S.M., S.V. and A.A.; supervision, S.M.Z. and M.A.I. funding acquisition, S.M.Z. and M.R.H. All authors have read and agreed to the published version of the manuscript.

Funding: This research was funded by the University Malaya Research Fund Assistance (BKP), grant number BK061-2015.

Institutional Review Board Statement: Not applicable.

Informed Consent Statement: Not applicable.

Data Availability Statement: The data presented in this study are available on request from the corresponding authors. The data are not publicly available due to confidentiality.

Conflicts of Interest: The authors declare no conflict of interest. The funders had no role in the design of the study; in the collection, analyses, or interpretation of data; in the writing of the manuscript, or in the decision to publish the results.

List of Abbreviation

Nomenclature

a	Indoor Air (Ambient)
CO_2	Carbon Dioxide Concentration (PPM)
T	Temperature (°C)
V	Air Velocity (m/s)
Acronyms	
ACEM	Association of Consulting Engineers of Malaysia
AHU	Air Handling Unit
ASHRAE	American Society of Heating, Refrigerating, and Air-Conditioning Engineers
BREEAM	Building Research Establishment Environmental Assessment Method
BUS	Building Use Studies
EE	Energy Efficiency
GBI	Green Building Index
HVAC	Heating, Ventilation, and Air Conditioning
IAQ	Indoor Air Quality
IEB	Industrial Existing Building
IEQ	Indoor Environment Quality
INC	Industrial New Construction
LEED	Leadership in Energy and Environmental Design
MGBC	Malaysia Green Building Confederation
MR	Materials and Resources
NRNC	Non-Residential New Construction
PAM	Persatuan Arkitek Malaysia
POE	Post-Occupancy Evaluation
PMV	Predicted Mean Vote
PPD	Predicted Percentage of Dissatisfied
RNC	Residential New Construction
SBS	Sick Building Syndrome
SM	Sustainable Site Planning and Management
SPL	Sound Pressure Level
TSV	Thermal Sensation vote
UFAD	Underfloor Air Distribution
VOCs	Volatile Organic Compounds
WE	Water Efficiency
WELL	WELL Building Standard
WWR	Window to Wall Ratio

References

1. American Society of Heating, Refrigerating and Air-Conditioning Engineers (ASHRAE). *Handbook Fundamentals*; American Society of Heating, Refrigerating and Air-Conditioning Engineers: Atlanta, GA, USA, 2013.
2. Fisk, W.J. How IEQ affects health, productivity. *ASHRAE J.* **2002**, *44*, 56.
3. Gawande, S.; Tiwari, R.R.; Narayanan, P.; Bhadri, A. Indoor air quality and sick building syndrome: Are green buildings better than conventional buildings? *Indian J. Occup. Environ. Med.* **2020**, *24*, 30.

4. Crook, B.; Burton, N.C. Indoor moulds, sick building syndrome and building related illness. *Fungal Biol. Rev.* **2010**, *24*, 106–113. [CrossRef]
5. Nicol, J.F.; Humphreys, M.A. Adaptive thermal comfort and sustainable thermal standards for buildings. *Energy Build.* **2002**, *34*, 563–572. [CrossRef]
6. Krüger, E.L.; Zannin, P.H. Acoustic, thermal and luminous comfort in classrooms. *Build. Environ.* **2004**, *39*, 1055–1063. [CrossRef]
7. Veitch, J.A.; Newsham, G.R. Lighting quality and energy-efficiency effects on task performance, mood, health, satisfaction, and comfort. *J. Illum. Eng. Soc.* **1998**, *27*, 107–129. [CrossRef]
8. Pellerin, N.; Candas, V. Effects of steady-state noise and temperature conditions on environmental perception and acceptability. *Indoor Air* **2004**, *14*, 129–136. [CrossRef]
9. Sadick, A.-M.; Kpamma, Z.E.; Agyefi-Mensah, S. Impact of indoor environmental quality on job satisfaction and self-reported productivity of university employees in a tropical African climate. *Build. Environ.* **2020**, *181*. [CrossRef]
10. Geng, Y.; Ji, W.; Lin, B.; Zhu, Y. The impact of thermal environment on occupant IEQ perception and productivity. *Build. Environ.* **2017**, *121*, 158–167. [CrossRef]
11. Lipczynska, A.; Schiavon, S.; Graham, L.T. Thermal comfort and self-reported productivity in an office with ceiling fans in the tropics. *Build. Environ.* **2018**, *135*, 202–212. [CrossRef]
12. Nematchoua, M.K.; Ricciardi, P.; Orosa, J.A.; Asadi, S.; Choudhary, R. Influence of indoor environmental quality on the self-estimated performance of office workers in the tropical wet and hot climate of Cameroon. *J. Build. Eng.* **2019**, *21*, 141–148. [CrossRef]
13. Abbaszadeh, S.; Zagreus, L.; Lehrer, D.; Huizenga, C. Occupant satisfaction with indoor environmental quality in green buildings. In Proceedings of the 8th International Conference and Exhibition on Healthy Buildings, Lisbon, Portugal, 4–8 June 2006; Volume III, pp. 365–370.
14. Liang, H.-H.; Chen, C.-P.; Hwang, R.-L.; Shih, W.-M.; Lo, S.-C.; Liao, H.-Y. Satisfaction of occupants toward indoor environment quality of certified green office buildings in Taiwan. *Build. Environ.* **2014**, *72*, 232–242. [CrossRef]
15. Lee, J.; Wargocki, P.; Chan, Y.; Chen, L.; Tham, K. Indoor environmental quality, occupant satisfaction, and acute building-related health symptoms in Green Mark-certified compared with non-certified office buildings. *Indoor Air* **2019**, *29*, 112–129. [CrossRef]
16. Paul, W.L.; Taylor, P.A. A comparison of occupant comfort and satisfaction between a green building and a conventional building. *Build. Environ.* **2008**, *43*, 1858–1870. [CrossRef]
17. MacNaughton, P.; Spengler, J.; Vallarino, J.; Santanam, S.; Satish, U.; Allen, J. Environmental perceptions and health before and after relocation to a green building. *Build. Environ.* **2016**, *104*, 138–144. [CrossRef]
18. Kats, G.; Capital, E. *The Costs and Financial Benefits of Green Buildings*; A Report to California's Sustainable Building Task Force; Massachusetts Technology Collaborative: Sacramento, CA, USA, 2003; p. 134.
19. Newsham, G.; Brand, J.; Donnelly, C.; Veitch, J.; Aries, M.; Charles, K. Linking indoor environment conditions to job satisfaction: A field study. *Build. Res. Inf.* **2009**, *37*, 129–147. [CrossRef]
20. Issa, M.; Rankin, J.; Attalla, M.; Christian, A. Absenteeism, performance and occupant satisfaction with the indoor environment of green Toronto schools. *Indoor Built Environ.* **2011**, *20*, 511–523. [CrossRef]
21. United Technologies Corporation (UTC) Climate, Controls & Security. *New Research Quantifies Value of Increased Productivity from Improved Indoor Air Quality*; PR Newswire: New York, NY, USA, 2015.
22. Olson, C. Wellness, wellbeing and productivity in the office. *Buildings* **2015**, *109*, 24–26.
23. Hoyt, T.; Lee, K.H.; Zhang, H.; Arens, E.; Webster, T. *Energy Savings from Extended Air Temperature Setpoints and Reductions in Room Air Mixing*; Center for the Built Environment: Berkeley, CA, USA, 2005.
24. Energy Efficiency. *Buildings Energy Data Book*; Department of Energy: Washington, DC, USA, 2009.
25. Koponen, I.K.; Asmi, A.; Keronen, P.; Puhto, K.; Kulmala, M. Indoor air measurement campaign in Helsinki, Finland 1999—The effect of outdoor air pollution on indoor air. *Atmos. Environ.* **2001**, *35*, 1465–1477. [CrossRef]
26. Damiati, S.A.; Zaki, S.A.; Rijal, H.B.; Wonorahardjo, S. Field study on adaptive thermal comfort in office buildings in Malaysia, Indonesia, Singapore, and Japan during hot and humid season. *Build. Environ.* **2016**, *109*, 208–223. [CrossRef]
27. Pellerin, N.; Candas, V. Combined effects of temperature and noise on human discomfort. *Physiol. Behav.* **2003**, *78*, 99–106. [CrossRef]
28. Navada, S.G.; Adiga, C.S.; Kini, S.G. A study on daylight integration with thermal comfort for energy conservation in a general office. *Int. J. Electr. Energy* **2013**, *1*, 18–22. [CrossRef]
29. Hirning, M.; Isoardi, G.; Cowling, I. Discomfort glare in open plan green buildings. *Energy Build.* **2014**, *70*, 427–440. [CrossRef]
30. Hwang, T.; Kim, J.T. Effects of indoor lighting on occupants' visual comfort and eye health in a green building. *Indoor Built Environ.* **2010**, *20*, 75–90. [CrossRef]
31. Li, P.; Froese, T.M.; Brager, G. Post-occupancy evaluation: State-of-the-art analysis and state-of-the-practice review. *Build. Environ.* **2018**, *133*, 187–202. [CrossRef]
32. Khalil, N.; Husin, H.N. Post occupancy evaluation towards indoor environment improvement in Malaysia's office buildings. *J. Sustain. Dev.* **2009**, *2*, 186. [CrossRef]
33. Ismail, A.S.; Mohidin, H.H.B.; Daud, M.M. A review on occupants' satisfaction and wellness level in low-cost housing in Malaysia. *Plan. Malays.* **2017**, *15*. [CrossRef]

34. Kamaruzzaman, S.N.; Egbu, C.O.; Mahyuddin, N.; Zawawi, E.M.A.; Chua, S.J.L.; Azmi, N.F. The impact of IEQ on occupants' satisfaction in Malaysian buildings. *Indoor Built Environ.* **2018**, *27*, 715–725. [CrossRef]
35. Mun, T.L. *The Development of GBI Malaysia (GBI)*; Pertubuhan Akitek Malaysia: Kuala Lumpur, Malaysia; Association of Consulting Engineers Malaysia: Kuala Lumpur, Malaysia, 2009.
36. MGBC. Malaysia Green Building Confederation. 2015. Available online: http://www.mgbc.org.my/about-mgbc/ (accessed on 3 March 2019).
37. Illankoon, I.C.S.; Tam, V.W.; Le, K.N.; Shen, L. Key credit criteria among international green building rating tools. *J. Clean. Prod.* **2017**, *164*, 209–220. [CrossRef]
38. Zian, O.B.; Fam, S.-F.; Liang, C.Z.; Wahjono, S.I.; Ying, T.Y. A Critical Research of Green Building Assessment Systems in Malaysia Context. *Int. J. Innov. Technol. Explor. Eng.* **2019**, *8*. [CrossRef]
39. Hadjri, K.; Crozier, C. Post-occupancy evaluation: Purpose, benefits and barriers. *Facilities* **2009**, *27*, 21–33. [CrossRef]
40. Ilesanmi, O.A. Post-occupancy evaluation and residents' satisfaction with public housing in Lagos, Nigeria. *J. Build. Apprais.* **2010**, *6*, 153–169. [CrossRef]
41. Seppänen, O.A.; Fisk, W. Some quantitative relations between indoor environmental quality and work performance or health. *HVAC R Res.* **2006**, *12*, 957–973. [CrossRef]
42. Preiser, W.F.E.; Vischer, J.C. *Assessing Building Performance*; Routledge: Milton Park, UK, 2006.
43. Altomonte, S.; Schiavon, S. Occupant satisfaction in LEED and non-LEED certified buildings. *Build. Environ.* **2013**, *68*, 66–76. [CrossRef]
44. Armitage, L.; Murugan, A.; Kato, H. Green offices in Australia: A user perception survey. *J. Corp. Real Estate* **2011**, *13*, 169–180. [CrossRef]
45. Birt, B.; Newsham, G.R. Post-occupancy evaluation of energy and indoor environment quality in green buildings: A review. In Proceedings of the 3rd CIB International Conference on Smart and Sustainable Built Environment, Delft, The Netherlands, 15–19 June 2009.
46. Preiser, W.; Rabinowitz, H.; White, E. *Post Occupancy Evaluation*; Van Nostrand Reinhold: New York, NY, USA, 1988.
47. Cheung, T.; Schiavon, S.; Graham, L.T.; Tham, K.W. Occupant satisfaction with the indoor environment in seven commercial buildings in Singapore. *Build. Environ.* **2021**, *188*. [CrossRef]
48. GBI. Green Building Index. 2021. Available online: http://www.greenbuildingindex.org/ (accessed on 4 April 2021).
49. Wong, L.; Mui, K.; Cheung, C.T.; Lee, M.; Lee, E. Performance impact of indoor environmental policy implementation for airside systems in Hong Kong Grade A offices. *Build. Serv. Eng. Res. Technol.* **2015**, *36*, 525–534.
50. Candanedo, L.M.; Feldheim, V. Accurate occupancy detection of an office room from light, temperature, humidity and CO2 measurements using statistical learning models. *Energy Build.* **2016**, *112*, 28–39.
51. Leaman, A. *The Building Use Studies (BUS) Occupant Survey: Origins and Approach Q&A*; Building Use Studies: London, UK, 2011.
52. BUS. BUS Methodology. 2021. Available online: https://busmethodology.org.uk/ (accessed on 4 April 2021).
53. Barbadilla-Martín, E.; Martín, J.G.; Lissén, J.M.S.; Ramos, J.S.; Domínguez, S.Á. Assessment of thermal comfort and energy savings in a field study on adaptive comfort with application for mixed mode offices. *Energy Build.* **2018**, *167*, 281–289.
54. Kántor, N.; Unger, J. The most problematic variable in the course of human-biometeorological comfort assessment—The mean radiant temperature. *Cent. Eur. J. Geosci.* **2011**, *3*, 90–100. [CrossRef]
55. Langner, M.; Scherber, K.; Endlicher, W.R. Indoor heat stress: An assessment of human bioclimate using the UTCI in different buildings in Berlin. *DIE ERDE J. Geogr. Soc. Berl.* **2013**, *144*, 260–273.
56. Matzarakis, A.; Amelung, B. Physiological equivalent temperature as indicator for impacts of climate change on thermal comfort of humans. In *Seasonal Forecasts, Climatic Change and Human Health*; Springer: Berlin/Heidelberg, Germany, 2008; pp. 161–172.
57. Tang, C.; Chin, N. *Building Energy Efficiency Technical Guideline for Passive Design*, 2nd ed.; Public Works Department: Kuala Lumpur, Malaysia, 2017.
58. International Organization for Standardization. *ISO 16000-1: Indoor Air–Part 1: General Aspects of Sampling Strategy*; International Organization for Standardization: Geneva, Switzerland, 2007.
59. Cochran, W.G. *Sampling Techniques*; Wiley: New York, NY, USA, 1977.
60. The Department of Standards Malaysia. *Energy Efficiency and Use of Renewable Energy for Non-Residential Buildings—Code of Practice*; MS 1525; The Department of Standards Malaysia: Cyberjaya, Malaysia, 2014.
61. Aziz, A.A.; Sumiyoshi, D.; Akashi, Y. Low cost humidity controlled air-conditioning system for building energy savings in tropical climate. *J. Build. Eng.* **2017**, *11*, 9–16. [CrossRef]
62. Mui, K.W.; Wong, L.T. Acceptable illumination levels for office occupants. *Archit. Sci. Rev.* **2006**, *49*, 116–119. [CrossRef]
63. Department of Public Works. *BSL, Design Standards and Guidelines*; Department of Public Works: Los Angeles, CA, USA, 2007.
64. Mui, K.W.; Wong, L.T. A method of assessing the acceptability of noise levels in air-conditioned offices. *Build. Serv. Eng. Res. Technol.* **2006**, *27*, 249–254. [CrossRef]
65. Ha, M.M. *Indoor Air Quality: Office Health, Safety and Well-Being*; University of Calgary (Canada): Calgary, AB, Canada, 1998; p. 128.
66. ASHRAE. *Standard 55, Thermal Environmental Conditions for Human Occupancy*; ASHRAE Inc.: Atlanta, GA, USA, 2017.
67. Fang, L. Impact of temperature and humidity on perception of indoor air quality during immediate and longer whole-body exposures. *Indoor Air* **1998**, *8*, 276–284. [CrossRef]

68. Melikov, A.K.; Kaczmarczyk, J. Air movement and perceived air quality. *Build. Environ.* **2012**, *47*, 400–409. [CrossRef]
69. Leaman, A.; Bordass, B. Productivity in buildings: The 'killer' variables. *Build. Res. Inf.* **1999**, *27*, 4–19. [CrossRef]
70. Bergs, J. Effect of healthy workplaces on well-being and productivity of office workers. In Proceedings of the International Plants for People Symposium, Amsterdam, The Netherlands, 3–6 June 2002.
71. Syahrul, N.K.; Ainur, M.A. Evaluation of occupants' well-being and perception towards indoor environmental quality in Malaysia affordable housing. *J. Facil. Manag.* **2019**, *17*, 90–106.
72. Colenberg, S.; Jylhä, T.; Arkesteijn, M. The relationship between interior office space and employee health and well-being—A literature review. *Build. Res. Inf.* **2020**, 1–15. [CrossRef]
73. Pandis, N. Comparison of 2 means (independent z test or independent t test). *Am. J. Orthod. Dentofacial. Orthop.* **2015**, *148*, 350–351. [CrossRef]
74. Geng, Y.; Ji, W.; Wang, Z.; Lin, B.; Zhu, Y. A review of operating performance in green buildings: Energy use, indoor environmental quality and occupant satisfaction. *Energy Build.* **2019**, *183*, 500–514. [CrossRef]
75. Buonocore, C.; de Vecchi, R.; Scalco, V.; Lamberts, R. Thermal preference and comfort assessment in air-conditioned and naturally-ventilated university classrooms under hot and humid conditions in Brazil. *Energy Build.* **2020**, *211*. [CrossRef]
76. Tan, C.K.; Ogawa, A.; Matsumura, T. *Innovative Climate Change Communication: Team Minus 6%*; Global Environment Information Centre (GEIC), United Nations University (UNU): Tokyo, Japan, 2008; pp. 53–70.
77. Aghniaey, S.; Lawrence, T.M. The impact of increased cooling setpoint temperature during demand response events on occupant thermal comfort in commercial buildings: A review. *Energy Build.* **2018**, *173*, 19–27. [CrossRef]
78. Ravindu, S.; Rameezdeen, R.; Zuo, J.; Zhou, Z.; Chandratilake, R. Indoor environment quality of green buildings: Case study of an LEED platinum certified factory in a warm humid tropical climate. *Build. Environ.* **2015**, *84*, 105–113. [CrossRef]
79. Banbury, S.; Berry, D. Office noise and employee concentration: Identifying causes of disruption and potential improvements. *Ergonomics* **2005**, *48*, 25–37. [CrossRef]
80. Menadue, V.; Soebarto, V.; Williamson, T. The effect of internal environmental quality on occupant satisfaction in commercial office buildings. *HVAC R Res.* **2013**, *19*, 1051–1062. [CrossRef]
81. Lee, Y.S. Office layout affecting privacy, interaction, and acoustic quality in LEED-certified buildings. *Build. Environ.* **2010**, *45*, 1594–1600. [CrossRef]
82. Passero, C.R.; Zannin, P.H. Acoustic evaluation and adjustment of an open-plan office through architectural design and noise control. *Appl. Ergon.* **2012**, *43*, 1066–1071. [CrossRef]
83. Virjonen, P.; Keränen, J.; Helenius, R.; Hakala, J.; Hongisto, O.V. Speech privacy between neighboring workstations in an open office-a laboratory study. *Acta Acust. United Acust.* **2007**, *93*, 771–782.
84. Hongisto, V.; Varjo, J.; Leppämäki, H.; Oliva, D.; Hyönä, J. Work performance in private office rooms: The effects of sound insulation and sound masking. *Build. Environ.* **2016**, *104*, 263–274. [CrossRef]
85. Lim, G.-H.; Keumala, N.; Ghafar, N.A. Energy saving potential and visual comfort of task light usage for offices in Malaysia. *Energy Build.* **2017**, *147*, 166–175. [CrossRef]
86. Park, J.; Loftness, V.; Aziz, A.; Wang, T.-H. Strategies to achieve optimum visual quality for maximum occupant satisfaction: Field study findings in office buildings. *Build. Environ.* **2020**, *195*. [CrossRef]
87. Kwong, Q.J. Light level, visual comfort and lighting energy savings potential in a green-certified high-rise building. *J. Build. Eng.* **2020**, *29*. [CrossRef]
88. Luo, Z.; Sun, C.; Dong, Q.; Yu, J. An innovative shading controller for blinds in an open-plan office using machine learning. *Build. Environ.* **2021**, *189*. [CrossRef]

energies

MDPI

Article

Acquiring the Foremost Window Allocation Strategy to Achieve the Best Trade-Off among Energy, Environmental, and Comfort Criteria in a Building

Seyedeh Farzaneh Mousavi Motlagh [1], Ali Sohani [2], Mohammad Djavad Saghafi [1], Hoseyn Sayyaadi [2] and Benedetto Nastasi [3,*]

1 Department of Architecture, Architecture School, College of Fine Arts, University of Tehran, Tehran 1415 564583, Iran; farzanehmousavim@gmail.com (S.F.M.M.); djsaghafifr@yahoo.fr (M.D.S.)
2 Lab of Optimization of Thermal Systems' Installations, Faculty of Mechanical Engineering Energy Division, K.N. Toosi University of Technology, Tehran 1999 143344, Iran; alisohany@yahoo.com (A.S.); sayyaadi@kntu.ac.ir (H.S.)
3 Department of Planning, Design and Technology of Architecture, Sapienza University of Rome, Via Flaminia 72, 00196 Rome, Italy
* Correspondence: benedetto.nastasi@outlook.com

Abstract: The purpose of this investigation is to propose a way for acquiring the foremost window allocation scheme to have the best trade-off among energy, environmental, and comfort criteria in a building. An advanced decision-making tool, named the technique for order preference by similarity to ideal solution (TOPSIS), is utilized to find the best building amongst different alternatives for having windows on the building façades. Three conditions, namely two parallel, two perpendicular, and three façades, considered as A, B, and C types, respectively, are investigated. For each type, four possible orientations are studied. Heating, cooling, and lighting energy demands in addition to carbon dioxide equivalent emission and thermal and visual comfort are taken into account as the investigated criteria, and they are all evaluated in a simulation environment. The results show that for the modular residential buildings chosen as the case study and located in Tehran, Iran, having windows on the north and east façades is the best scheme. This alternative, which belongs to the B type, has about 40% and 37% lower heating and cooling energy demands than the C type's foremost alternative. It is also able to provide about 10% better CO_2 equivalent emission and 28% higher thermal comfort.

Keywords: building performance simulation; CO_2 emission; energy saving; occupant's comfort; window allocation

Citation: Mousavi Motlagh, S.F.; Sohani, A.; Djavad Saghafi, M.; Sayyaadi, H.; Nastasi, B. Acquiring the Foremost Window Allocation Strategy to Achieve the Best Trade-Off among Energy, Environmental, and Comfort Criteria in a Building. *Energies* **2021**, *14*, 3962. https://doi.org/10.3390/en14133962

Academic Editor: Alberto-Jesus Perea-Moreno

Received: 10 May 2021
Accepted: 24 June 2021
Published: 1 July 2021

1. Introduction

As the concern about energy and environmental crises increases, presenting solutions and methods to cope with such issues becomes more and more important to researchers from different fields. Since the building sector is recognized as having a huge contribution to these crises, researchers in this field have also come up with some solutions to achieve energy and environmental improvements, and they are trying hard to find more effective methods as their crucial mission. Table 1 presents a list of the recent investigations in the field as the literature in which window-related parameters in a building were evaluated, considering the topic of this work. As the mentioned items in that table show, the previously conducted studies can be investigated from different viewpoints, including employed software programs and studied building aspects. The literature is reviewed from the mentioned viewpoints in the remainder of this section.

Table 1. A list of the recent works which have been carried out in the field of improving building performance.

Study	Year	Discussed in Section 1.1.		Discussed in Section 1.2.	
		Employed Software Program	**Considered Building Aspects**	**The Stage at Which the Improvements Have Been Applied**	**Considered Window Parameters**
Delgarm et al. [1]	2016	EnergyPlus	Energy consumption Thermal comfort	Early design stage	Window size and glazing material
Manigandan et al. [2]	2016	Computational Fluid Dynamics approach	Natural ventilation	N.A.	Orientation and size of windows
Manigandan et al. [3]	2017	Computational Fluid Dynamics approach	Natural ventilation	N.A.	Orientation and size of windows
Misiopecki et al. [4]	2018	THERM	Energy consumption	N.A.	Window position in the window opening
Azmy and Ashmawy [5]	2018	EnergyPlus	Energy consumption Visual comfort	Early design stage	Window position in the building envelope
Jafari and Valentin [6]	2018	eQuest	N.A.	Retrofitting stage	Replacement of windows
Selen Solmaz et al. [7]	2018	EnergyPlus	Energy consumption Economic benefits	Retrofitting stage	Window properties and configuration of shading system
Xue et al. [8]	2019	EnergyPlus, Radiance	Energy consumption Visual comfort	Early design stage	Window-to wall ratio (WWR) and sunshade configuration
Zhai et al. [9]	2019	EnergyPlus	Energy consumption Thermal comfort Visual comfort	Early design stage	WWR and glazing material
Troup et al. [10]	2019	N.A.	Energy consumption	Early design stage	WWR
Feng et al. [11]	2019	Autodesk® Dynamo and Revit	Environmental impact	Early design stage	Window size
Ashrafian and Moazzen [12]	2019	DIALux Evo, DesignBuilder, and EnergyPlus	Energy consumption Occupants' comfort	Early design stage	WWR and window configuration
Hart et al. [13]	2019	EnergyPlus	Energy consumption	Retrofitting stage	Replacement of windows
Rizal et al. [14]	2020	N.A.	Visual comfort	N.A.	Window position
Kunwar and Bhandari [15]	2020	EnergyPlus, DIVA-for-Rhino	Energy consumption Visual comfort	N.A.	Window shading systems and control strategies
Elghamry and Hassan [16]	2020	DesignBuilder	Energy consumption Thermal comfort Economic benefits Environmental impact	Early design stage	WWR and position on the wall
Kaasalainen et al. [17]	2020	IDA Indoor Climate and Energy (IDA ICE)	Energy consumption	Early design stage	Window area, proportions, horizontal position, external shading, and glazing properties
Al-Saggaf et al. [18]	2020	Revit and Ecotect	Energy consumption	Early design stage	Glazing area and percentage
Ascione et al. [19]	2020	EnergyPlus	Energy consumption Economic benefits	Retrofitting stage	Replacement of windows
Zhao and Du [20]	2020	DesignBuilder	Energy consumption Thermal comfort	Early design stage	Configuration of windows and shading system
The current work	2021	EnergyPlus	Energy consumption Environmental impact Thermal comfort Visual comfort	Selection stage	Number and combination of building façades

1.1. Considered Building Aspects and Employed Software Programs in the Literature

In a building, several aspects are involved, and researchers investigate a building from different perspectives. Each of these aspects can be then evaluated by some quantitative indicators that have been considered as the objective functions in the literature [21]. In order to investigate these objectives in a building, depending on the considered objectives, researchers usually use a variety of building simulation software programs, such as EnegyPlus (funded by the U.S. Department of Energy (DOE)), DesignBuilder (Stroud, Gloucestershire, UK), Radiance (developed by Greg Ward, Berkeley, CA, USA) etc.

For instance, Zhao and Du [20] presented an optimum design for an office building using DesignBuilder in which thermal comfort and energy consumption indicators were considered as the objectives. In another study, Delgarm et al. [1] considered the same two building aspects as [20] and used the EnergyPlus building simulation tool to evaluate them. Moreover, in some other works, [8,15], the impact of some design strategies, such as window-to-wall ratio (WWR) and window shading system, on energy demand and daylighting have been analyzed by employing EnergyPlus for energy simulation and Radiance and DAYSIM (developed by the National Research Council Canada and the Fraunhofer Institute for Solar Energy Systems) for daylighting analysis.

Another example that can be given is the study of Zhai et al. [9]. In this investigation, the authors presented a three-objective optimization approach to evaluate the effect of window design on energy demand and thermal and visual comfort by combining an optimization algorithm with EnergyPlus. Elghamry and Hassan [16] also analyzed the impact of window parameters on energy consumption and thermal comfort, in addition to cost and environment. Moreover, some recent works have only considered the impact of some design strategies on energy demand (e.g., [5,10,17]).

Reviewing the studies based on software program reveals that EnergyPlus has been the most popular building simulation tool. In addition, analyzing the investigated building aspects shows that, to the best of the authors' knowledge, in a large part of the research works, only some of the important building performance criteria have been investigated, and others have been neglected. Since there is a trade-off among different performance criteria in a building, not considering this interaction will lead to obtaining unfavorable results from some viewpoints. It can be said that for having a favorable condition, different building aspects should be considered at the same time, and not considering one will lead to an unfavorable condition from other perspectives.

1.2. The Stages at Which the Evaluation Was Done in the Literature

Reviewing the literature demonstrates that the evaluation of the building performance has been done at different stages during a building's lifespan, including the early design stage and retrofitting stage.

In the early design stage, building parameters that are not possible to change later have been analyzed. The results help the architects to choose the appropriate building variables in the design process.

Within this framework, Feng et al. [11] implemented a parametric design method to improve the environmental performance of buildings in the early design phase. Moreover, Al-Saggaf et al. [18] developed a system to analyze the impact of architectural design features on energy consumption. The proposed system was implemented in three different building design alternatives in a hot climate, and their impact on cooling energy demand was assessed.

Window parameters, as a group of early design features that have a great impact on improving building performance, have been widely considered in the previously conducted studies. For instance, Ashrafian and Moazzen [12] studied the impact of WWR and orientation on energy demand and occupants' comfort. Moreover, Misiopecki et al. [4] investigated different window-to-wall connections to find the most energy-efficient position. Azmy and Ashmawy [5] also considered WWR, window position, and orientation as different variables to optimize energy consumption.

Inappropriate conditions in an existing building caused by inefficient design strategies can be changed in the retrofitting stage. Despite the huge amount of cost and effort it takes, it can sometimes be beneficial in low-performance buildings. As a few examples, Asclonea et al. [19], Jafari and Valentin [6], Hart et al. [13], and Selen Solmaz et al. [7] addressed the building retrofitting phase in their works.

Reviewing the investigations according to the stages at which they have been carried out reveals that most of the studies have proposed steps to be taken in the early design stage, and a few of them have suggested solutions for retrofitting plans. However, to the best of the authors' knowledge, a framework for selecting the best building among several alternatives has not been proposed. Presenting such a method in the selection stage can help a customer choose the best building to buy among all the existing ones.

Moreover, in the conducted studies, the analysis has usually been done by considering a constant number for the façades on which windows are installed. This means that the impact of changing the number and combination of building façades on the building performance criteria and objective functions is still missing in the literature.

1.3. The Novelty of the Current Work

Despite being valuable, the literature reveals three gaps, as discussed in the final paragraph of Section 1.1 and two last paragraphs of Section 1.2. As a result, the current study aims to cover the gaps by considering the following items as novelties:

- This paper provides a comparative method to find the best building considering the interaction between the top four most used building aspects, including energy consumption, environmental impact, and thermal and visual comfort. The best building is selected based on a systematic decision-making method called the technique for order preference by similarity to the ideal solution, also known as TOPSIS.
- The framework is presented for the selection stage, in which instead of one, a number of buildings are evaluated, and the best of them is introduced as the winning alternative for a customer.
- A number of buildings with the same characteristics, such as floor area and wall construction, but different orientations, numbers, and combinations of façades have been taken into account as the alternatives. The best one is then found in a comprehensive comparative analysis. In this way, the foremost window allocation strategy for a building is found.

Expressed in question format, this study aims to find answers for the following questions as some gaps and areas of concern in the literature:

- If a customer is going to buy a new building among a set of alternatives with the same architectural plans but different window allocation strategies, which building will provide the most benefits to them? Is having the highest number of façades, which is usually thought to be the best condition, really the optimal condition in terms of different building aspects?
- Is the optimal orientation the same for different buildings with the same architectural plans and located in the same climatic region, but with different window allocation strategies? Or does the optimal orientation vary as the number and combination of façades changes in buildings?

The following structure is chosen for this paper. After this part, i.e., the Introduction, the employed methodology is presented in Section 2. Then, the details of the case study and results are given in Sections 3 and 4, respectively. Finally, the conclusions are proposed in Section 5.

2. Methodology

The methodology employed in this study will be described in this section. Initially, the working principle of the proposed method is given in Section 2.1. Then, the details about the EnergyPlus and decision-making method are given in Sections 2.2 and 2.3

respectively. As an important point, it should be noted that in this study, a commercially developed software is used. In such a condition, and especially for well-known software programs such as those employed in this study, as a widely accepted assumption, it has been considered that the results of the simulation have been validated by the software developers, and for that reason, no further validation is done.

2.1. Working Principle Description

The research presents a comparative method, which is carried out in the selection stage of buildings. The selection stage refers to a phase in which the building alternatives are already designed and ready to be occupied. The method aims to select the winning alternative among all the existing building choices based on their overall performance. In order to observe the interaction between different aspects that define a building's performance, the four most important building aspects, including energy consumption, environmental impact, thermal comfort, and visual comfort, are taken into account in this study. These different aspects are assessed by some quantitative indicators and analyzed with the aid of the EnergyPlus building simulation tool. The final optimal building with the highest performance is selected using the TOPSIS decision-making method.

As shown in Figure 1, the overall framework of the proposed method consists of the following steps:

1. Some building alternatives are selected among the existing choices. Some factors, including the customer's preferences and budget, also play a significant role in choosing the appropriate alternatives.
2. Depending on the differences between the considered alternatives, some decision criteria are taken into account.
3. A set of objective functions that contribute to the four building aspects are specified to conduct the comparison based on them.
4. The EnergyPlus building simulation tool is used to evaluate the performance of each alternative based on the considered objectives.
5. The Decision-making method is developed to prioritize the alternatives according to their similarity to the ideal performance.
6. The final winning alternative is presented.

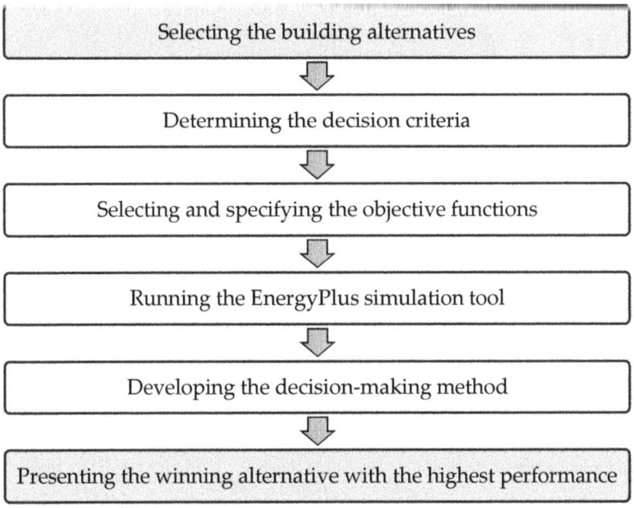

Figure 1. Schematic workflow of the proposed method.

To evaluate the proficiency of the developed method, it is applied to select the final optimal building among 12 residential building alternatives in a small residential town. These buildings have the same architectural plans but different orientations, numbers, and combinations of façades. The mentioned case studies are located in Tehran, which is in the mild climatic region of Iran. It should be underlined that other building functions with some other decision criteria can be also assessed using the presented method in this paper.

2.2. EnergyPlus Simulation Tool

EnergyPlus is a building performance analysis software developed by the U.S. Department of Energy (DOE) [22]. The findings of the review by Mousavi Motlagh et al. [21] indicate that taking advantage of EnergyPlus to simulate building performance has been very popular in the recent studies. Thus, this software is used in this paper to evaluate the performance of building alternatives in terms of different objective functions. Since EnergyPlus is a text-based file format interface, the OpenStudio SketchUp Plug-in is also implemented as a graphical user interface to model the geometry of the buildings.

2.3. Decision-Making Method

Once the values of objective functions are calculated, a decision-making method is performed to find the winning alternative. As the most efficient and popular decision-making method for selecting the final optimal building, TOPSIS is used in this study. This method was first introduced by Hwang and Yoon in 1981 [23]. Based on this approach, the winning alternative is the one that has the shortest distance to the ideal condition and the longest distance to the nonideal condition [24]. In this work, the ideal condition is a situation that minimizes energy consumption and environmental impact while maximizes thermal and visual comfort. In contrast, the nonideal condition is a situation that maximizes energy consumption and environmental impact while minimizing thermal and visual comfort.

Before starting the calculations, due to the different dimensions of objective functions, they should be normalized via Equation (1) [25]:

$$F_{ij} = \frac{Obj_{ij}}{\sqrt{\sum_{i=1}^{Num_i} (Obj_{ij})^2}} \tag{1}$$

where i and j are the number of alternatives and objective functions, respectively. Moreover, F is the normalized objective function, and Obj is the actual value of the objective function.

In the second step, the parameters d^+ and d^- are calculated for each building alternative using the Equations (2) and (3), respectively [25]. These two parameters correspond to the spatial distance of each alternative from the ideal and nonideal conditions, respectively [26].

$$d_i^+ = \sqrt{\sum_{i=1}^{Num_{Obj}} \left(F_{ij} - F_j^{ideal}\right)^2} \tag{2}$$

$$d_i^- = \sqrt{\sum_{i=1}^{Num_{Obj}} \left(F_{ij} - F_j^{non-ideal}\right)^2} \tag{3}$$

Finally, the parameter Cl is defined for each alternative using Equation (4) [25]. This parameter is used to rank all the existing choices. The final optimal building is the one with the highest value of Cl.

$$Cl_i = \frac{d_i^-}{d_i^- + d_i^+} \tag{4}$$

3. Case Study

The information about the considered case study is given here. It contains an introduction of the plans and location of the buildings, considered the decision criteria, and definitions of the objective functions. Sections 3.1–3.4 provide information about each of the mentioned items, respectively.

3.1. Description of the Case Study

In this section, the proposed method is applied to select the optimal building among all the existing buildings in a small residential town, which is already designed and ready to be occupied. The town is located in Tehran, Iran. Figure 2 shows different configurations of the urban blocks in this town. All these blocks consist of some three-story buildings with the same architectural plans, which is also demonstrated in Figure 2. Moreover, as shown in Table 2, the total area of spaces with a controlled thermal condition, known as net conditioned area, is the same in all these buildings. Other characteristics of the residential apartments in this town are also reported in Table 2. It should be underlined that the material properties reported in Table 2 are obtained based on the Iranian National Building Regulations [27]. Packaged terminal heat pump (PTHP) air conditioning systems are provided for all these apartments. The COP of the systems for the cooling and heating operations are 3 and 2.75, respectively [28]. Moreover, the set points of the systems are 22 °C for heating and 26 °C for cooling [27]. It is worth mentioning that this study considers a case study with all the obstructions previous works have taken into account.

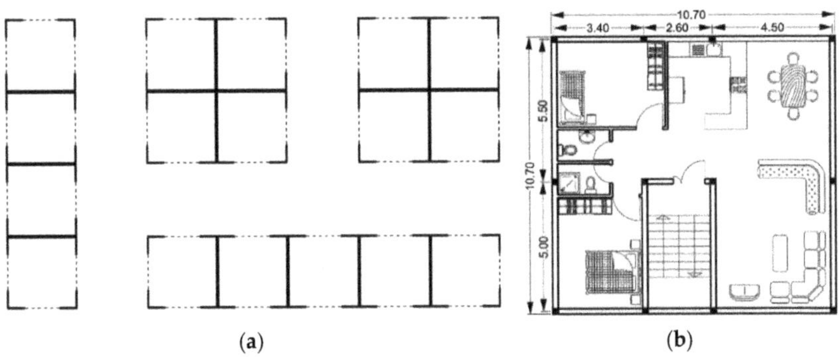

Figure 2. The considered case study: (**a**) different configurations of the urban blocks; (**b**) architectural plan of the buildings.

Table 2. Characteristics of the residential apartments in the investigated town.

Parameter	Value	Unit
Net conditioned building area	86.69	m^2
Building height	2.8	m
WWR	30 for each zone's external walls	%
External walls' U-value	0.7	W/m^2K
Internal walls' U-value	2.58	W/m^2K
Floors/ceilings' U-value	1.45	W/m^2K
Double-glazed windows' U-value	2.67	W/m^2K

3.2. Location of the Considered Case Study

Tehran is the capital of Iran and is located in the mild climatic region of this country. Due to the increasing demand for residential apartments in this densely populated city, making any improvement to the performance of the residential buildings will avoid a significant proportion of the energy and environmental issues occurring as an impact of the population growth. Given this, a small residential town in Tehran is investigated in this study. The climatic properties of this city are introduced in Table 3. Moreover, the temperature range in different months of the year for Tehran is shown in Figure 3. This chart is obtained using the EnergyPlus weather data for Tehran.

Table 3. Climatic properties of Tehran (reproduced with permission from Abbasi et al., Applied Thermal Engineering; published by Elsevier, 2018 [29]).

City	Climatic Type	Dry Bulb Temperature (°C)		Wet Bulb Temperature (°C)	Latitude (°N)	Elevation (m)
		Summer	Winter	Summer		
Tehran	Hot Semidesert	37.8	−4.4	19.4	35.68	1190.0

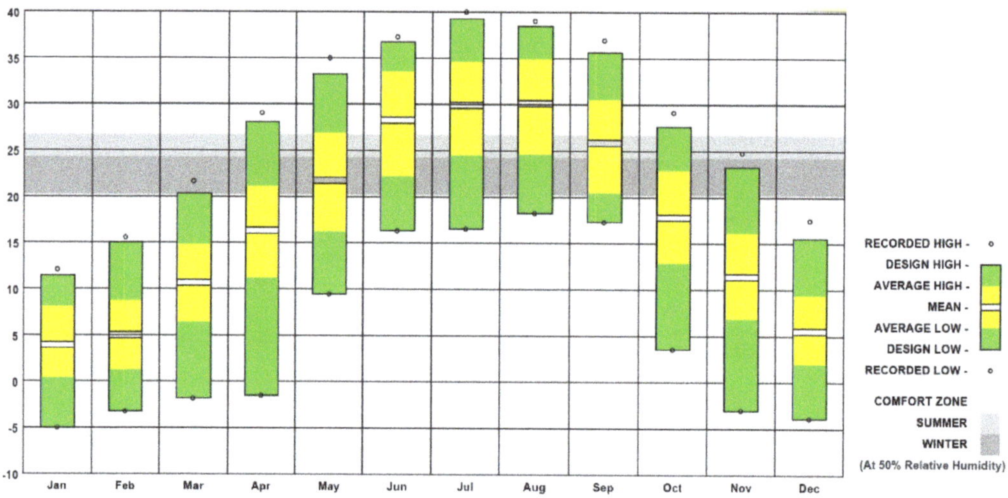

Figure 3. Temperature range in different months of a year for Tehran.

3.3. Decision Criteria

To consider all the possible choices for a customer in the selection phase, 12 different building alternatives located in the second story with the same plans but different orientations, numbers, and combinations of façades are taken into account. The mentioned parameters that distinguish the different alternatives are considered as the decision criteria in this study.

Three variations of number and combination of façades are found in the existing buildings, including two parallel façades, two perpendicular façades, and three façades. The classification of the considered alternatives based on these three variations is shown in Figure 4. Sketchup 3D modeling software (developed by Trimble Inc.) is used to model the buildings. Then, the EnergyPlus building simulation tool is employed to analyze the models in different orientations based on the four objectives explained in Section 3.4.

3.4. Definition of the Considered Objectives

In order to find the winning alternative, the effect of the decision criteria on energy consumption, environmental impact, thermal comfort, and visual performance is taken into account. These four aspects are defined by some indicators considered as the objective functions, which are described in the following.

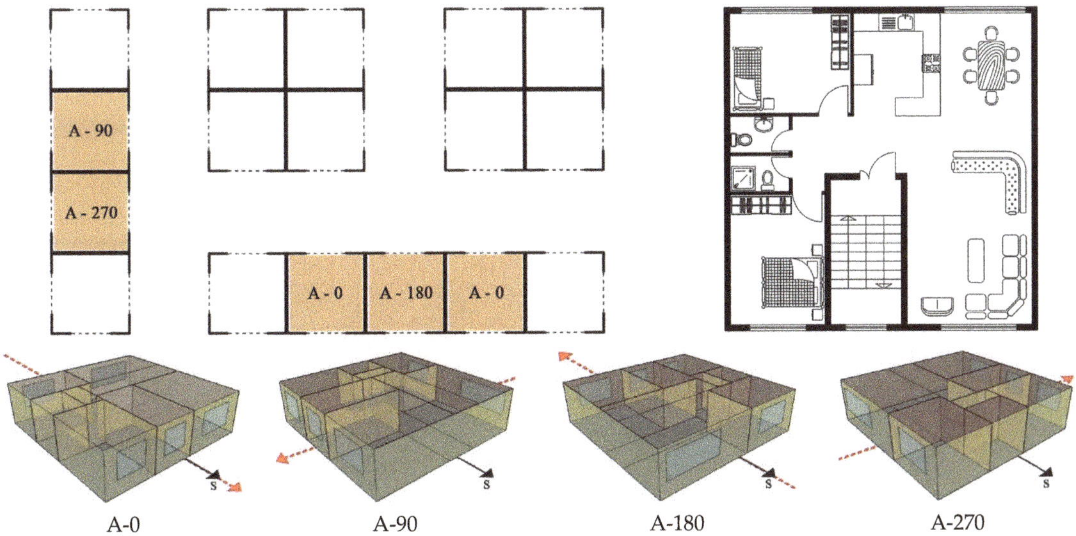

Type (A): Two Parallel Façades

(**a**)

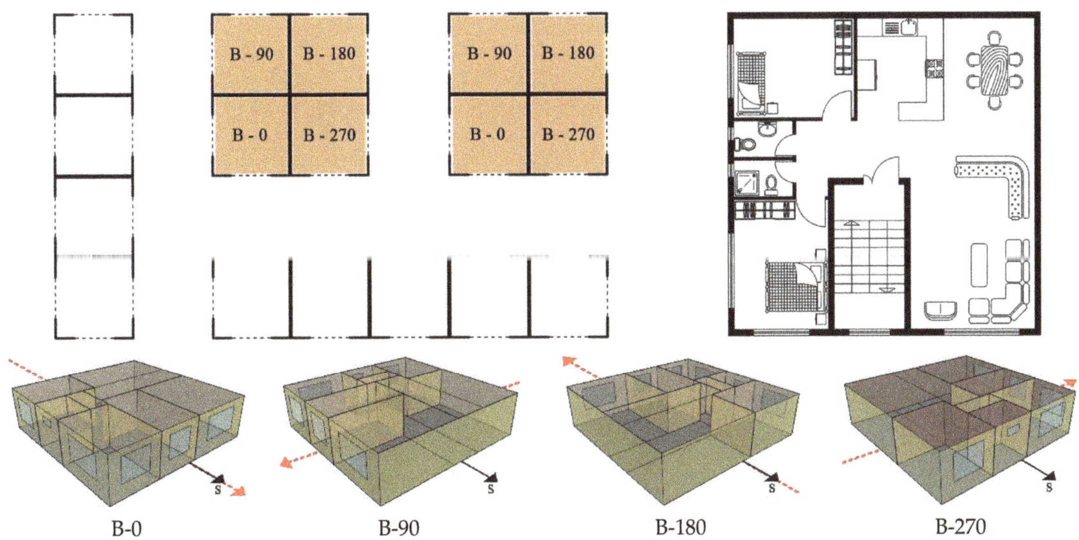

Type (B): Two Perpendicular Façades

(**b**)

Figure 4. *Cont.*

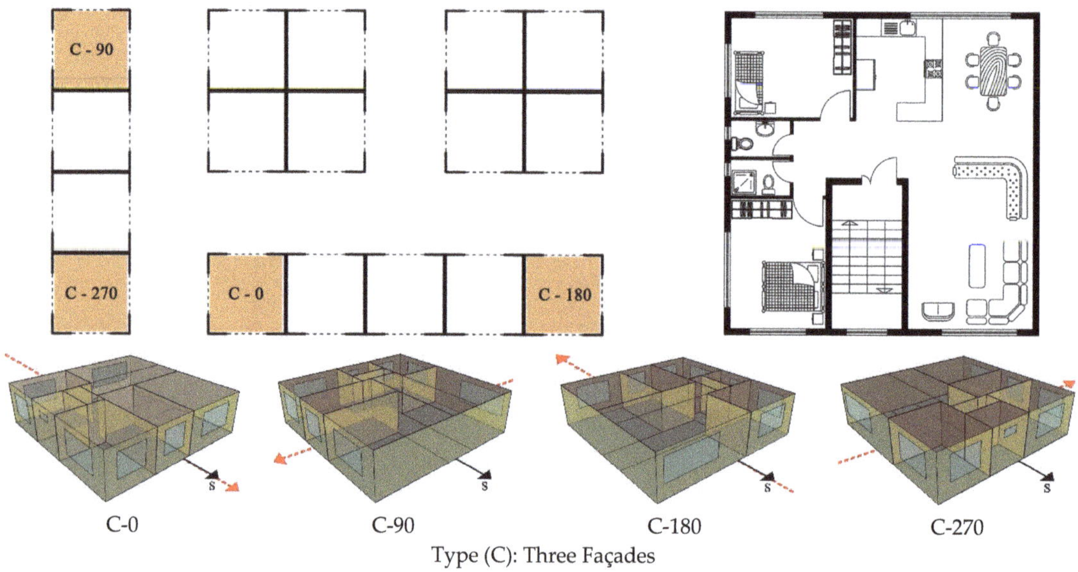

(c)

Figure 4. Classification of the considered alternatives: (**a**) type (A) with two parallel façades; (**b**) type (B) with two perpendicular façades; (**c**) type (C) with three façades.

3.4.1. Energy Consumption

In many studies (e.g., [9,30,31]), the total energy consumption has been considered as an indicator for investigating the energy performance of a building. This indicator is usually composed of cooling, heating, and lighting energy demand and calculated as follows:

$$TEC = Q_c + Q_h + Q_l \tag{5}$$

where Q_c is the annual cooling energy demand, Q_h is the annual heating energy demand, and Q_l is the annual lighting load of a building.

In this paper, to present a more comprehensive method, cooling, heating, and lighting energy consumptions are taken into account separately as three independent indicators.

3.4.2. Environmental Impact

Reviewing the literature, it can be recognized that different indicators have been used to assess the environmental impact in a building. For instance, Sohani et al. [32] presented a multi-objective optimization method, and considered annual carbon dioxide emission as one of the objective functions. In another work [33], life cycle emissions have been minimized as an environmental impact metric in addition to economic and thermal comfort indicators. As another considered indicator in the literature (e.g., [34]), carbon dioxide equivalent (CO_2-eq) emission is taken into account in this study.

The electricity consumption in a building is consumed as one of the sources for CO_2-eq. The produced amount of CO_2-eq related to heating, cooling, and lighting energy consumption is obtained as follows [34]:

$$CO_2 - eq = \frac{Q \cdot EF}{\eta} \tag{6}$$

where Q is the total annual energy use, EF is the primary greenhouse gas factor, and η is the average annual efficiency of the system.

3.4.3. Thermal Comfort

As the standard of living increases, designing a building that provides occupants comfort becomes more and more important. It can be said that comfort in a building is a condition in which occupants feel satisfied thermally and visually [35]. The considered thermal and visual comfort metrics are described in this section and Section 3.4.4, respectively.

In this paper, the Fanger model is developed to investigate the percentage of people dissatisfied (PPD), which is a metric to assess thermal comfort in a building. To calculate this metric, the predicted mean vote (PMV) should first be calculated using the following equations [36]:

$$\begin{aligned} PMV = {} & [0.303\times exp(-0.036\times M)+0.028]\times\{(M-EW) \\ & -3.05\times10^{-3}\times[5733-6.99\times(M-EW)-P_a]-0.42\times[(M-EW)-58.15] \\ & -1.7\times10^{-5}\times M\times(5867-P_a)-0.0014\times M\times(34-T_{air}) \\ & -3.96\times10^{-8}\times f_{cl}\times[(T_{cl}+273)^4-(T_r+273)^4]-f_{cl}\times h_c\times(T_{cl}-T_{air}) \end{aligned} \tag{7}$$

where

$$\begin{aligned} T_{cl} = {} & 33.7-0.028\times(M-EW) \\ & -0.155\times I_{cl} \\ & \times\{3.96\times10^{-8}\times f_{c;}\times[(T_{cl}+273)^4-(T_r+273)^4]+f_{cl}\times h_{cl}\times(T_{cl}-T_{air})\} \end{aligned} \tag{8}$$

$$h_c=\begin{cases} 2.38\times|T_{cl}-T_{air}|_{0.25} & for \quad 2.38\times|T_{cl}-T_{air}|_{0.25}>12.1\times\sqrt{V_{rel}} \\ 12.1\times\sqrt{V_{rel}} & for \quad 2.38\times|T_{cl}-T_{air}|_{0.25}<12.1\times\sqrt{V_{rel}} \end{cases} \tag{9}$$

$$f_{cl}=\begin{cases} 1.00+1.290\times I_{cl} & for \quad I_{cl}\le 0.078 \quad m^2\cdot K\cdot W^{-1} \\ 1.05+0.645\times I_{cl} & for \quad I_{cl}>0.078 \quad m^2\cdot K\cdot W^{-1} \end{cases} \tag{10}$$

In Equations (7) to (10), the parameters f_{cl}, I_{cl}, T_{cl}, T_r, and M are the area of clothing surface factor, clothing insulation, clothing surface temperature, mean radiant temperature, and the metabolic rate, respectively. Moreover, V_{rel}, h_c, T_{air}, and P_a refer to the relative air velocity, convective heat transfer coefficient, air temperature, and the partial pressure of water vapor, respectively. External work is introduced by EW, which is also another parameter in Equations (7) to (10), and is related to the system.

Finally, PPD can be obtained from Equation (11) [36].

$$PPD=100-\left[95\exp\left(-0.03353PMV^4-0.2179PMV^2\right)\right] \tag{11}$$

To compare different building alternatives, the annual average PPD ($AAPPD$) and the monthly average PPD ($MAPPD$) of the three conditioned zones, shown in Figure 5, are used and calculated as follows [30]:

$$AAPPD=\frac{1}{n}\sum_{i=1}^{n}\sum_{t=1}^{12}PPD_{i,\,t} \tag{12}$$

$$MAPPD=\frac{1}{n}\sum_{i=1}^{n}\sum_{j=1}^{m}PPD_{i,\,j} \tag{13}$$

where $PPD_{i,t}$ is the PPD of the conditioned zone i in the tth month of a year, and n is the total number of the conditioned zones. In addition, $PPD_{i,j}$ is the PPD of the conditioned zone i in the tth day of a month, and m is the total number of days in a month.

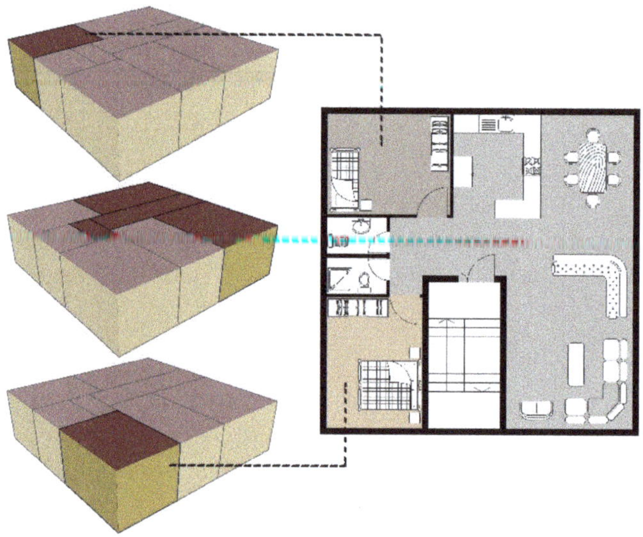

Figure 5. The conditioned zones.

3.4.4. Visual Comfort

Visual performance as another aspect that demonstrates occupants' comfort is considered in this study. To evaluate this aspect, the level of daylight illuminance in the four control points shown in Figure 6 is analyzed. These points are placed 0.8 m above the floor with a distance of 3 m from the external walls. The goal of this paper is to minimize the ratio of hours in a year that the level of daylight illuminance falls out of the comfort range. This metric is introduced as $UDI_{Discomfort}$ in the study done by Carlucci et al. [37] and is calculated as follows:

$$UDI_{Discomfort} = UDI_{Underlit} + UDI_{Overlit} \tag{14}$$

$$UDI = \frac{\sum_{i=1}^{8760} v_i}{8760} \tag{15}$$

$$\begin{cases} UDI_{Underlit} \quad with \quad v_i = \begin{cases} 1. & E_{Daylight} < E_{Lower\ limit} \\ 0. & E_{Daylight} \geq E_{Lower\ limit} \end{cases} \\ UDI_{Overlit} \quad with \quad v_i = \begin{cases} 1. & E_{Daylight} > E_{Upper\ limit} \\ 0. & E_{Daylight} \leq E_{Upper\ limit} \end{cases} \end{cases} \tag{16}$$

To describe the overall visual performance of the buildings with a single factor, the average value of $UDI_{Discomfort}$ of P_1, P_2, P_{31}, and P_{32} is calculated using Equation (17).

$$AUDI_{Discomfort} = \frac{1}{m} \sum_{i=1}^{m} UDI_{Discomfort} \tag{17}$$

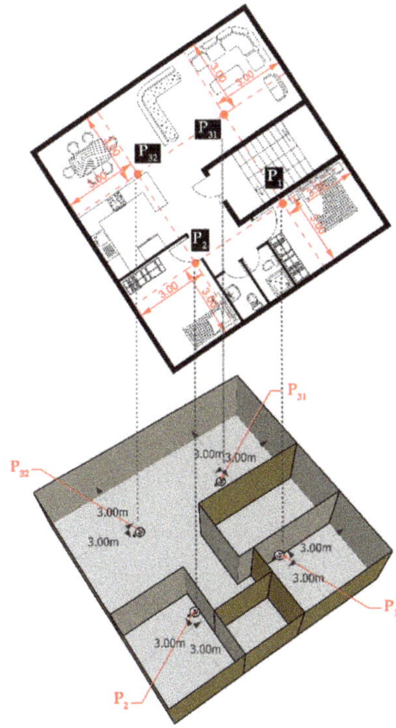

Figure 6. Placement of the control points.

4. Results

In this section, the results of the proposed comparative method, which contributes to the selection phase, is presented based on the following structure. First, different orientations of the three building types, including, type A, type B, and type C, are analyzed based on the considered objective functions, and the optimal orientation of each group is selected by the TOPSIS decision-making method. Then, the final winning alternative is chosen among the selected optimal orientations of the three mentioned groups.

4.1. Optimal Orientation of the Buildings Type A

The annual performance of the building alternatives with two parallel façades, classified as building type A, in four different orientations is analyzed in this part. The analysis is done based on the six considered objective functions. The findings demonstrated in Figure 7 indicate that there is not a single building in this group with minimum values of all the objective functions. For instance, A-0 with the lowest annual cooling energy demand is also one of the highest energy consumers for heating. This happens due to the high heat loss from the large net area of the northern façade of A-0. To better understand the interaction between these two objectives, in A-90 the annual heating energy consumption is the lowest, while its cooling energy demand is about 23% higher than the lowest value, which is a considerable amount. Reviewing the results, it is clear that the heating energy demand is linked to the heat loss from the north rather than the heat gain from the south, meaning that the lower the heat loss from the north, the less energy is consumed for heating. In contrast, the cooling energy consumption is lower when the heat loss from the north is higher.

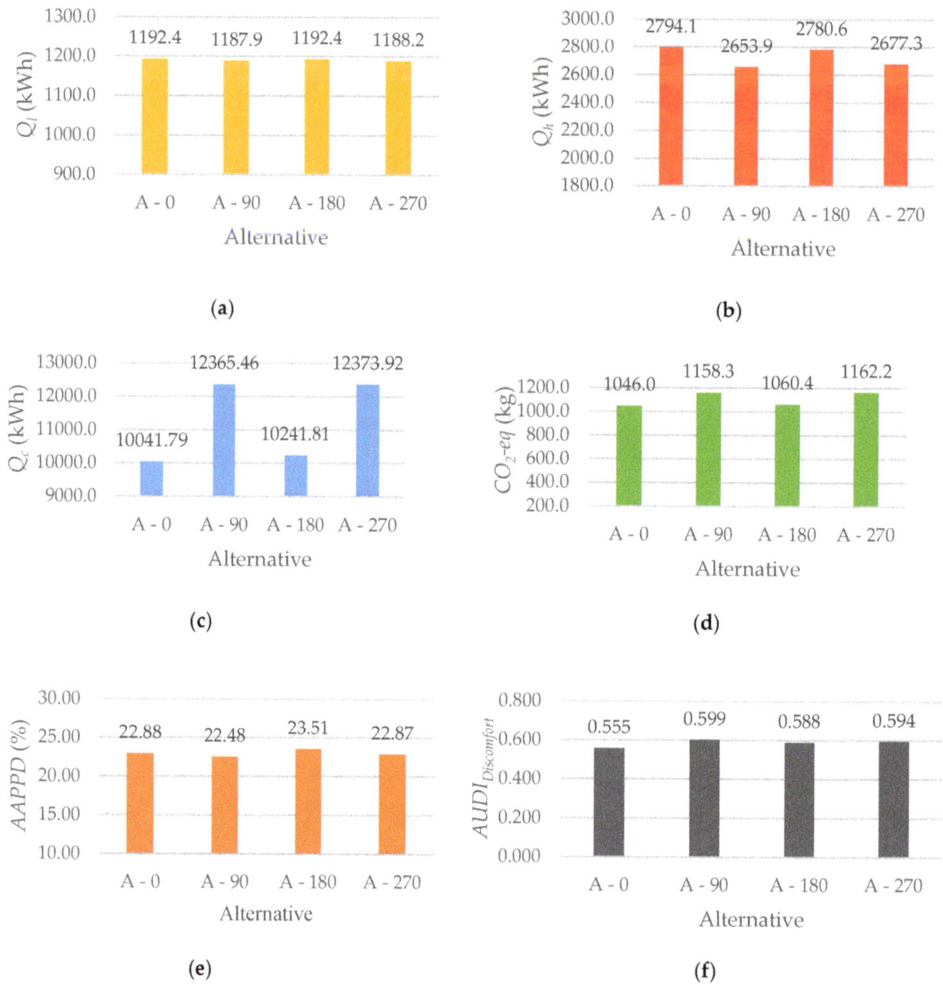

Figure 7. Annual profiles of the considered objective functions for different orientations of the building type A: (**a**) annual profiles of Q_l; (**b**) annual profiles of Q_h; (**c**) annual profiles of Q_c; (**d**) annual profiles of CO_2-*eq*; (**e**) annual profiles of *AAPPD*; (**f**) annual profiles of $AUDI_{Discomfort}$.

Since the values of annual lighting energy demand and $AUDI_{Discomfort}$ change very slightly from one direction to another, it can be said that they are not notably affected by the building orientation. Moreover, the results reveal that there is less visual discomfort in A-0 and A-180 with southern and northern façades compared to A-90 and A-270 with east–west orientations. On the contrary, the lighting energy consumption increases in the south–north orientations in comparison with those of east–west.

The greater amount of *AAPPD* in A-0 and A-180 compared to A-90 and A-270 is a result of the discomfort caused by the high solar radiation from the southern windows in summer. This can be due to the lack of a designed overhang for the southern windows of A-0 and A-180, which causes overheating in the summer. It should be mentioned that the difference between the least and the greatest values of *AAPPD* in the four orientations is only about 1%.

Given that the electricity consumption is the main source of CO_2-*eq*, its values in different alternatives are in line with the amount of the total energy demand. This means that A-0, A-180, A-90, and A-270 arranged from the lowest CO_2-*eq* to the highest value also vary in the same order for total energy consumption.

Based on the results of the TOPSIS decision-making method reported in Table 4, A-0 is found to be the optimal alternative in this group of buildings. It is considered to have the highest performance with the minimum values of cooling energy demand, CO_2-*eq*, and $AUDI_{Discomfort}$, and its values of *AAPPD* and lighting and heating energy consumption are relatively close to the ideal situation.

Table 4. The TOPSIS results for different orientations of the building type A.

Alternative	Normalized Objective Functions (*F*)						d^+	d^-	*CI*	Rank
	Q_l	Q_h	Q_c	CO_2-*eq*	*AAPPD*	$AUDI_{Discomfort}$				
A-0	0.5009	0.5123	0.4439	0.4720	0.4987	0.4750	0.0272	0.1224	0.8181	1
A-90	0.4990	0.4866	0.5466	0.5227	0.4900	0.5126	0.1206	0.0342	0.2211	3
A-180	0.5009	0.5098	0.4527	0.4785	0.5125	0.5032	0.0443	0.1053	0.7037	2
A-270	0.4991	0.4908	0.5470	0.5244	0.4985	0.5083	0.1207	0.0260	0.1771	4

4.2. Optimal Orientation of the Buildings Type B

The annual values of the considered objectives for building type B in different orientations are reported in Figure 8. As shown in this figure, B-0 and B-270 with south-facing glazing on one side consume less energy for heating than the two other cases. This happens because of the higher solar radiation and heat gain in the south. Accordingly, A-270 with the greatest area of southern glazing is recognized to demand the least heating energy. It should be underlined that the high solar radiation in the buildings from the southern windows is considered disadvantageous in terms of *AAPPD*. Given this, *AAPPD* is increased in B-0 and B-270 compared to B-90 and B-180.

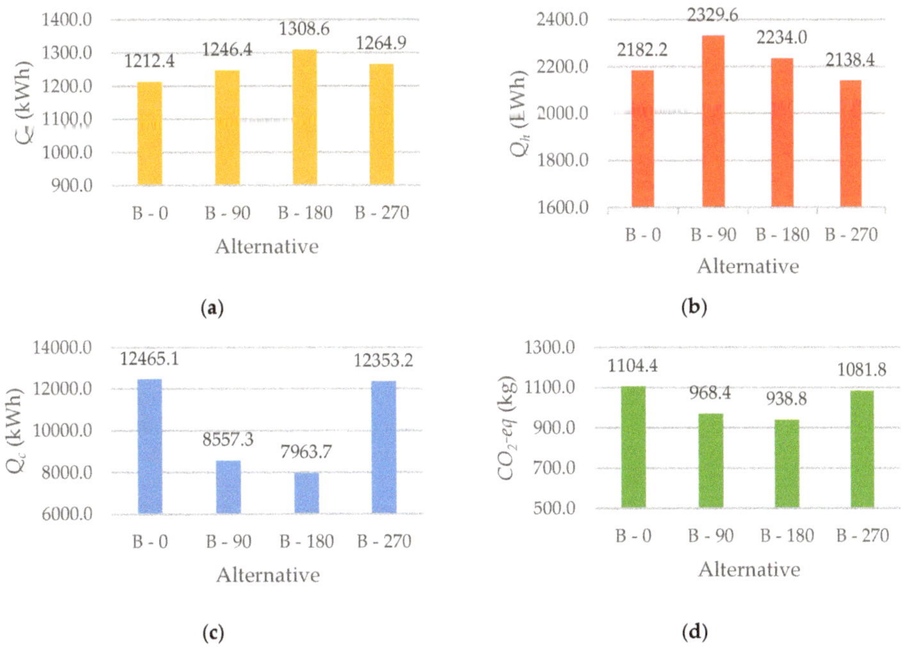

Figure 8. *Cont.*

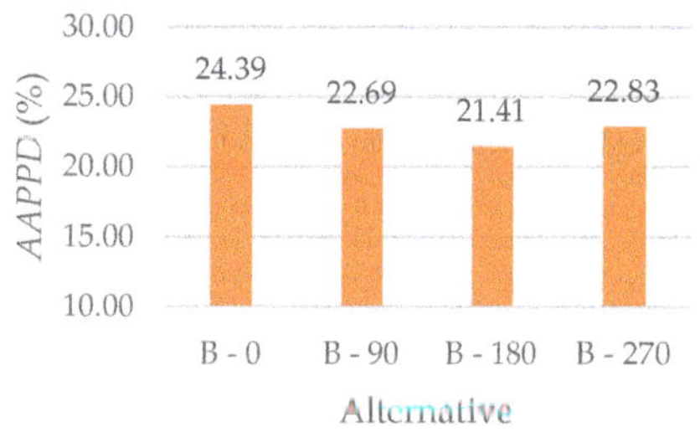

(e)

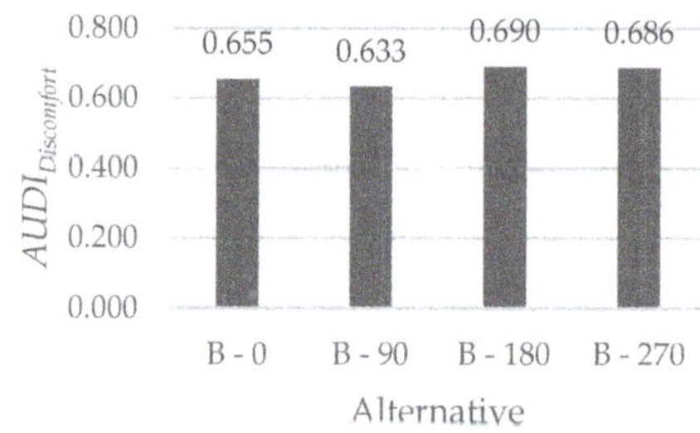

(f)

Figure 8. Annual profiles of the considered objective functions for different orientations of the building type B: (**a**) annual profiles of Q_l; (**b**) annual profiles of Q_h; (**c**) annual profiles of Q_c; (**d**) annual profiles of CO_2-eq; (**e**) annual profiles of *AAPPD*; (**f**) annual profiles of $AUDI_{Discomfort}$.

Moreover, the overheating in B-0 and B-270 and the heat loss from the northern façade in B-90 and B-180 result in a huge difference in the cooling energy demand of these cases. The lowest and the highest values of this objective are seen in B-180 and B-0, respectively.

Since CO_2-eq is a function of the total energy consumption, its annual value is the highest in B-0 due to its greatest amount of total energy demand, and it is about 18% higher than the minimum value in B-180.

Analyzing the results of $AUDI_{Discomfort}$ and lighting energy demand reveal that in B-0 and B-90 with a western façade in both, the values of these objective functions are less than B-180 and B-270, which both have an eastern façade.

As shown in Table 5, the results of the TOPSIS decision-making method point out that B-180 is introduced as the optimal orientation. Despite a slight difference in its $AUDI_{Discomfort}$ and lighting and heating energy consumption with the lowest values, it has the best performance in terms of CO_2-eq, *AAPPD*, and cooling energy demand.

Table 5. The TOPSIS results for different orientations of the building type B.

Alternative	Normalized Objective Functions (*F*)						d^+	d^-	*CI*	Rank
	Q_l	Q_h	Q_c	CO_2-eq	*AAPPD*	$AUDI_{Discomfort}$				
B-0	0.4817	0.4910	0.5911	0.5383	0.5335	0.4912	0.2381	0.0570	0.1932	4
B-90	0.4952	0.5242	0.4058	0.4720	0.4965	0.4751	0.0619	0.2062	0.7692	2
B-180	0.5199	0.5027	0.3777	0.4576	0.4683	0.5175	0.0610	0.2383	0.7962	1
B-270	0.5025	0.4811	0.5858	0.5273	0.4995	0.5150	0.2263	0.0589	0.2065	3

4.3. Optimal Orientation of the Buildings Type C

Buildings categorized as type C are the ones with three façades such that their configuration is a combination of type A and B. Here, four alternatives of this group are analyzed. Figure 9 shows that in buildings type C, as in types A and B, the greatest range of variation in the values of the considered objective functions is contributed to the cooling energy demand. The maximum annual cooling energy in C-270 is due to its highest heat gain from the south and zero heat loss from the north. In contrast, the least annual cooling energy demand in C-90 is a result of the highest heat loss from the north and zero heat gain from the south. Moreover, the value of CO_2-eq is also affected by this huge range of variation in cooling energy; thus, its minimum and maximum amounts are reached in C-90 and C-270, respectively.

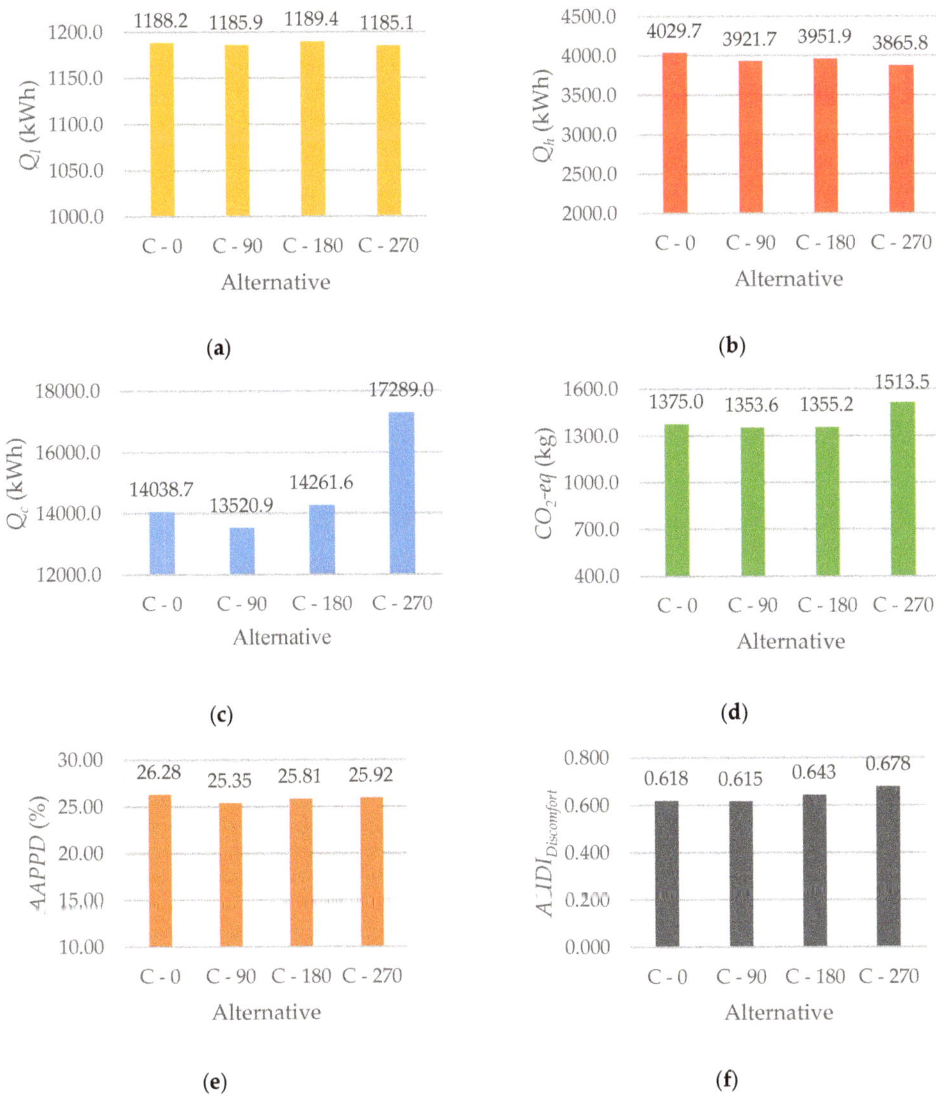

Figure 9. Annual profiles of the considered objective functions for different orientations of the building type C: (**a**) annual profiles of Q_l; (**b**) annual profiles of Q_h; (**c**) annual profiles of Q_c; (**d**) annual profiles of CO_2-*eq*; (**e**) annual profiles of *AAPPD*; (**f**) annual profiles of $AUDI_{Discomfort}$.

The findings also indicate that C-90 provides more thermal and visual comfort compared to other cases, which means that the lowest values of *AAPPD* and $AUDI_{Discomfort}$ are achieved in this case. In terms of lighting and heating energy consumption, the minimum value is reached in C-270.

Considering the conflicting relationship between the investigated objective functions, C-90 is selected as the optimal orientation by the TOPSIS decision-making method, as demonstrated in Table 6. Even though the heating and lighting energy demand in C-90 are a little higher than the least values, respectively, it has the minimum amounts of cooling energy consumption, CO_2-*eq*, *AAPPD*, and $AUDI_{Discomfort}$.

Table 6. The TOPSIS results for different orientations of the building type C.

Alternative	Normalized Objective Functions (F)						d^+	d^-	CI	Rank
	Q_l	Q_h	Q_c	CO_2-eq	$AAPPD$	$AUDI_{Discomfort}$				
C-0	0.5005	0.5110	0.4727	0.4907	0.5085	0.4838	0.0336	0.1288	0.7934	2
C-90	0.4995	0.4973	0.4552	0.4831	0.4905	0.4812	0.0071	0.1493	0.9546	1
C-180	0.5010	0.5012	0.4802	0.4837	0.4994	0.5030	0.0361	0.1205	0.7697	3
C-270	0.4991	0.4902	0.5821	0.5402	0.5015	0.5304	0.1480	0.0220	0.1295	4

4.4. The Winning Alternative

The monthly profiles of A-0, B-180, and C-90 as the optimal orientations of each building type are compared in Figure 10. According to the results presented in Figure 10, increasing the number of façades (larger area of external walls), as in C-90, is only favorable to the values of lighting energy consumption, which is due to the increased daylight availability. Moreover, in terms of visual discomfort, the values of $AUDI_{Discomfort}$ in C-90 are the lowest only at the end of autumn and the beginning of winter. In the rest of the year, because of the great range of variations in the monthly amount of $AUDI_{Discomfort}$ in A-0, the lowest values are seen in this case.

Due to the large area of external walls and windows in C-90, it is more affected by the weather conditions. Subsequently, the range of variations in the monthly cooling energy consumption, in this case, is very large. As a result, the values of this objective are the maximum in C-90 in the hot months; however, in November, December, January, and February, its values, in this case, are close to the minimum amounts. For the same reason, the worst condition in terms of thermal comfort also happens in C-90 in all months of the year except for August and September. In the mentioned months, the values of *MAPPD* in A-0 exceed the values in C-90 because of the overheating in the south.

Since the increase in the heating energy demand is in a linear relationship with the buildings' heat loss from the external walls and windows, its monthly maximum and minimum values are reported in C-90 and B-180, respectively. Moreover, the minimum values of CO_2-eq are also obtained in B-180 during the whole year except for summer. In summer, the lowest values of this objective function are seen in A-0, which is affected by its lowest cooling energy consumption in this season. It should be underlined that the increase in the values of CO_2-eq in both summer and winter is due to the peak energy usage for cooling and heating energy demand in these two seasons, respectively. Furthermore, since the values of lighting energy consumption in different alternatives vary in a small range, it is not considered as an effective factor in the values of CO_2-eq.

Finally, based on the results of the TOPSIS decision-making method shown in Table 7, first, second, and third place go to B-180, A-0, and C-90, respectively, and B-180 with two perpendicular façades facing north and east is selected as the winning alternative among all the existing buildings. Even though B-180 is not the optimal alternative in terms of lighting energy consumption and visual discomfort, the highest performance is achieved in this case considering the trade-off between all the objective functions. As it was argued, it has the lowest values of heating energy demand during the whole year. The cooling energy consumption and CO_2-eq in B-180 are a bit higher than the lowest values in May, June, July, and August only. Moreover, the minimum values of *MAPPD* are observed in this case in all seasons except for spring. Overall, comparing the annual results in Figures 7–9, there is a significant decrease of about 40%, 37%, 28%, and 10% in the values of Q_h, Q_c, CO_2-eq, and *AAPPD* for B-180 compared to the highest values observed in C-90, respectively. However, its annual values of Q_l and $AUDI_{Discomfort}$ are only about 7% and 14% higher than the lowest values obtained in C-90 and A-0, respectively.

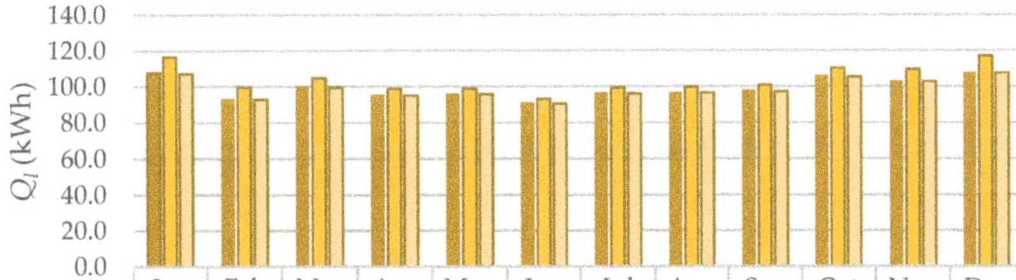

	Jan	Feb	Mar	Apr	May	Jun	Jul	Aug	Sep	Oct	Nov	Dec
A - 0	107.7	93.1	100.1	95.4	96.3	91.3	96.7	97.0	97.9	106.0	103.1	107.8
B - 180	116.3	99.4	104.5	98.6	98.7	93.0	99.1	99.6	100.7	110.1	109.5	116.9
C - 90	107.1	92.8	99.4	94.9	95.8	90.6	96.0	96.6	97.1	105.3	102.7	107.4

(**a**)

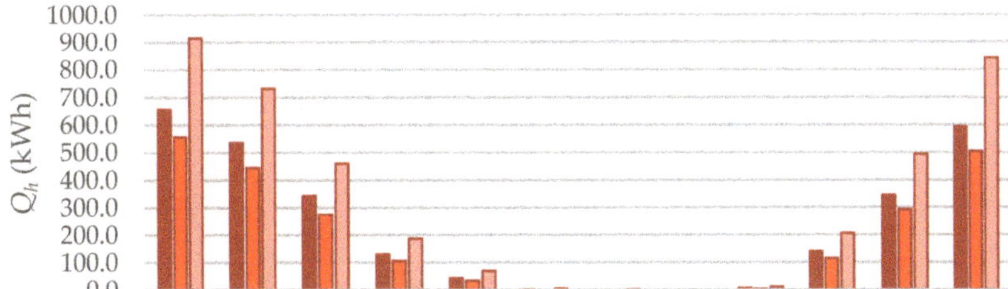

	Jan	Feb	Mar	Apr	May	Jun	Jul	Aug	Sep	Oct	Nov	Dec
A - 0	656.0	534.6	342.6	130.0	43.4	2.4	1.0	0.7	6.9	140.1	343.8	592.6
B - 180	556.5	443.7	274.9	105.8	34.5	1.0	0.5	0.3	3.5	114.2	291.5	503.0
C - 90	916.0	732.2	459.6	185.8	69.2	4.2	1.8	1.3	11.3	204.9	493.5	841.9

(**b**)

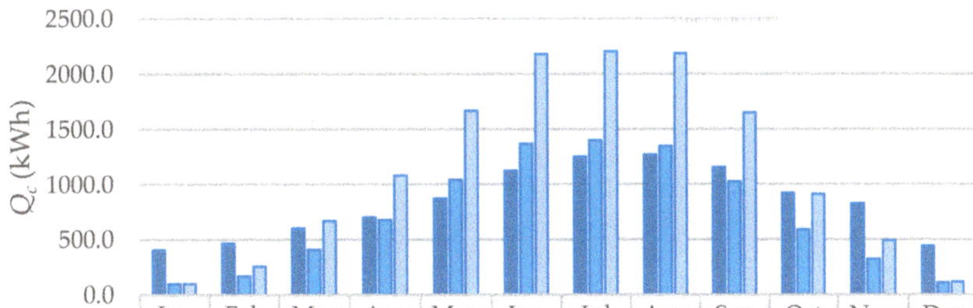

	Jan	Feb	Mar	Apr	May	Jun	Jul	Aug	Sep	Oct	Nov	Dec
A - 0	405.4	465.5	602.8	703.4	872.2	1123.	1251.	1271.	1155.	919.8	827.7	442.1
B - 180	98.9	169.9	409.7	679.3	1042.	1368.	1399.	1346.	1022.	588.0	322.7	108.8
C - 90	101.5	258.9	671.6	1079.	1668.	2180.	2204.	2187.	1651.	911.1	489.1	116.8

(**c**)

Figure 10. *Cont.*

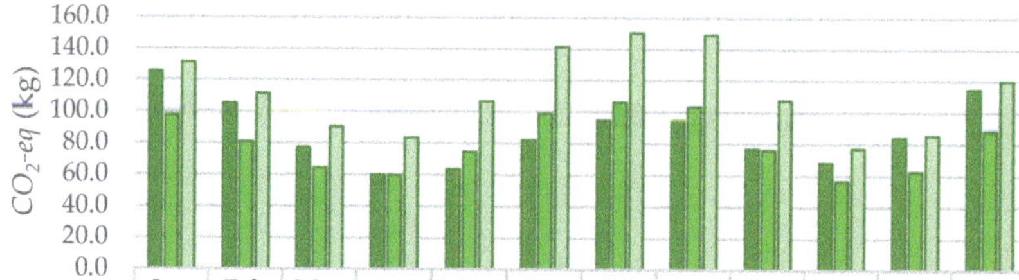

	Jan	Feb	Mar	Apr	May	Jun	Jul	Aug	Sep	Oct	Nov	Dec
A - 0	125.3	105.1	77.1	59.8	63.5	82.2	94.6	94.7	77.0	68.0	84.0	114.5
B - 180	97.3	80.9	64.5	59.6	74.7	98.9	106.2	103.3	76.0	56.5	62.4	88.2
C - 90	131.0	111.4	90.4	83.6	106.5	141.4	150.4	149.1	107.5	77.1	85.4	119.9

(d)

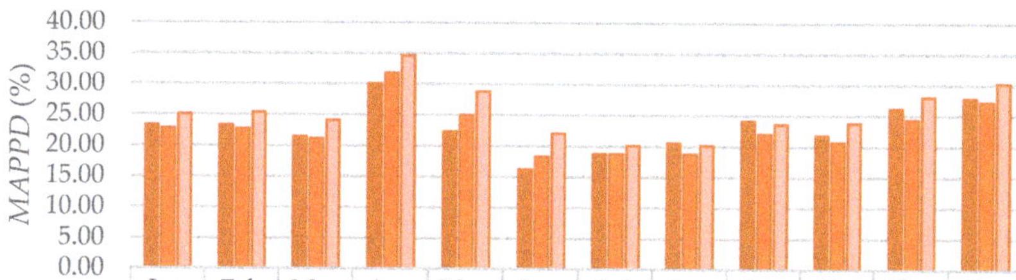

	Jan	Feb	Mar	Apr	May	Jun	Jul	Aug	Sep	Oct	Nov	Dec
A - 0	23.27	23.30	21.38	30.00	22.30	16.22	18.69	20.55	24.13	21.77	26.12	27.81
B - 180	22.74	22.67	21.03	31.85	24.96	18.26	18.69	18.69	22.02	20.69	24.31	27.25
C - 90	25.04	25.34	24.06	34.63	28.78	22.03	20.09	20.11	23.60	23.78	28.04	30.26

(e)

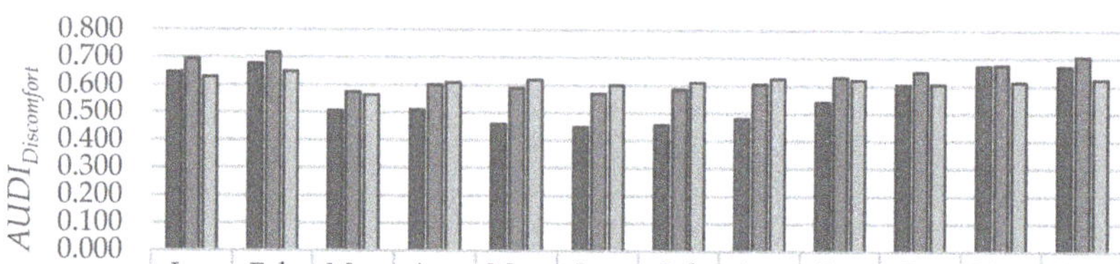

	Jan	Feb	Mar	Apr	May	Jun	Jul	Aug	Sep	Oct	Nov	Dec
A - 0	0.646	0.677	0.508	0.510	0.458	0.449	0.457	0.481	0.542	0.605	0.673	0.671
B - 180	0.695	0.717	0.575	0.601	0.590	0.571	0.587	0.607	0.631	0.651	0.676	0.706
C - 90	0.630	0.650	0.565	0.610	0.620	0.602	0.612	0.626	0.622	0.606	0.614	0.624

(f)

Figure 10. Monthly results of the objective functions for the optimal orientations of the three different building types, including, types A, B, and C: (**a**) monthly results of Q_l; (**b**) monthly results of Q_h; (**c**) monthly results of Q_c; (**d**) monthly results of CO_2-eq; (**e**) monthly results of $MAPPD$; (**f**) monthly results of $AUDI_{Discomfort}$.

Table 7. The TOPSIS results for the optimal orientations of the three different building types, including, types A, B, and C.

Alternative	Normalized Objective Functions (F)						d^+	d^-	CI	Rank
	Q_l	Q_h	Q_c	CO_2-eq	AAPPD	$AUDI_{Discomfort}$				
A-0	0.5596	0.5264	0.5390	0.5361	0.5677	0.5149	0.1672	0.3567	0.6809	2
B-180	0.6141	0.4208	0.4275	0.4811	0.5311	0.6399	0.1377	0.4948	0.7824	1
C-90	0.5565	0.7388	0.7258	0.6937	0.6290	0.5704	0.4979	0.0903	0.1535	3

5. Conclusions

The results revealed some very important conclusions. First of all, the best orientation is not necessarily the same for all the buildings located in a specific climatic region. Many parameters can affect the selection of the optimal orientation, including the number and combination of façades in a building. As the results show, A-0, B-180, and C-90 were selected as the best orientations of A, B, and C building types, respectively. Moreover, as another remarkable outcome, it is found that, despite what a customer usually selects, when all the important building aspects are involved, the best window allocation strategy is not having windows on the greatest number of façades, i.e., a member of C plans. Even though this strategy enjoys a lower lighting energy demand, other criteria are not put in a good position. Based on the TOPSIS decision-making method results and the in-depth conducted analysis, which included a monthly comparison of the performance indicators, B-180 was the winning alternative among the best of each type. In B-180, the annual values of Q_h, Q_c, CO_2-eq, and AAPPD were about 40%, 37%, 28%, and 10% lower than the highest values observed in C-90, respectively, while Q_l and $AUDI_{Discomfort}$ were about 7% and 14% higher than the lowest value achieved in C-90 and A-0, respectively. This highlighted that to select the most appropriate building for a customer, the trade-off between all the important performance criteria should be taken into account simultaneously.

Ranking existing buildings in the selection stage by implementing the proposed framework can create a competitive environment among architects to apply the optimization methods presented in the literature in the early design stages. Furthermore, the proposed method can be also used by architects in the predesign phase to compare different design strategies and select the best one.

A software program could be designed and developed based on the method, the development of which could be followed up on in future works. In this software, the plans of alternatives and the climatic conditions could be given, and the rank of alternatives in addition to the values of important performance criteria could be provided as the output. Moreover, employing the presented method in this paper to other building functions, such as offices and schools that are not occupied during the whole day, can be taken into account in future investigations. The results of such studies can create a new perspective for selecting the optimal buildings for the mentioned functions. As another suggestion for future works, the optimal direction for different plans could be selected, and the best direction of various plans with different window setups could be evaluated.

Author Contributions: Conceptualization: S.F.M.M., A.S. Formal analysis: S.F.M.M. Investigation: S.F.M.M., Methodology: S.F.M.M., A.S. Software: S.F.M.M., Supervision: A.S., M.D.S., H.S., B.N. Validation: S.F.M.M., Visualization: S.F.M.M., Writing-original draft: S.F.M.M., Writing-review and editing: A.S., B.N. All authors have read and agreed to the published version of the manuscript.

Funding: This research received no external funding.

Institutional Review Board Statement: Not applicable.

Informed Consent Statement: Not applicable.

Data Availability Statement: Not applicable.

Acknowledgments: Not applicable.

Conflicts of Interest: The authors declare no conflict of interest.

Nomenclature

d^{+}_{i}	the ith answer in Pareto optimal front distance from the ideal point
d^{-}_{i}	the ith answer in Pareto optimal front distance from the nonideal point
F	normalized objective functions
Obj	objective functions
Num	number
Cl_i	decision making parameter in TOPSIS method for the ith answer in Pareto optimal front
Q_c	cooling energy demand (kWh)
Q_h	heating energy demand (kWh)
Q_l	lighting energy demand (kWh)
Q	total annual energy use (kWh)
EF	primary greenhouse gas emission factor ($kg_{CO_2}\cdot(kWh)^{-1}$)
η	average annual efficiency of the system
f_{cl}	area of clothing surface factor
I_{cl}	clothing insulation ($(W\cdot m^{-2}\cdot K^{-1})^{-1}$)
T_{cl}	clothing surface temperature (K)
T_r	mean radiant temperature (K)
M	metabolic rate
V_{rel}	relative air velocity ($m\cdot s^{-1}$)
h_c	convective heat transfer coefficient ($W\cdot m^{-2}\cdot K^{-1}$)
T_{air}	air temperature (K)
P_a	the partial pressure of water vapor (Pa)

Abbreviations

TOPSIS	technique for order preference by similarity to the ideal solution
DOE	department of energy
PTHP	packaged terminal heat pump
COP	coefficient of performance
WWR	window-to-wall ratio
TEC	total energy consumption
CO_2-*eq*	carbon dioxide equivalent
PMV	predicted mean vote
PPD	percentage of people dissatisfied
EW	external work
AAPPD	annual average PPD
MAPPD	monthly average PPD
UDI	useful daylight illuminance
AUDI	average UDI
Scripts	
c	cooling
h	heating
l	lighting
air	air
cl	clothing
rel	relative
Discomfort	discomfort
Underlit	underlit
Upperlit	upperlit
Daylight	daylight
Lower limit	lower limit
Upper limit	upper limit
Superscripts	
Ideal	ideal
Nonideal	nonideal

References

1. Delgarm, N.; Sajadi, B.; Delgarm, S. Multi-objective optimization of building energy performance and indoor thermal comfort: A new method using artificial bee colony (ABC). *Energy Build.* **2016**, *131*, 42–53. [CrossRef]
2. Manigandan, S.; Gunasekar, P.; Devipriya, J.; Saravanan, W. Reduction of greenhouse gases by the effect of window position and its size in isolated building. *J. Chem. Pharm. Sci.* **2016**, *9*, 3304–3307.
3. Manigandan, S.; Gunasekar, P.; Devipriya, J.; Anderson, A.; Nithya, S. Energy-saving potential by changing window position and size in an isolated building. *Int. J. Ambient Energy* **2018**, *39*, 462–466. [CrossRef]
4. Misiopecki, C.; Bouquin, M.; Gustavsen, A.; Jelle, B.P. Thermal modeling and investigation of the most energy-efficient window position. *Energy Build.* **2018**, *158*, 1079–1086. [CrossRef]
5. Azmy, N.Y.; Ashmawy, R.E. Effect of the window position in the building envelope on energy consumption. *Int. J. Eng. Technol* **2018**, *7*, 1861. [CrossRef]
6. Jafari, A.; Valentin, V. Selection of optimization objectives for decision-making in building energy retrofits. *Build. Environ.* **2018**, *130*, 94–103. [CrossRef]
7. Senel Solmaz, A.; Halicioglu, F.H.; Gunhan, S. An approach for making optimal decisions in building energy efficiency retrofit projects. *Indoor Built Environ.* **2018**, *27*, 348–368. [CrossRef]
8. Xue, P.; Li, Q.; Xie, J.; Zhao, M.; Liu, J. Optimization of window-to-wall ratio with sunshades in China low latitude region considering daylighting and energy saving requirements. *Appl. Energy* **2019**, *233*, 62–70. [CrossRef]
9. Zhai, Y.; Wang, Y.; Huang, Y.; Meng, X. A multi-objective optimization methodology for window design considering energy consumption, thermal environment and visual performance. *Renew. Energy* **2019**, *134*, 1190–1199. [CrossRef]
10. Troup, L.; Phillips, R.; Eckelman, M.J.; Fannon, D. Effect of window-to-wall ratio on measured energy consumption in US office buildings. *Energy Build.* **2019**, *203*, 109434. [CrossRef]
11. Feng, K.; Lu, W.; Wang, Y. Assessing environmental performance in early building design stage: An integrated parametric design and machine learning method. *Sustain. Cities Soc.* **2019**, *50*, 101596. [CrossRef]
12. Ashrafian, T.; Moazzen, N. The impact of glazing ratio and window configuration on occupants' comfort and energy demand: The case study of a school building in Eskisehir, Turkey. *Sustain. Cities Soc.* **2019**, *47*, 101483. [CrossRef]
13. Hart, R.; Selkowitz, S.; Curcija, C. Thermal performance and potential annual energy impact of retrofit thin-glass triple-pane glazing in US residential buildings. *Build. Simul.* **2019**, *12*, 79–86. [CrossRef]
14. Rizal, Y.; Robandi, I.; Yuniarno, E.M. Optimization of daylight factor distribution using standard deviations based on shifting window position. *J. Ilm. Kursor* **2020**, 10. [CrossRef]
15. Kunwar, N.; Bhandari, M. A Comprehensive analysis of energy and daylighting impact of window shading systems and control strategies on commercial buildings in the United States. *Energies* **2020**, *13*, 2401. [CrossRef]
16. Elghamry, R.; Hassan, H. Impact of window parameters on the building envelope on the thermal comfort, energy consumption and cost and environment. *Int. J. Vent.* **2020**, *19*, 233–259. [CrossRef]
17. Kaasalainen, T.; Mäkinen, A.; Lehtinen, T.; Moisio, M.; Vinha, J. Architectural window design and energy efficiency: Impacts on heating, cooling and lighting needs in Finnish climates. *J. Build. Eng.* **2020**, *27*, 100996. [CrossRef]
18. Al-Saggaf, A.; Nasir, H.; Taha, M. Quantitative approach for evaluating the building design features impact on cooling energy consumption in hot climates. *Energy Build.* **2020**, *211*, 109002. [CrossRef]
19. Ascione, F.; Bianco, N.; Iovane, T.; Mauro, G.M.; Napolitano, D.F.; Ruggiano, A.; Viscido, L. A real industrial building: Modeling, calibration and Pareto optimization of energy retrofit. *J. Build. Eng.* **2020**, *29*, 101186. [CrossRef]
20. Zhao, J.; Du, Y. Multi-objective optimization design for windows and shading configuration considering energy consumption and thermal comfort: A case study for office building in different climatic regions of China. *Sol. Energy* **2020**, *206*, 997–1017. [CrossRef]
21. Mousavi Motlagh, S.F.F.M.; Sohani, A.; Djavad Saghafi, M.D.; Sayyaadi, H.; Nastasi, B. The Road to developing economically feasible plans for green, comfortable and energy efficient buildings. *Energies* **2021**, *14*, 636. [CrossRef]
22. Crawley, D.B.; Lawrie, L.K.; Pedersen, C.O.; Winkelmann, F.C. Energy plus: Energy simulation program. *ASHRAE J.* **2000**, *42*, 49–56.
23. Hwang, C.-L.; Yoon, K. Methods for multiple attribute decision making. In *Multiple Attribute Decision Making*; Springer: Berlin/Heidelberg, Germany, 1981; pp. 58–191.
24. Assari, A.; Mahesh, T.; Assari, E. Role of public participation in sustainability of historical city: Usage of TOPSIS method. *Indian J. Sci. Technol.* **2012**, *5*, 2289–2294. [CrossRef]
25. Sohani, A.; Sayyaadi, H. Design and retrofit optimization of the cellulose evaporative cooling pad systems at diverse climatic conditions. *Appl. Therm. Eng.* **2017**, *123*, 1396–1418. [CrossRef]
26. Sohani, A.; Sayyaadi, H.; Hoseinpoori, S. Modeling and multi-objective optimization of an M-cycle cross-flow indirect evaporative cooler using the GMDH type neural network. *Int. J. Refrig.* **2016**, *69*, 186–204. [CrossRef]
27. Naderi, E.; Sajadi, B.; Behabadi, M.A.; Naderi, E. Multi-objective simulation-based optimization of controlled blind specifications to reduce energy consumption, and thermal and visual discomfort: Case studies in Iran. *Build. Environ.* **2020**, *169*, 106570. [CrossRef]
28. Hasani Balyani, H.; Sohani, A.; Sayyaadi, H.; Karami, R. Acquiring the best cooling strategy based on thermal comfort and 3E analyses for small scale residential buildings at diverse climatic conditions. *Int. J. Refrig.* **2015**, *57*, 112–137. [CrossRef]
29. Abbasi, M.H.; Sayyaadi, H.; Tahmasbzadebaie, M. A methodology to obtain the foremost type and optimal size of the prime mover of a CCHP system for a large-scale residential application. *Appl. Therm. Eng.* **2018**, *135*, 389–405. [CrossRef]

30. Si, B.; Wang, J.; Yao, X.; Shi, X.; Jin, X.; Zhou, X. Multi-objective optimization design of a complex building based on an artificial neural network and performance evaluation of algorithms. *Adv. Eng. Inform.* **2019**, *40*, 93–109. [CrossRef]
31. Li, Z.; Chen, H.; Lin, B.; Zhu, Y. Fast bidirectional building performance optimization at the early design stage. *Build. Simul.* **2018**, *11*, 647–661. [CrossRef]
32. Sohani, A.; Sayyaadi, H.; Azimi, M. Employing static and dynamic optimization approaches on a desiccant-enhanced indirect evaporative cooling system. *Energy Convers. Manag.* **2019**, *199*, 112017. [CrossRef]
33. Mostavi, E.; Asadi, S.; Boussaa, D. Development of a new methodology to optimize building life cycle cost, environmental impacts, and occupant satisfaction. *Energy* **2017**, *121*, 606–615. [CrossRef]
34. Hamdy, M.; Mauro, G.M. Multi-objective optimization of building energy design to reconcile collective and private perspectives: CO2-eq vs. Discounted payback time. *Energies* **2017**, *10*, 1016. [CrossRef]
35. Sohani, A.; Sayyaadi, H.; Mohammadhosseini, N. Comparative study of the conventional types of heat and mass exchangers to achieve the best design of dew point evaporative coolers at diverse climatic conditions. *Energy Convers. Manag.* **2018**, *158*, 327–345. [CrossRef]
36. Sohani, A.; Sayyaadi, H.; Zeraatpisheh, M. Optimization strategy by a general approach to enhance improving potential of dew-point evaporative coolers. *Energy Convers. Manag.* **2019**, *188*, 177–213. [CrossRef]
37. Carlucci, S.; Cattarin, G.; Causone, F.; Pagliano, L. Multi-objective optimization of a nearly zero-energy building based on thermal and visual discomfort minimization using a non-dominated sorting genetic algorithm (NSGA-II). *Energy Build.* **2015**, *104*, 378–394. [CrossRef]

energies

Article

The Impact Assessment of Climate Change on Building Energy Consumption in Poland

Hassan Bazazzadeh [1,*], Peiman Pilechiha [2], Adam Nadolny [1], Mohammadjavad Mahdavinejad [2] and Seyedeh sara Hashemi safaei [3]

[1] Faculty of Architecture, Poznan University of Technology, 61-131 Poznan, Poland; Adam.nadolny@put.poznan.pl
[2] Faculty of Architecture, Tarbiat Modares University, Tehran 14115-111, Iran; P.pilechiha@modares.ac.ir (P.P.); mahdavinejad@modares.ac.ir (M.M.)
[3] Faculty of Architecture, Jundi-Shapur University of Technology, Dezful 334-64615, Iran; Sarahashemii92@gmail.com
* Correspondence: Hassan.bazazzadeh@doctorate.put.poznan.pl; Tel.: +48-503-945-736

Abstract: A substantial share of the building sector in global energy demand has attracted scholars to focus on the energy efficiency of the building sector. The building's energy consumption has been projected to increase due to mass urbanization, high living comfort standards, and, more importantly, climate change. While climate change has potential impacts on the rate of energy consumption in buildings, several studies have shown that these impacts differ from one region to another. In response, this paper aimed to investigate the impact of climate change on the heating and cooling energy demands of buildings as influential variables in building energy consumption in the city of Poznan, Poland. In this sense, through the statistical downscaling method and considering the most recent Typical Meteorological Year (2004–2018) as the baseline, the future weather data for 2050 and 2080 of the city of Poznan were produced according to the HadCM3 and A2 GHG scenario. These generated files were then used to simulate the energy demands in 16 building prototypes of the ASHRAE 90.1 standard. The results indicate an average increase in cooling load and a decrease in heating load at 135% and 40%, respectively, by 2080. Due to the higher share of heating load, the total thermal load of the buildings decreased within the study period. Therefore, while the total thermal load is currently under the decrease, to avoid its rise in the future, serious measures should be taken to control the increased cooling demand and, consequently, thermal load and GHG emissions.

Keywords: climate change; energy consumption; building energy load; thermal load; future weather

Citation: Bazazzadeh, H.; Pilechiha, P.; Nadolny, A.; Mahdavinejad, M.; Hashemi safaei, S.s. The Impact Assessment of Climate Change on Building Energy Consumption in Poland. *Energies* **2021**, *14*, 4084. https://doi.org/10.3390/en14144084

Academic Editors: Andrea Mauri and Benedetto Nastasi

Received: 18 May 2021
Accepted: 28 June 2021
Published: 6 July 2021

1. Introduction

The growth of the urban population due to economic and industrial development has sharply raised the demand for urban infrastructures such as energy systems and housing. These changes, alongside the enhancement of life quality, have led to higher greenhouse gas emissions. Increasing greenhouse gas (GHG) emissions are among the significant causes of climate change, and their impacts include changing weather patterns, extreme weather conditions, and global warming [1]. The Fifth Assessment Report of IPCC (AR5) concluded that as a consequence of these changes, the global mean surface temperature would rise around 2.5–4.5 °C by the end of the 21st century [2]. That is to say, the outdoor climate condition is among the main factors that substantially affect the energy consumption of buildings [3]. This sector, which has a considerably high share in the global energy consumption and total GHG emissions, according to EIA, is a crucial player in the energy context as it accounts for 67% of energy demand worldwide [4] and up to 40% in the E.U. and the U.S. in 2019 [5,6]. Furthermore, it is reported that the total energy-related CO_2 emissions of the building sector have risen in recent years after flattening between 2013 and 2016, reaching an all-time high of 10 $GtCO_2$ in 2019 [7]. Therefore, various scholars have tried to address this issue from their own discipline [8].

At the same time, the evidence of climate change indicates its impact on the different aspects of urban life such as health, water resources, energy, economy, and politics. This is why adaptation to climate change is one of the vital challenges of the 21st century to mitigate its adverse effects. Several international committees have attempted to use various methods to study this subject [9]. According to recent studies, the impact of climate change on the building sector can be categorized into "HVAC system", "heating and cooling demand", and "power peak demand" [10].

To evaluate the categories above-mentioned, building performance simulation (BPS) helps designers to have a comprehensive energy consumption assessment of a design or an existing building about what it would face during its life cycle. However, studies have indicated that thermal load is a crucial factor in assessing the impact of climate change on a building's energy consumption as a high proportion of the building's energy consumption is usually dedicated to heating and cooling systems [11].

To be more precise, the thermal behavior of a building is related to three primary attributes: building physics, the microclimate of the outside environment, and the required thermal comfort inside the building [12]. Therefore, due to the significant impact of weather parameters on building energy demand, an appropriate portion of energy consumption in buildings can be controlled by evaluating only the heating and cooling space demand.

In the E.U., the energy consumption and, more specifically, thermal load in buildings has been decreasing since 2008. In detail, the average annual energy consumption per m^2 in the E.U. for all building types in 2013 was around 180 kWh/m^2 with different rates for each country (e.g., from 55 kWh/m^2 in Malta to 300 kWh/m^2 in Romania) according to the illustration of the energy consumption of buildings per m^2 (Figure 1). Nevertheless, even for countries with a similar climate, notable differences have been reported (e.g., 200 kWh/m^2 in Sweden, 18% lower than Finland). Climatic conditions can mainly explain such discrepancies, which is why there is a crucial need to focus on each country to analyze the impact of climate change on the building's energy consumption by considering their current and future climatic conditions.

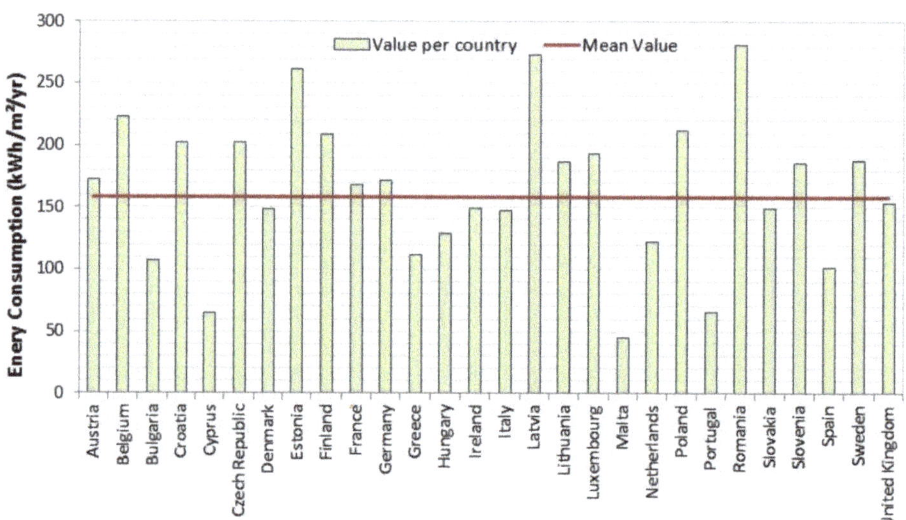

Figure 1. Annual energy consumption of buildings per m^2 on average [13].

Poland, as one of the top five energy consumers in buildings per m^2 in the E.U. (Figure 1) with more than 200 $kWh/m^2/yr$, has experienced substantial growth in energy efficiency by reducing at least 50% of its energy intensity over the last decades [14] thanks to Poland's Thermo-modernization and Rehabilitation Fund program. However, despite this progress, coal still dominates the power sector. The draft energy policy of Poland has placed a particular emphasis on reducing greenhouse gas (GHG) emissions and increasing the energy efficiency of the building sector. Thus, the Polish government has formed the "Polish National Strategy for Adaptation to Climate Change (NAS 2020)" to focus on an adaptation plan for sectors vulnerable to climate change including the building industry. Among all energy-consuming sectors, the building industry ranked the highest in 2019, with around one-third of total energy consumption. More precisely, residential buildings were responsible for 29% of Poland's TFC in 2014, and commercial buildings account for roughly 17%. They are also responsible for around 18% of CO_2 emissions in Poland since 1990 [15]. The share of the construction sector in Poland's total energy consumption in recent years is notably high (Figure 2), which indicates the importance of this sector. Furthermore, around three-quarters of Poland's buildings have either low or deficient energy efficiency standards, according to a survey by "The Buildings Performance Institute Europe (BPIE)" in 2016 [16]. In this sense, climate change can affect Poland's energy consumption rate, which is why the impact assessment of climate change for the building sector in Poland is quite vital.

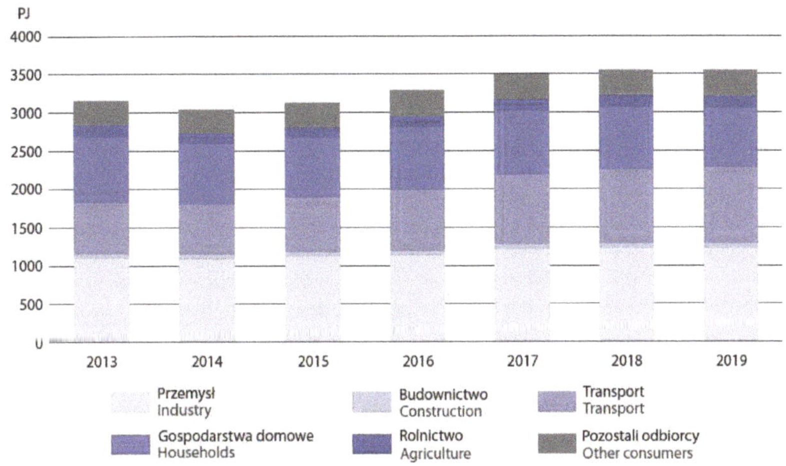

Figure 2. Poland's energy consumption by sectors [17].

There is an increasing tendency in Poland to enhance energy efficiency, and fortunately, the building sector has always been one of the focal points of this improvement plan. However, limited information and unreliable data about building energy use allow for provisions to circumvent the system [14]. In this respect, solid national cooperation, led by the 'build desk', has been formed to address this issue by conducting comprehensive studies to analyze the energy consumption in the building sector in each region. Subsequently, several studies have been carried out in Poznan due to its rapid urban growth and its significance in Poland's energy sector (Figure 3). Poznan is among the oldest Polish cities in the western region of this country and the fifth-largest city with a population of more than 534,813 according to the 2019 census [18].

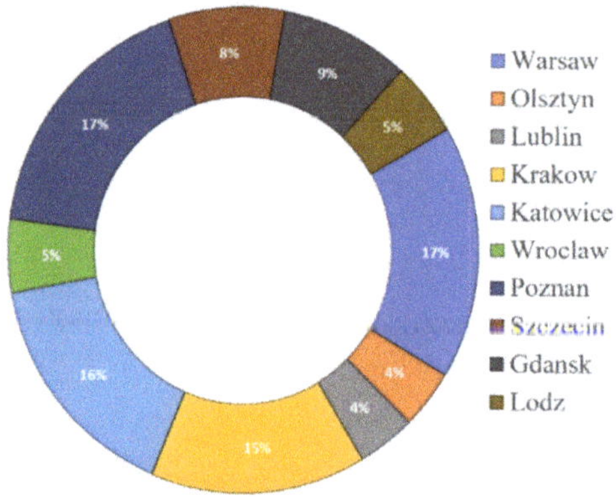

Figure 3. Share of each region in Poland in a comprehensive study to analyze energy consumption in the building sector [19].

In Poznan, in line with the E.U.'s Clean Energy Package goals, an EU-Horizon 2020 project entitled "Energy island communities for energy transition" has started to integrate and demonstrate solutions that will foster a substantial increase in energy efficiency with a total budget of €6,694,000 and one of the energy islands that is going to be studied is "Warta Campus in Poznan" [20]. However, the need for a thorough study of energy performance in different sectors, especially buildings, by considering the current (Cfb according to the Köppen climate classification [20]) and future conditions and uncertainties such as climate change is still crucial (see Table 1). Cfb climate zone, according to the Köppen climate classification, refers to locations where the oceanic climate is temperate; the coldest month averages above 0 °C or − 3 °C, all months had average temperatures below 22 °C, and at least four months averaging above 10 °C. [20]. Therefore, as a response to this need and the significance of the building energy consumption in Poland and, more specifically, in Poznan, this study aimed to analyze the impact of climate change on the thermal load of the building sector in Poznan as a critical node of the E.U. energy-efficient research program.

Table 1. A short description of the research case study.

Country	City	Latitude	Longitude	Time Zone	KGC
Poland	Poznan	52.42N	16.83E	1.00	Cfb
Population	Elevation	Weather Data	Heating DB 99.6%	Cooling DB 0.4%	Cooling 0.4% MCWB
533,830	302 (m)	2004–2018 (TMY)	−14.27 °C	29.77 °C	19.22 °C

2. Literature Review

2.1. Projecting Energy Consumption of Buildings

Due to the significance of energy demand, prediction of buildings' energy demand (more specifically thermal load) has been one of goals of researchers from different perspectives [21–23] to reach sustainability [24,25] through optimizing energy consumption [26–28]. This knowledge, indeed, can lead to integrated management of buildings [29]. Evaluate and projecting the impact of climate change on building energy consumption and, more specifically, the thermal load, are among the challenging areas for scholars. Rosenthal et al. [30] were among the pioneer scholars who pointed out that the impacts of global warming do not necessarily lead to increased energy expenditure, particularly for reasonably cold regions. They estimated that with 1.0 °C global warming at the end of 2010, the U.S. will save more than $5.5 billion (1991 USD) by reducing energy consumption. Their claim, indeed, is in contrast to earlier studies [31,32] in which an increase in U.S.

energy consumption had been projected. In 2006, Nik estimated a sharp decrease in heating load and a more considerable increase of cooling load in Stockholm [33] in 1961–2100 according to uncertainty factors. Christenson and his colleagues [34] reached similar points for Switzerland until 2085. Similarly, Hosseini and her colleagues [35] predicted an average reduction in heating load and an increase in cooling load in Montreal in the period of 2020–2080 by studying a one-storey commercial building.

Cellura [36] studied all scenarios and concluded that we would experience an overall increase in thermal load with a relative decrease in heating and an increase in cooling demand. He stated this by studying 15 cities in southern Europe and generating future weather data by downscaling different general circulation models (GCMs) through the morphing method. To be more precise, he projected that the thermal load of buildings in their study areas increased in a range of 50–119%.

The projections of buildings' thermal loads influenced by climate change have not always been consistent. That is to say, Crawley [37] concluded that regions with the predominant heating load usually experience a decrease in thermal load. Similarly, Triana, Lamberts, and Sassi asserted that since cooling consumption will significantly increase, the thermal load of social housing will also increase by analyzing the performance of social housing in Sao Paulo and Salvador [38]. Several scholars have reported that this figure could also be increased in a significant number of cases [39,40]. For instance, Shen [41], through the impact assessment of climate change in 2040–2069 for four cities in the U.S. under the IPCC SRES A2 scenario for GHG emissions, reported an average increase and a decrease in annual energy use was reported for residential buildings and office buildings, respectively.

In the most recent studies, Moazami, Nik, and their colleagues [1], by generating several future weather data for Geneva, clarified the importance of considering extreme weather condition for improving the reliability of future weather data. They also indicated that the cooling load would show a more considerable increase (20%) compared to typical conditions. Berardi and Jafarpur [3] also used the statistical downscaling method to analyze the future energy consumption of buildings in Toronto (Dfb zone according to the Köppen—Geiger climate classification [20,42]) by 2070. They discussed that Toronto's cold climate would see a low magnitude decrease in heating needs and a considerable increase in cooling loads. Additionally, Velashjerdi Farahani et al., to assess the impact of using different passive measures to reduce the risk of overheating in an old and a new apartment in southern Finland, generated future weather data under two scenarios 2050 [43].

2.2. Weather Data

One of the most common kinds of weather data to assess a building's energy performance and its emission is Typical Meteorological Year (TMY) files. These files consist of 8760 hourly values of selected climatic parameters in which each typical month is selected from different years over a long-term weather dataset [44]. TMY files can be generated through a different method, but the most common method of generating them is using the Finkelstein–Schafer (FS) statistics method, which was developed by Hall et al. [45] and has been widely used by scholars [46]. TMY files are usually generated by considering the data of a long time period as it should represent the typical weather of the intended location. For instance, Petrakis et al., by considering data of seven years, generated TMY for Nicosia in Cyprus [47]. There are also other types of typical meteorological files such as TRY (Typical Reference Year) and DSY (Design Summer Year), which are indeed introduced, analyzed, and modified through various efforts for creating weather data [48,49].

For generating TMY files, according to the definition of these types of weather data, a complete set of historical data is needed to assess before using an acceptable method to choose each month from the years of a dataset. Therefore, the availability of a historical dataset is highly critical. As it is limited for Poznan, which was the case study of this research, the authors, by admitting the fact that different historical weather data may result in some slight differences in the results, used a historical weather dataset of Poznan belonging to 2004–2018, which is relatively recent compared to the usually available TMY files (Table 2).

Table 2. Selected years of each month in the Poznan Typical Meteorological Year for the period of 2004–2018.

Location	Jan	Feb	Mar	Apr	May	Jun	Jul	Aug	Sep	Oct	Nov	Dec
Ławica Airport	2016	2017	2009	2005	2007	2005	2013	2017	2015	2008	2012	2014

2.3. Climate Models and Projection

In terms of research method, although GCMs have been widely used for projecting future climate files, they are not quite suitable for building performance simulation purposes as they provide monthly or daily data instead of hourly data needed for BPSs. To consider deviating results of different climate models because of internal climate variability and differences in model formulations, some scholars have used different climate models according to their availability for their study area. For instance, Berardi and Jafarpur, in one of the most recent studies in this area [3], apart from using the HadCM3 model for statistical downscaling, used Hadley Regional Model 3 (HRM3) coupled with Hadley Climate Model 3 (HadCM3) to generate future weather files through dynamical downscaling, which was indeed drawing on their previous work for generating future weather data [50].

More specifically, for climate projection in Poland, the Polish–Norwegian CHASE-PL project helps authors to have a clear overview of Poland's future climatic conditions influenced by climate change. This projection was obtained through the downscaling of GCM simulations for future conditions. Indeed, model-based projection analyses of CHASEPL were carried out with the ensemble of climate projections comprising nine RCMs outputs stemming from the EURO-CORDEX ensemble for two time periods: 2021–2050 and 2071–2100 [51]. Mezghani et al. illustrated the outcomes of CHASE-PL climate projections [52] and the 5 km (CPLCP-GDPT5) dataset [53]. The projection showed that mean annual temperature is expected to increase by 1 °C until 2021–2050 and by 2 °C until 2071–2100, under the RCP4.5 (an intermediate scenario in which emissions peak around 2040, then decline [54]). However, by considering the RCP8.5 (the worst-case climate change scenario in which emissions continue to rise throughout the 21st century [55]), the same variable was projected to increase to 4 °C by 2071–2100 [56].

Several methods have been widely used to integrate the climate change impacts into weather data according to the current state of this field [57]. These methods can be categorized into two groups: the first group mainly relies on historical weather data and includes the imposed offset method, extrapolating statistical method, and the stochastic weather model. However, the second group relies on numeric climate models instead of historical weather data such as using GCMs to generate local future weather data through downscaling or RCMs. As one of the most widely used methods, downscaling can be performed in two ways: statistical and dynamic downscaling. The dynamic one is a computationally-intensive method that uses the regional scale forcing combined with the lateral boundary condition to generate regional climate models (RCMs) from a GCM.

On the other hand, statistical downscaling is less demanding because its required computations consist of two main stages: the statistical relationship development between large-scale and local climate variables. Second, it uses the previous relationship for the simulation of local climate conditions [58]. By assessing the existing downscaling method, Stephen Belcher introduced a new method of statistical downscaling called "morphing" in 2005 [59]. He discussed dynamic downscaling as being computationally expensive, unlike practical implementation on building design projects. In his study, the stochastic weather generation method that uses empirically derived statistics [60,61] was not considered an efficient way to generate future weather data. Although it is computationally cheap, it requires large datasets to train the model to have a reliable model, and the weather series it produces may not always be meteorologically consistent.

Although several scholars such as Nik [33] are still employing other methods such as regional climate models (RCMs) to downscale the results of the GCMs dynamically, they mainly need special requirements for the generated data. For instance, Nik's work better represented topography and mesoscale processes than the statistical downscaling employed by authors to use other methods rather than morphing. However, in general, statistical downscaling and, more specifically, morphing has recently been one of the main preferred methods chosen by scholars for its relatively low computational requirements and fast calculations [3,62–67].

3. Methodology

For the research case study, a set of historical data in the form of a Typical Meteorological Year (TMY file) was used as it was very recent compared to the typical TMY files available for most locations in Poland. In this research, as the first step, a weather file from the Ławica Airport weather station in Poznan was used as the baseline of projection.

These data were obtained from GCMs, and subsequently used as an input in the Climate Change World Weather Generator (CCWorldWeatherGen; the Sustainable energy research group (SERG) at Southampton university, Southampton, England) tool for projection. Initially developed by Jentsch [68,69], the Microsoft® Excel-based tool called the 'Climate Change World Weather Generator', commonly referred to as CCWorldWeatherGen, uses the output data of HadCM3 (Met Office Hadley Centre for Climate Science and Services; 2010), forced with the IPCC A2 emission scenario to generate future weather files by applying the morphing method. Jentsch et al., after assessing 23 GCMs under AR4 and six GCMs under AR3 around the world, showed that HadCM3 for the A2 emission scenario seems like a suitable GCM for use in the morphing technique [69]. This emission scenario put forward by the IPCC AR3 refers to a 'business as usual' condition in which the global population would continuously increase, and regionally oriented economic growth would grow. Therefore, it is evident that using HadCM3 for the A2 emission scenario for this tool, which uses a morphing method, is appropriate. Thus, this tool helps scholars have a fast projection about the future of climate change [70]. In detail, by adding the TMY file into CCWorldWeatherGen, the future weather data for 2050 and 2080 were created under the A2 scenario through the morphing procedure. The data obtained from GCMs were subsequently used as an input in CCWorldWeatherGen to statistically downscale the baseline and generate future weather data in the second stage. In this research, the life cycles of buildings were assumed to be at least 60 years, which means that up to the end of the analysis period, there would be no changes in the performance of building components in terms of thermal load.

Finally, current and future weather files were used to perform building energy modeling using EnergyPlus 9.0.1. To simulate the impact of climate change on the buildings' thermal load for the city of Poznan, 16 building prototypes of the ASHRAE standard 90.1 were chosen to be used in this study [71] to assess the impact of climate change on a building's thermal load. These prototypes were initially obtained from DOE's Commercial Reference Building Models with modifications from the Advanced Energy Design Guide series and the ASHRAE 90.1 committee. Their detailed descriptions and modeling strategies are accessible in the Pacific Northwest National Laboratory's (PNNL) reports [72,73]. Having realistic building characteristics, these 16 prototypes (Figure 4) were simulated with the current and future weather data to compare the relative impact of climate change on their energy performance. Technical descriptions of the prototypes' envelope components are presented as given in Table 3.

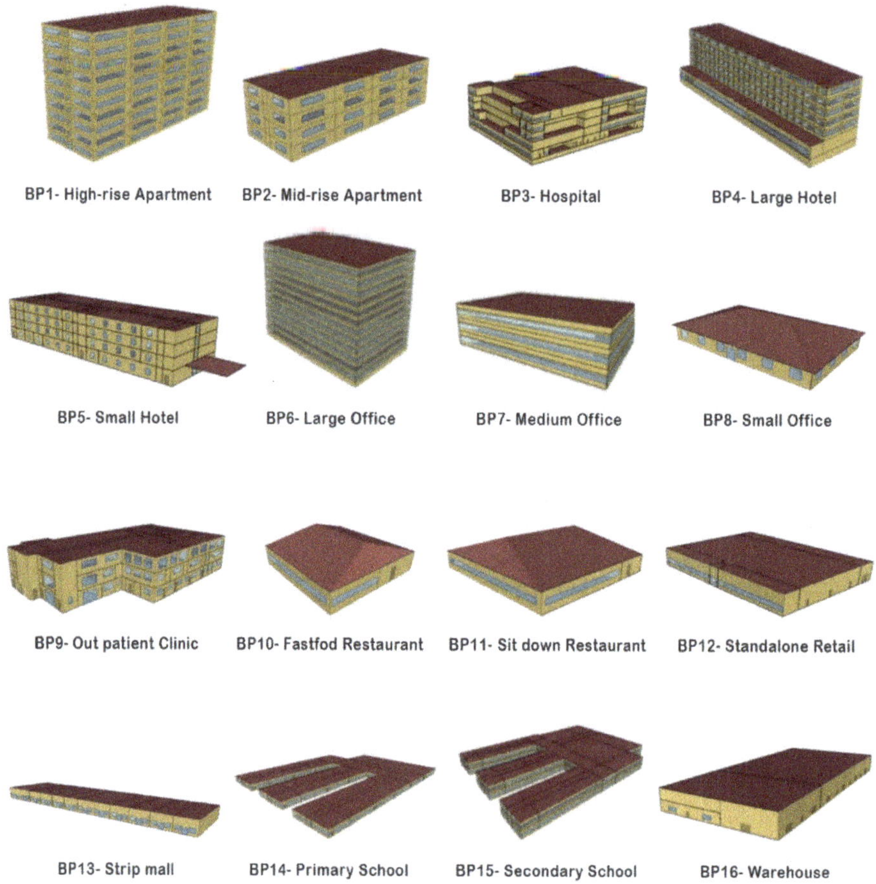

Figure 4. Building prototypes from the ASHRAE 90.1 standard [74].

Table 3. Technical description of the prototype envelope components.

Building Types		U-Factor (W/m²K)				Solar Heat Gain Coefficient (SHGC)	
		Roof	External Wall	Glazing		Glazing	
				Window	Skylight	Window	Window
Apartment	High-rise (BP1)	0.18	0.31	2.65	-	0.43	-
	Mid-rise (BP2)	0.18	0.31	ti2.65	-	0.43	-
Hotel	Large (BP4)	0.18	0.45, 0.51	2.65	-	0.43	-
	Small (BP5)	0.18	0.31	2.65, 2.85	-	0.43, 0.29	-
Office	Large (BP6)	0.18	0.51	2.65	-	0.43	-
	Medium (BP7)	0.18	0.31	2.65	-	0.43	-
	Small (BP8)	0.15	0.29	2.65	-	0.43	-
Health	Hospital (BP3)	0.18	0.45, 0.51	2.65	-	0.43	-
	Outpatient (BP9)	0.18	0.31	2.65	-	0.43	-
Restaurant	Fast food (BP10)	0.15	0.29	2.65	-	0.43	-
	Sit-down (BP11)	0.15	0.31	2.65	-	0.43	-
Retail	Stand-alone (BP12)	0.18	0.51	2.65	2.96	0.43	0.34
	Strip-mall (BP13)	0.18	0.31	2.65	-	0.43	-
School	Primary (BP14)	0.18	0.31	2.65	-	0.43	-
	Secondary (BP15)	0.18	0.31	2.65	2.96	0.43	0.34
Warehouse	(BP16)	0.21, 0.53	0.28, 0.47	2.65	2.96	0.43	0.34

After performing the energy simulation for each prototype using current and future climate weather files, the comparative analysis showed the impact of climate change on the thermal load of buildings, which was the primary goal of this paper (Figure 5).

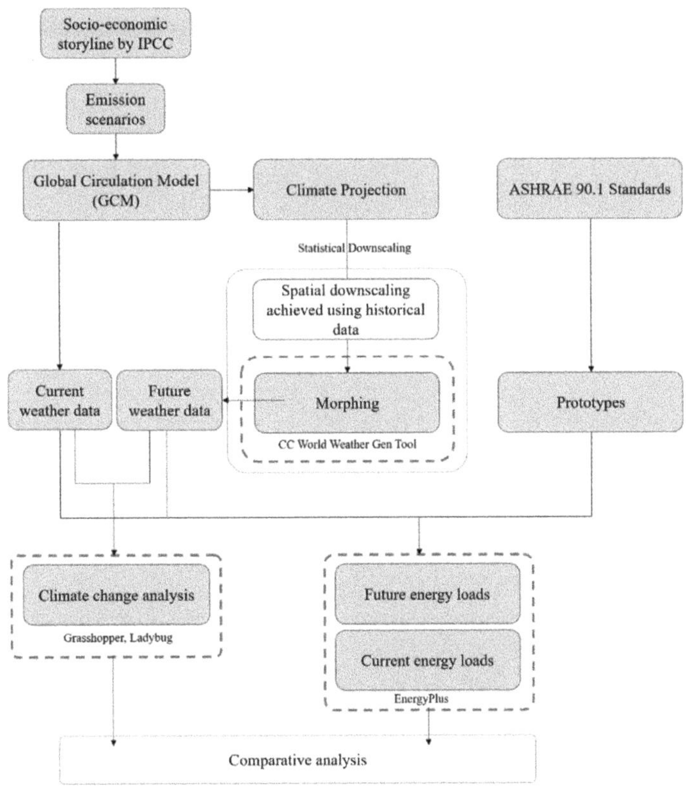

Figure 5. Research workflow.

4. Results

This section initially presents the results of the weather projection and the comparison between projected values and the current conditions to provide an overview of the future trend of climatic condition in Poznan. This section is followed by employing each weather data to build a prototype to assess the impact of climate change on their thermal load.

Comparing current and generated future weather data for the next 30 and 60 years showed the future trend discussed in detail. There were 13 influential variables that this study took into account, namely: dry bulb temperature, dew point temperature, relative humidity, direct normal radiation, global horizontal radiation, diffuse horizontal radiation, horizontal infrared radiation, direct normal illuminance, global horizontal illuminance, diffuse horizontal illuminance, ground temperature, total sky cover, and atmospheric pressure.

The future weather files forecast a notably warmer period with more direct illuminance and radiation and less humidity with lower clouds in the sky (Table 4). In general, the temperature-related variables showed an average increase of 4 °C in this period, while this figure for the radiation and illuminances group was around 14.3 Wh/m^2 and 463 lux, respectively. On the other hand, although the change rate for total sky cover was considerable with an almost 15% decrease, rates for atmospheric pressure were negligible. To be more precise, the highest increase belonged to direct normal illuminance at around one third, while the figure for atmospheric pressure ranked last. In contrast, total sky cover had the highest decrease (−13%), and the lowest decrease belonged to diffuse horizontal illuminance less than 1%.

Table 4. Relative changes of weather parameters for 2050 and 2080 compared to 2020.

Weather Parameters	2020	2050	2080
Dew point temperature (°C)	4.2	6.0	7.2
Average ground temperature (°C)	8.0	10.8	12.6
Dry bulb temperature (°C)	8.2	10.9	12.6
Direct normal radiation (Wh/m^2)	59.1	81.3	86.4
Diffuse horizontal radiation (Wh/m^2)	75.9	72.5	71.4
Global horizontal radiation (Wh/m^2)	109.7	113.0	115.7
Horizontal infrared radiation (Wh/m^2)	303.0	322.1	331.5
Direct normal illuminance (lux)	7015	7385	7875
Diffuse horizontal illuminance (lux)	8578.5	8609.7	8527.7
Global horizontal illuminance (lux)	12,157	12,434	12,739
Atmospheric pressure (Pa)	101,325	101,272	102,249
Relative humidity (%)	78.1	74.6	72.7
Total sky cover (0–10)	4.5	4.2	4.3

The next step to assess the impact of climate change on the building's energy consumption, heating and cooling use intensity (EUI), and thermal load for all 16 prototypes were calculated, and the results are as follows. After the simulation process, the values for heating and cooling EUI were mapped in scatterplot graphs, and trendlines of these variables for each prototype were drawn. As expected, the heating load for most cases saw a steady fall while the cooling load showed a sharp rise in the study period.

In terms of heating EUI, a blatant downward trend with different decrease rates, depending on building prototypes with an average decrease of more than 41 kWh/m^2, was witnessed (Figure 6a). The heating load average value in 2020 was around 114 kWh/m^2, while for 2050 and 2080, it was more than 87 kWh/m^2 and 72 kWh/m^2 with a roughly 25% and 40% decrease, respectively. To be more specific, the heating load of the fast-food restaurant building (BP10) significantly decreased (324 kWh/m^2) while in contrast, in the large office building (BP06), the decrease was not considerable (around 6 kWh/m^2). On the other hand, while heating load changed averagely near 40% in building prototypes, the maximum decrease belonged to the small office building (BP08) with around 57%, and the minimum was for the hospital building (BP03) at roughly 13%.

As far as the cooling load is concerned, the results indicated a clear, stable upward trend in the cooling load with an average increase of 13 kWh/m^2 (Figure 6b). Indeed, while the average cooling load in 2020 was 13 kWh/m^2, it increased 1.5 and 2 times in 2050 and 2080, respectively. Among all BPS, the cooling load of the fast-food restaurant building (BP10) increased markedly (46 kWh/m^2) while the warehouse building (BP16) increased slightly (0.7 kWh/m^2). All prototypes saw the average increase rate of cooling load at 135%, while the maximum increase rate belonged to the warehouse building (BP16) with 371% and the minimum one was for the large office building (BP06) with 20%.

Comparing the heating and cooling from 2020 to 2080, a considerable difference was noticeable. The results suggest that increasing the cooling load was certain, and the heating load decreased for all cases. It is evident that although the average rate absolute change for the cooling load was 13 kWh/m^2, which was notably lower than that of the heating load with more than 41 kWh/m^2, the percentages of this change were roughly 135% and 40%, respectively. This difference indicates the relatively sharp rise of the cooling load and negligible decrease in heating load for Poznan in the study period. The highest relative difference change rate of the heating load was witnessed in the small office (BP08) with around a 60% (around 10 kWh/m^2) decrease. In addition, the relative change rate of the cooling load was reported to be the highest in the warehouse building (BP16) at 372%, where the absolute figure of this change was the lowest (0.7 kWh/m^2).

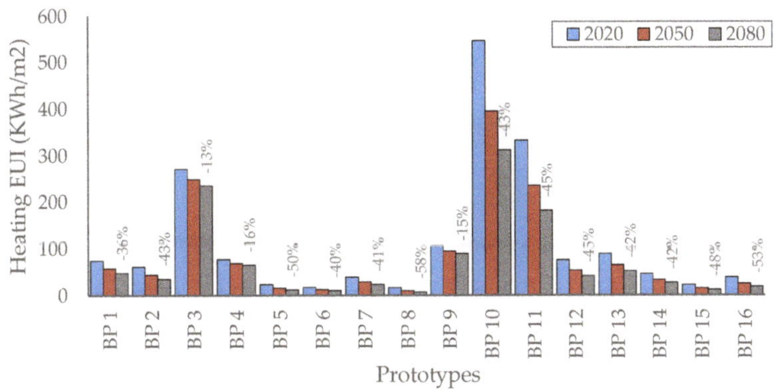

(a)

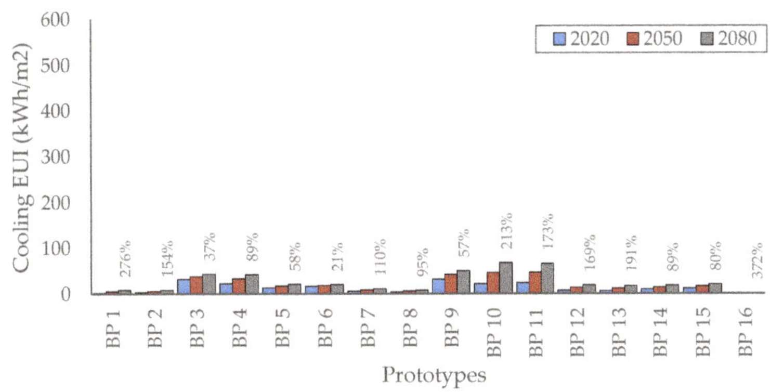

(b)

Figure 6. Heating EUI (**a**) and cooling EUI (**b**) in building prototypes from 2020 to 2080; the numbers demonstrate relative changes of values for 2080 compared to 2020.

On the other hand, the lowest relative change rate of the heating and cooling load was for the hospital (BP03) with a 13% decrease and the large office building (BP06) with a 58% increase. Due to global warming, which has already started, the need for cooling has increased relatively higher. Therefore, the demand for heating energy would decrease. Indeed, according to this figure, the role of cooling load, which is not as important as heating load due to its small share in the thermal load within the study period, would probably be more significant as this share is changing.

5. Discussion

Required data availability and quality are essential for successful scenario projections, especially to assess the built environment where accurate energy data are very critical [75]. Admittedly, using a different historical weather dataset for different periods could provide an overview of the accuracy of the results by comparing their outcomes. However, the unavailability of historic weather datasets is among the limitations of studies, which is why only one set of historical weather data was used in this study. Moreover, in contrast to megacities, there are various climate projections according to different climate models; Poznan's only available climate model for projection was the HadCM3 and IPCC A2 emission scenario.

It should be noted that solely comparing the relative changing rate of heating and cooling demand cannot give us a clear image of the thermal load's future as the initial magnitude of these figures was highly distinguished. In other words, although the relative changing rate of cooling demand was about three times higher than that of the heating load, however, as the share of cooling load in 2020 was considered negligible, this change would not change the pattern of total thermal load extensively. As shown in Figure 7, most BPs showed a relatively downward trend with different rates, with an average decrease of roughly 28 kWh/m^2 in thermal loads. To be more specific, while the average rate of changing thermal load was 20% (decrease), the minimum value was observed in the outpatient health care building (BP09) with 1% positive change, and the maximum value was for the warehouse building (BP16) with more than 50% negative change. Therefore, it can be concluded that the decrease rate in thermal loads in most cases was slight. The decrease rate also shows that despite the increasing trend of cooling load, its contribution in thermal load was negligible such as in the large hotel building (BP04) and the outpatient health care building (BP09) where the drop in the heating load was not sharp like in other cases, where the thermal load increased. The results predicted that while the average thermal load of cases was around 127 kWh/m^2, this figure changed for 2080 at around 98 kWh/m^2 with a more than 22% decrease. It was also reported that the slightest decrease in thermal load associated with the secondary school building (BP15) with 1.3 kWh/m^2, and the maximum value was for the fast-food restaurant building (BP10) with 188 kWh/m^2.

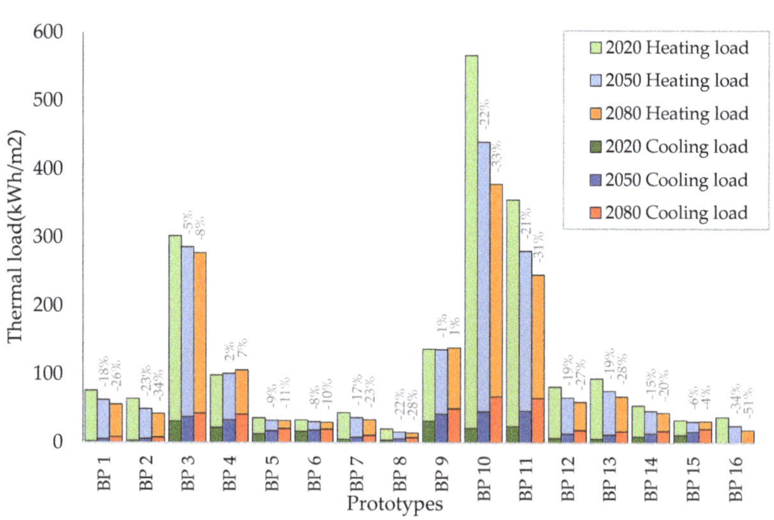

Figure 7. Change in the thermal load of building prototypes in 2020–2080 categorized by heating and cooling share; numbers demonstrate relative changes of values for 2050 and 2080 compared to 2020.

Despite the upward trend in the cooling loads, their contributions in thermal load were still lower than heating loads in most BPs. In a more detailed report (Figure 8), in 25% of prototypes (BP05, BP06, BP08, and BP15), the contribution of cooling and heating load in 2080 was reversed in another group with the same size (BP01, BP02, BP03, and BP16). Therefore, a considerable difference between the contribution of heating and cooling load has remained. Therefore, it can be concluded that the closer the heating and cooling contributions were, the more likely the contributions were to reverse.

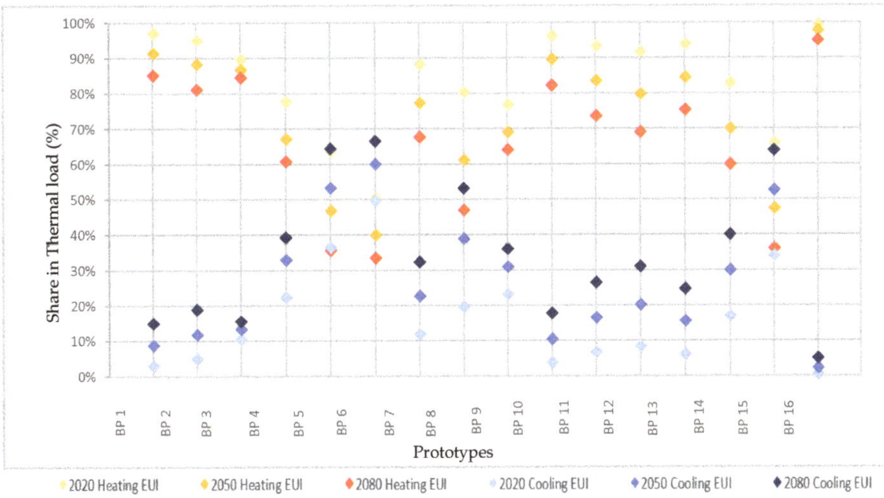

Figure 8. The share of thermal load of building prototypes in 2020, 2050, and 2080.

Therefore, considering the earlier discussions, it can be concluded that, in general, climate change heavily changed the energy performance of all building prototypes through increasing cooling and decreasing heating load. In this sense, as the contribution of cooling was more petite than heating in most cases, the total thermal load of most cases was reduced due to the decreased heating load. Furthermore, in one quarter of cases, the cooling increment altered heating and cooling contributions in thermal load. Finally, whenever increased heating load was reported, the total thermal load showed an upward trend in the study period.

6. Conclusions

These days, we are witnessing that the share of the building sector in the total world energy consumption has been reported to be critically high, and it has also been predicted to increase for several reasons. These changes may lead to higher GHG emissions, one of the significant reasons behind climate change that can potentially affect the buildings' thermal load. As the impacts of climate change on a building's energy consumption have been different from one region to another according to several studies, this paper presents the impact assessment of climate change on the energy performance buildings for Poznan city in Poland. Furthermore, the current state literature established the importance of considering future weather data to evaluate the buildings' energy performance. This is why generating future weather data based on statistical downscaling in the initial part of this paper enabled the authors to project the energy demand of prototypes in the future. Thus, one of the prevailing methods for generating future weather data, using the HadCM3 climate model and considering IPCC SRES A2 GHG scenario through statistical downscaling future weather data for Poznan for 2020–2080, was generated.

According to the developed future weather data, Poznan will see a notable increase in average temperature in the following 60 years. In this sense, changing the climate zone of Poznan would be a strong possibility. The results indicated that the cooling load of prototypes was in the mid-range, which is why the despite the negligible rate of cooling load in total thermal load compared to the share of heating load, almost all prototypes saw a sharp rise in cooling load and a slight fall in heating load at around 135% and 40% by 2080, respectively. This change led to altering the most significant share in total thermal load from heating load to cooling load in 25% of prototypes, but for other cases where heating load remained the highest share, the total thermal loads saw a steady decrease in the study period. Similar to other studies in this area, in general, it can be concluded that as the average temperature of Poznan will rise from 2020 to 2080, the climate zone of this region will probably shift as other scholars have had such predictions for cities with similar climatical conditions. As a result, the heating demand will decrease, and the cooling demand will substantially increase. However, what has been missing in the studies is the impact on the total thermal load. According to the results, the total thermal load of more than 85% of prototypes decreased because of the sharp decrease in heating loads. Therefore, by comparing the change rate of total thermal load and heating load, it is evident that the more heating load a prototype consumes, the more likely it will see a decrease in total thermal load in the future. It should also be noted that considering only the HadCM3 climate model for the A2 emission scenario was one of the limitations of this research.

The impacts of predicted climate change were seen in mild cooling seasons and warmer heating seasons. Buildings with high heating loads will likely decrease the total thermal load for the next 60 years for a cold Polish climate. Therefore, buildings that tend to sustain for decades should evaluate how their cooling load can be reduced as increasing it will lead to more GHG emissions and more extreme weather conditions. This evaluation means that to mitigate the impact of climate change on the energy load of buildings, it is highly critical to investigate ways to understand the energy resiliency of buildings. Future studies should consider a broader range of scenarios and use other climate models to increase the study's accuracy. Furthermore, conducting a comprehensive study about all major cities of Poland would be a real boon for Polish decision-makers.

Author Contributions: Conceptualization, H.B., P.P., A.N., and M.M.; Methodology, H.B., P.P., and S.H.; Software, H.B. and S.H.; Validation, H.B., P.P., and M.M.; Formal analysis, H.B., P.P., and S.H.; Investigation, H.B.; Resources, H.B. and S.H.; Data curation, H.B. and S.H.; Writing—original draft preparation, H.B. and S.H.; Writing—review and editing, P.P., A.N., and M.M.; Visualization, H.B. and S.H.; Supervision, P.P., A.N., and M.M.; Project administration, H.B. and A.N.; Funding acquisition, H.B. and A.N. All authors have read and agreed to the published version of the manuscript.

Funding: This research was funded by Poznan University of Technology, within the framework of the research project entitled "Mapping of architectural space, the history, theory, practice, contemporaneity II", grant number 0112/SBAD/0185.

Institutional Review Board Statement: Not applicable.

Informed Consent Statement: Not applicable.

Data Availability Statement: Not applicable.

Conflicts of Interest: The authors declare no conflict of interest.

References

1. Moazami, A.; Nik, V.M.; Carlucci, S.; Geving, S. Impacts of future weather data typology on building energy performance—Investigating long-term patterns of climate change and extreme weather conditions. *Appl. Energy* **2019**, *238*, 696–720. [CrossRef]
2. IPCC. *Synthesis Report Contribution of Working Groups I, II and III to the Fifth Assessment Report of the Intergovernmental Panel on Climate Change*; IPCC: Geneva, Switzerland, 2014.
3. Berardi, U.; Jafarpur, P. Assessing the impact of climate change on building heating and cooling energy demand in Canada. *Renew. Sustain. Energy Rev.* **2020**, *121*, 109681. [CrossRef]
4. Madlener, R.; Sunak, Y. Impacts of urbanization on urban structures and energy demand: What can we learn for urban energy planning and urbanization management? *Sustain. Cities Soc.* **2011**, *1*, 45–53. [CrossRef]
5. Cao, X.; Dai, X.; Liu, J. Building energy-consumption status worldwide and the state-of-the-art technologies for zero-energy buildings during the past decade. *Energy Build.* **2016**, *128*, 198–213. [CrossRef]
6. Bazazzadeh, H.; Nadolny, A.; Hashemi Safaei, S.S. Climate Change and Building Energy Consumption: A Review of the Impact of Weather Parameters Influenced by Climate Change on Household Heating and Cooling Demands of Buildings. *Eur. J. Sustain. Dev.* **2021**, *10*, 1. [CrossRef]
7. IEA. *Tracking Buildings 2020*; IEA: Paris, France, 2020.
8. Tabrizikahou, A.; Nowotarski, P. Mitigating the Energy Consumption and the Carbon Emission in the Building Structures by Optimization of the Construction Processes. *Energies* **2021**, *14*, 3287. [CrossRef]

9. Mancini, F.; Lo Basso, G. How Climate Change Affects the Building Energy Consumptions Due to Cooling, Heating, and Electricity Demands of Italian Residential Sector. *Energies* **2020**, *13*, 410. [CrossRef]
10. Yau, Y.H.; Hasbi, S. A review of climate change impacts on commercial buildings and their technical services in the tropics. *Renew. Sustain. Energy Rev.* **2013**, *18*, 430–441. [CrossRef]
11. Kheiri, F. A review on optimization methods applied in energy-efficient building geometry and envelope design. *Renew. Sustain. Energy Rev.* **2018**, *92*, 897–920. [CrossRef]
12. Gero, J.S.; D'Cruz, N.; Radford, A.D. Energy in context: A multicriteria model for building design. *Build. Environ.* **1983**, *18*, 99–107. [CrossRef]
13. European Commission (EC). *Energy Consumption of Buildings per m^2*; European Commission, 2013. Available online: https://ec.europa.eu/energy/content/energy-consumption-m%C2%B2-2_en (accessed on 30 May 2021).
14. Stuggins, G.; Sharabaroff, A.; Semikolenova, Y. *Energy Efficiency; Lessons Learned from Success Stories*; The World Bank: Washington, DC, USA, 2013. [CrossRef]
15. IEA. *Energy Policies of IEA Countries: Poland 2016 Review*; IEA: Paris, France, 2017.
16. Staniaszek, D.; Firląg, S. *Financing Building Energy Performance Improvement in Poland*; BPIE: Brussels, Belgium, 2016.
17. BuildDeskPolska. *Energy Condition of Buildings in Poland*; BuildDeskPolska: Cigacice, Poland, 2011.
18. Parysek, J.J.; Mierzejewska, L. Poznań. *Cities* **2006**, *23*, 291–305. [CrossRef]
19. BuildDeskPolska. *BuildDesk Analytics Report; Energy Condition of Buildings in Poland*; BuildDeskPolska: Cigacice, Poland, 2019.
20. European Union (EU). Energy Island Communities for Energy Transition. Available online: https://cordis.europa.eu/project/id/957845/pl (accessed on 17 February 2021).
21. Seyedzadeh, S.; Pour Rahimian, F.; Oliver, S.; Rodriguez, S.; Glesk, I. Machine learning modelling for predicting non-domestic buildings energy performance: A model to support deep energy retrofit decision-making. *Appl. Energy* **2020**, *279*, 115908. [CrossRef]
22. Eslamirad, N.; Malekpour Kolbadinejad, S.; Mahdavinejad, M.; Mehranrad, M. Thermal comfort prediction by applying supervised machine learning in green sidewalks of Tehran. *Smart Sustain. Built Environ.* **2020**, *9*, 361–374. [CrossRef]
23. Ghaffour, W.; Ouissi, M.N.; Velay Dabat, M.A. Analysis of urban thermal environments based on the perception and simulation of the microclimate in the historic city of Tlemcen. *Smart Sustain. Built Environ.* **2021**, *10*, 141–168. [CrossRef]
24. Thanu, H.P.; Rajasekaran, C.; Deepak, M.D. Developing a building performance score model for assessing the sustainability of buildings. *Smart Sustain. Built Environ.* **2020**. ahead-of-print. [CrossRef]
25. Rahmouni, S.; Smail, R. A design approach towards sustainable buildings in Algeria. *Smart Sustain. Built Environ.* **2020**, *9*, 229–245. [CrossRef]
26. Dell'Anna, F.; Bottero, M.; Becchio, C.; Corgnati, S.P.; Mondini, G. Designing a decision support system to evaluate the environmental and extra-economic performances of a nearly zero-energy building. *Smart Sustain. Built Environ.* **2020**, *9*, 413–442. [CrossRef]
27. Seyedzadeh, S.; Pour Rahimian, F. Building Energy Data-Driven Model Improved by Multi-objective Optimisation. In *Data-Driven Modelling of Non-Domestic Buildings Energy Performance: Supporting Building Retrofit Planning*; Seyedzadeh, S., Pour Rahimian, F., Eds.; Springer International Publishing: Cham, The Netherlands, 2021; pp. 99–109. [CrossRef]
28. Weiss, T. Energy flexibility and shiftable heating power of building components and technologies. *Smart Sustain. Built Environ.* **2020**. ahead-of-print. [CrossRef]
29. Sheikhkhoshkar, M.; Pour Rahimian, F.; Kaveh, M.H.; Hosseini, M.R.; Edwards, D.J. Automated planning of concrete joint layouts with 4D-BIM. *Autom. Constr.* **2019**, *107*, 102943. [CrossRef]
30. Kottek, M.; Grieser, J.; Beck, C.; Rudolf, B.; Rubel, F. World Map of the Köppen-Geiger climate classification updated. *Meteorol. Z.* **2006**, *15*, 5. [CrossRef]
31. Rosenthal, D.H.; Gruenspecht, H.K.; Moran, E.A. Effects of Global Warming on Energy Use for Space Heating and Cooling in the United States. *Energy J.* **1995**, *16*, 77–96. [CrossRef]
32. Nordhaus, W.D. To slow or not to slow: The economics of the greenhouse effect. *Econ. J.* **1991**, *101*, 17. [CrossRef]
33. Cline, W.R. *The Economics of Global Warming*; Institute for International Economics: Washington, DC, USA, 1992.
34. Nik, V.M.; Sasic Kalagasidis, A. Impact study of the climate change on the energy performance of the building stock in Stockholm considering four climate uncertainties. *Build. Environ.* **2013**, *60*, 291–304. [CrossRef]
35. Christenson, M.; Manz, H.; Gyalistras, D. Climate warming impact on degree-days and building energy demand in Switzerland. *Energy Convers. Manag.* **2006**, *47*, 671–686. [CrossRef]
36. Hosseini, M.; Tardy, F.; Lee, B. Cooling and heating energy performance of a building with a variety of roof designs; the effects of future weather data in a cold climate. *J. Build. Eng.* **2018**, *17*, 107–114. [CrossRef]
37. Cellura, M.; Guarino, F.; Longo, S.; Tumminia, G. Climate change and the building sector: Modelling and energy implications to an office building in southern Europe. *Energy Sustain. Dev.* **2018**, *45*, 46–65. [CrossRef]
38. Crawley, D.B. Estimating the impacts of climate change and urbanization on building performance. *J. Build. Perform. Simul.* **2008**, *1*, 91–115. [CrossRef]
39. Triana, M.A.; Lamberts, R.; Sassi, P. Should we consider climate change for Brazilian social housing? Assessment of energy efficiency adaptation measures. *Energy Build.* **2018**, *158*, 1379–1392. [CrossRef]

40. Wang, H.; Chen, Q. Impact of climate change heating and cooling energy use in buildings in the United States. *Energy Build.* **2014**, *82*, 428–436. [CrossRef]
41. Cox, R.A.; Drews, M.; Rode, C.; Nielsen, S.B. Simple future weather files for estimating heating and cooling demand. *Build. Environ.* **2015**, *83*, 104–114. [CrossRef]
42. Shen, P. Impacts of climate change on U.S. building energy use by using downscaled hourly future weather data. *Energy Build.* **2017**, *134*, 61–70. [CrossRef]
43. Rubel, F.; Brugger, K.; Klaus, H.; Auer, I. The climate of the European Alps: Shift of very high resolution Köppen-Geiger climate zones 1800–2100. *Meteorol. Z.* **2017**, *26*, 11. [CrossRef]
44. Velashjerdi Farahani, A.; Jokisalo, J.; Korhonen, N.; Jylhä, K.; Ruosteenoja, K.; Kosonen, R. Overheating Risk and Energy Demand of Nordic Old and New Apartment Buildings during Average and Extreme Weather Conditions under a Changing Climate. *Appl. Sci.* **2021**, *11*, 3972. [CrossRef]
45. Herrera, M.; Natarajan, S.; Coley, D.A.; Kershaw, T.; Ramallo-González, A.P.; Eames, M.; Fosas, D.; Wood, M. A review of current and future weather data for building simulation. *Build. Serv. Eng. Res. Technol.* **2017**, *38*, 602–627. [CrossRef]
46. Hall, I.J.; Prairie, R.R.; Anderson, H.E.; Boes, E.C. Generation of a typical meteorological year. In Proceedings of the 1978 Annual Meeting of American Section of ISES, Denver, CO, USA, 12–15 August 1978.
47. Lam, J.C.; Hui, S.C.M.; Chan, A.L.S. A Statistical Approach to the Development of a Typical Meteorological Year for Hong Kong. *Archit. Sci. Rev.* **1996**, *39*, 201–209. [CrossRef]
48. Petrakis, M.; Kambezidis, H.D.; Lykoudis, S.; Adamopoulos, A.D.; Kassomenos, P.; Michaelides, I.M.; Kalogirou, S.A.; Roditis, G.; Chrysis, I.; Hadjigianni, A. Generation of a "typical meteorological year" for Nicosia, Cyprus. *Renew. Energy* **1998**, *13*, 381–388. [CrossRef]
49. Bilbao, J.; Miguel, A.; Franco, J.; Ayuso, A. Test Reference Year Generation and Evaluation Methods in the Continental Mediterranean Area. *J. Appl. Meteorol.* **2004**, *43*, 390–400. [CrossRef]
50. Chan, A.L.S.; Chow, T.T.; Fong, S.K.F.; Lin, J.Z. Generation of a typical meteorological year for Hong Kong. *Energy Convers. Manag.* **2006**, *47*, 87–96. [CrossRef]
51. Jafarpur, P.; Berardi, U. Building energy demand within a climate change perspective. *IOP Conf. Series Mater. Sci. Eng.* **2019**, *609*, 072037. [CrossRef]
52. Kundzewicz, Z.W.; Piniewski, M.; Mezghani, A.; Okruszko, T.; Pińskwar, I.; Kardel, I.; Hov, Ø.; Szcześniak, M.; Szwed, M.; Benestad, R.E.; et al. Assessment of climate change and associated impact on selected sectors in Poland. *Acta Geophys.* **2018**, *66*, 1509–1523. [CrossRef]
53. Mezghani, A.; Dobler, A.; Haugen, J.E.; Benestad, R.E.; Parding, K.M.; Piniewski, M.; Kardel, I.; Kundzewicz, Z.W. CHASE-PL Climate Projection dataset over Poland—Bias adjustment of EURO-CORDEX simulations. *Earth Syst. Sci. Data* **2017**, *9*, 905–925. [CrossRef]
54. Mezghani, A.; Dobler, A.; Haugen, J.H. *CHASE-PL Climate Projections: 5-km Gridded Daily Precipitation & Temperature Dataset (CPLCP-GDPT5)*; Norwegian Meteorological Institute: Oslo, Norway, 2016. [CrossRef]
55. Thomson, A.M.; Calvin, K.V.; Smith, S.J.; Kyle, G.P.; Volke, A.; Patel, P.; Delgado-Arias, S.; Bond-Lamberty, B.; Wise, M.A.; Clarke, L.E.; et al. RCP4.5: A pathway for stabilization of radiative forcing by 2100. *Clim. Chang.* **2011**, *109*, 77. [CrossRef]
56. Riahi, K.; Rao, S.; Krey, V.; Cho, C.; Chirkov, V.; Fischer, G.; Kindermann, G.; Nakicenovic, N.; Rafaj, P. RCP 8.5—A scenario of comparatively high greenhouse gas emissions. *Clim. Chang.* **2011**, *109*, 33. [CrossRef]
57. Horton, B.P.; Khan, N.S.; Cahill, N.; Lee, J.S.H.; Shaw, T.A.; Garner, A.J.; Kemp, A.C.; Engelhart, S.E.; Rahmstorf, S. Estimating global mean sea-level rise and its uncertainties by 2100 and 2300 from an expert survey. *NPJ Clim. Atmos. Sci.* **2020**, *3*, 18. [CrossRef]
58. Guan, L. Preparation of future weather data to study the impact of climate change on buildings. *Build. Environ.* **2009**, *44*, 793–800. [CrossRef]
59. Laflamme, E.M.; Linder, E.; Pan, Y. Statistical downscaling of regional climate model output to achieve projections of precipitation extremes. *Weather Clim. Extrem.* **2016**, *12*, 15–23. [CrossRef]
60. Belcher, S.E.; Hacker, J.N.; Powell, D.S. Constructing design weather data for future climates. *Build. Serv. Eng. Res. Technol.* **2005**, *26*, 49–61. [CrossRef]
61. Wilby, R.L.; Wigley, T.M.L. Downscaling general circulation model output: A review of methods and limitations. *Prog. Phys. Geogr. Earth Environ.* **1997**, *21*, 530–548. [CrossRef]
62. van Paassen, A.H.; Luo, Q.X. Weather data generator to study climate change on buildings. *Build. Serv. Eng. Res. Technol.* **2002**, *23*, 251–258. [CrossRef]
63. Zhang, L.; Xu, Y.; Meng, C.; Li, X.; Liu, H.; Wang, C. Comparison of Statistical and Dynamic Downscaling Techniques in Generating High-Resolution Temperatures in China from CMIP5 GCMs. *J. Appl. Meteorol. Climatol.* **2020**, *59*, 207–235. [CrossRef]
64. Tootkaboni, M.P.; Ballarini, I.; Zinzi, M.; Corrado, V. A Comparative Analysis of Different Future Weather Data for Building Energy Performance Simulation. *Climate* **2021**, *9*, 37. [CrossRef]
65. Liu, S.; Kwok, Y.T.; Lau, K.K.-L.; Tong, H.W.; Chan, P.W.; Ng, E. Development and application of future design weather data for evaluating the building thermal-energy performance in subtropical Hong Kong. *Energy Build.* **2020**, *209*, 109696. [CrossRef]
66. Hosseini, M.; Bigtashi, A.; Lee, B. Generating future weather files under climate change scenarios to support building energy simulation—A machine learning approach. *Energy Build.* **2021**, *230*, 110543. [CrossRef]

67. Yassaghi, H.; Gurian, P.L.; Hoque, S. Propagating downscaled future weather file uncertainties into building energy use. *Appl. Energy* **2020**, *278*, 115655. [CrossRef]
68. Fiorillo, D.; Kapelan, Z.; Xenochristou, M.; De Paola, F.; Giugni, M. Assessing the Impact of Climate Change on Future Water Demand using Weather Data. *Water Resour. Manag.* **2021**. [CrossRef]
69. Jentsch, M.F. *Climate Change Weather File Generators. Technical Reference Manual for the CCWeatherGen and CCWorldWeatherGen Tools;* GB. University of Southampton Press: Southampton, UK, 2010.
70. Jentsch, M.F.; James, P.A.B.; Bourikas, L.; Bahaj, A.S. Transforming existing weather data for worldwide locations to enable energy and building performance simulation under future climates. *Renew. Energy* **2013**, *55*, 514–524. [CrossRef]
71. *CCWorldWeatherGen Tool;* University of Southampton press: Southampton, UK, 2009.
72. Deru, M.; Field, K.; Studer, D.; Benne, K.; Griffith, B.; Torcellini, P.; Liu, B.; Halverson, M.; Winiarski, D.; Rosenberg, M. *US Department of Energy Commercial Reference Building Models of the National Building Stock;* US Department of Energy Press: Washington, DC, USA, 2011.
73. Zhang, J.; Athalye, R.A.; Hart, P.R.; Rosenberg, M.I.; Xie, Y.; Goel, S.; Mendon, V.V.; Liu, B. *Energy and Energy Cost Savings Analysis of the IECC for Commercial Buildings;* Pacific Northwest National Lab. (PNNL): Richland, WA, USA, 2013.
74. Halverson, R.; Hart, R.; Athalye, W. *ANSI/ASHRAE/IES Standard 90.1-2013 Determination of Energy Savings: Qualitative Analysis;* Pacific Northwest National Lab. (PNNL): Richland, WA, USA, 2014.
75. US Department of Education (U.S.D.o.E.). *Commercial Building Prototype Models;* US Department of Energy Press: Washington, DC, USA, 2018.

energies

MDPI

Article

Analysis of Daily Energy Demand for Cooling in Buildings with Different Comfort Categories—Case Study

Imre Csáky

Faculty of Engineering, University of Debrecen, Ótemető Str., 2-4, 4028 Debrecen, Hungary; imrecsaky@eng.unideb.hu

Abstract: Climate change has a potential impact on the number of hot and torrid days in the summer period. Due to the occupants' comfort needs, and because of the high heat loads during the summer period, in several European countries, the energy used for air conditioning in buildings increased. With multiple environmental monitoring systems (Testo Saveris) in two similar offices, having west and east orientation of glazing, 1920 data (internal air, mean radiant temperature) related to operative temperature were collected in order to show the differences between heat loads of rooms with similar geometry in the same building. Data were measured in a 15 min interval. The diffuse and direct solar radiation had been determined for the horizontal and vertical surfaces, using the measured hourly global radiation (Debrecen, Hungary) data for the analyzed days (summer, hot, and torrid days). The local climatic results were compared with other climatic days used in different national standards. The daily energy need for cooling for different building comfort categories was also determined in the case of the representative days. The maximum daily energy need for cooling can be even 2.3 times higher for east orientation in comparison to the west orientation of the facades.

Keywords: operative temperature; cooling load; daily energy need; solar gains

Citation: Csáky, I. Analysis of Daily Energy Demand for Cooling in Buildings with Different Comfort Categories—Case Study. *Energies* **2021**, *14*, 4694. https://doi.org/10.3390/en14154694

Academic Editors: Andrea Mauri and Benedetto Nastasi

Received: 7 July 2021
Accepted: 27 July 2021
Published: 2 August 2021

1. Introduction

Extreme weather events, just like the summer period's hot and torrid days, are challenging the inhabitant's health and comfort. Numerous research had been performed, and plenty of scientific papers studying thermal comfort, indoor operative temperature, indoor air quality were published [1–3].

Majewsky et al. analyze the indoor environment quality in two different intelligent buildings, which are located in Poland. The authors stated that energy consumption worldwide is constantly rising. Most of the energy is used by the building sector, mainly for the daily operation of modern buildings, intelligent buildings for offices, and also for public use [4].

Increasing the airtightness of the buildings' envelope, special attention should be paid to the CO_2 concentration level in the indoor air. It was shown that the CO_2 concentration may exceed even 4000 ppm in the case of two persons performing office work if aeration of the room is not performed properly [5]. Bullová et al. have shown that the air change rate can be reduced to even 0.1–0.3 h^{-1} in the case of refurbished buildings [6]. In the last decades, the need for thermal comfort increased considerably. Several air conditioning systems were mounted in the existing office buildings, while new office buildings are equipped with cooling systems.

Thus, the energy use for cooling in the buildings increased substantially in the last years [7,8]. Mousavi Motlagh et al. have shown that most researchers analyzed the energy use in residential and office buildings [9].

In the European Commission Report related to energy performance in buildings, the annual final energy consumption per square meter in the residential sector in Hungary was 150 kWh/m^2 year in 2013 [10]. Hungarian Energy and Public Utility Regulatory Authority, in the yearly report, presents the energy consumption for cooling used by air conditioning systems in the summer period [11]. It was shown that the energy consumption for cooling increased from 271 TJ (2015) to 570 TJ (2019). Bazazzadeh et al., in their research, conclude that the cooling load will increase by 135% by 2080 in Poland [12]. The proper architectural design of buildings can mitigate the energy need for cooling. Long et al. and Reilly et al., in their research, present the thermal mass and heat storage capacity effects for the heat load of buildings [13,14].

The present paper focuses on the importance of operative temperature for different office building comfort categories and energy needs for cooling. In order to determine the operative temperature and the daily energy need for cooling in buildings, according to standards, the first step is to determine the local specific meteorological data, such as global diffuse radiation, diffuse and direct radiation on vertical surfaces, mean outdoor temperature, and so on. Duties of the Meteorological Observatory (Debrecen, Hungary, GPS coordinates: 47.53253, 21.62537) are the collection and measurement of global hourly radiation data and mean hourly outdoor temperatures.

In order to obtain an appropriate overview of the building's cooling needs, it is essential to review different standards on the topic. According to ISO-52000-1:2017, "Building: construction as a whole, including the fabric and all technical building systems, where energy may be used to condition the indoor environment" [15]. Cooling energy needs can be determined using the algorithm given by ISO 13790:2008 standard, as a room or enclosure, which is specified to be cooled to a given setpoint temperature. According to ISO standard 13790:2008, energy needs for cooling in a conditioned space have to maintain the expected temperature conditions until a given period of time. Solar heat gains are defined as heat that is provided by solar radiation, entering directly or indirectly through openings, opaque walls, and roofs, or by passive solar devices such as sunspace, transparent insulation, and solar walls [16]. The calculation methodology presented in ISO 13790:2008 standard was previously validated, and the results were published in different articles [17,18].

The aim of the present research is to discuss and analyze through a case study the effect of different building structures, size, and orientation of glazed areas on the indoor operative temperature for different comfort categories and on the daily cooling energy need in the case of hot and torrid days (selected from the 11 analyzed years).

2. Materials and Methods

In order to see the effects of the orientation of transparent surfaces, building structure, and glazing ratio on the operative temperature, measurements and simulations were simultaneously performed.

2.1. Methods for Measurements of the Operative Temperature

A multiple environmental monitoring system provided the opportunity for complex research related to building performance on different aspects, including thermal comfort, energy use, and indoor air quality.

With the expansion of wireless communication, data storage, small electronic devices, etc., will be able to transfer bigger data for increasing the building performance [19].

Firstly, it was identified a western and an eastern-orientation office in a 3-story building with heavy structure. The selected offices are located on the upper building level (similar shading) and have similar geometry. In these offices, a Testo Saveris measurement system was placed (Figure 1a,b). This measurement system can be used for monitoring building internal air and mean radiant temperature.

Very long distances can be bridged using a converter (5), which converts the radio signals of the probe or router and then transmits this measurement data to the base via an Ethernet cable. The Ethernet probes (6) can also be connected to the base using an Ethernet cable. NTC sensor measuring range of the Saveris T2/T2D is from −35 to +50 °C, and the accuracy is ±0.2 °C (−25 to +70 °C) [20].

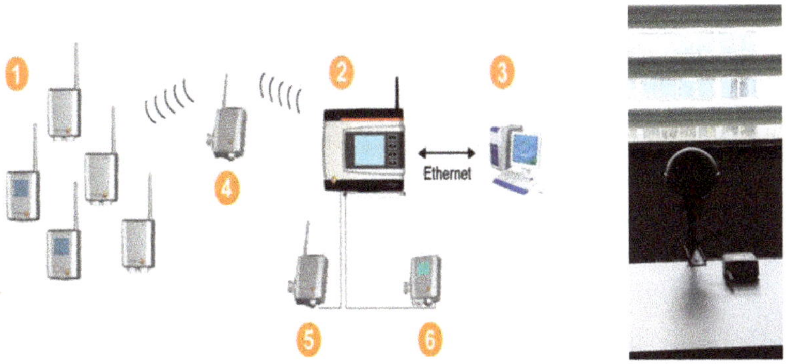

(a) (b)

Figure 1. The Testo Saveris measurement system [20]. (**a**) Testo Saveris base system (**b**) globe sensorWith this measurement system, ambient or process data for temperature and humidity in sealed rooms and/or during transportation is measured and recorded using probes (1). These measured values are transmitted by radio to the Saveris base (2) and saved. A router (4) can be used to optimize the radio signal in the event of difficult structural conditions. The data is then called up from a computer (3) by the Saveris base and saved to a database [20].

The dimension of the analyzed offices is 7.5 × 6 × 4.5 m (Figure A1 in Appendix A). The windows dimension is 1.32 × 5.44 m. The overall heat transfer coefficients of external building elements were U_{wall} = 0.28 W/m^2 K; U_{window} = 1.1 W/m^2 K; and $U_{flatroof}$ = 0.2 W/m^2 K. The operative temperature was measured on the working places, which are located at 1.5 m from the window, similar for both offices (Figure 2a,b). Measurements were carried out in the summer period when no occupants were in the rooms (holiday 3 August 2018–7 August 2018). The HVAC systems were switched off.

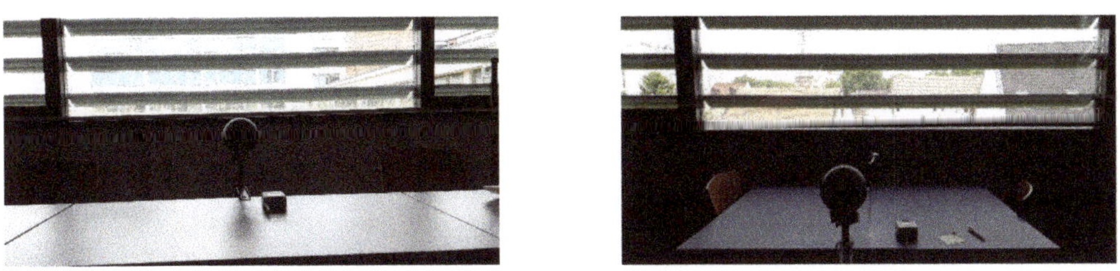

(a) (b)

Figure 2. Measurements of the Operative temperature (**a**) East-orientation office; (**b**) west-orientation office.

2.2. Methods for Calculation of the Operative Temperature and Daily Energy Need for Cooling

The elements of a cooling system are chosen depending on the setpoint operative temperature and the maximum value of the operative temperature in a room. In order to determine the daily variation of the operative temperature, different calculation methods can be used. In the following, the method given by standard EN 13790 was used. The validation of the calculation model was performed and presented in previously published papers [17,18]. Furthermore, the day for which the calculation is performed has to be chosen carefully. This has to be a torrid day, but for energy need for a day or a whole summer period, the hot days and summer days are important as well.

The calculation method uses the RC network of the heat flows, as is shown in Figure 3 [16].

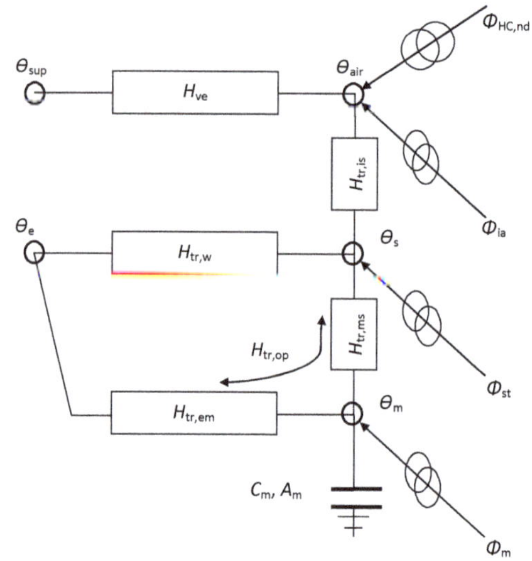

Figure 3. RC network heat flows [16].

The internal heat capacity of the building zone, C_m, (J/K), was calculated for a maximum thickness of 10 cm.

The operative temperature is obtained using Equation (1), [16]:

$$\theta_{op} = 0.3\theta_{air} + 0.7\theta_s \tag{1}$$

The cooling need is [16]:

$$\phi_{HC,nd,un} = \frac{\phi_{HC,nd10}(\theta_{air,set} - \theta_{air,0})}{(\theta_{air,10} - \theta_{air,0})} \tag{2}$$

where:

$$\phi_{HC,nd10} = 10A_f \tag{3}$$

The ASHRAE 90.1 standard presents the prototype buildings. In the standard 16 prototype buildings are presented: apartment, hospital, hotel, office (large, medium, small), restaurant, school. The room dimensions used for calculus in the present paper is a model room for each prototype building defined in the ASHRAE 90.1 standard [21].

In buildings with various functions, rooms have different dimensions. In the case of office buildings, the geometrical sizes of rooms depends on the number of persons working in the office. Usually, the dimensions of a room result from a grid. Generally, in the case of a suspended ceiling, the dimension of one element of the grid is 600 × 600 mm. In the following, for cooling load calculus, a room with a 4.8 × 4.8 × 3.0 m dimension is taken into account.

The overall heat transfer coefficients of external building elements were U_{wall} = 0.226 W/m^2 K; U_{window} = 1.0 W/m^2 K; and $U_{flatroof}$ = 0.166 W/m^2 K.

The room is located in the corner of the attic of a three-story building. The window has no shading, the solar energy transmittance: g = 0.7, and the glazed ratio is 85% (in all cases, 15% is frame). The air change rate was considered ACH = 0.5 h^{-1}.

The window glazed of analyzed room ratio is 85% (for example: exterior glass wall, window wall).

According to Hungarian regulation related to the energy performance of buildings [22], the building is considered light structured if the specific thermal mass is lower or equal to

400 kg/m^2, and the building is considered heavy structured if the specific thermal mass exceeds 400 kg/m^2.

Four different cases were analyzed for each window orientation.

(a) Heavy construction, external wall with 50% window;
(b) Light construction, external wall with 50% window;
(c) Heavy construction, external wall with 100% window;
(d) Light construction, external wall with 100% window.

3. Input Data for Calculation of the Operative Temperature and Daily Energy Need for Cooling

In the last decades, the heat waves are longer lasting, and their number increases. During the heatwaves, besides health problems [23], the electrical energy consumption in the building sector caused by air conditioners operation at maximum capacity raises substantially.

The local design standards for energy efficiency do not contain the input local climatic data, which are: hourly values of the external air temperature and solar radiation intensity.

Hourly global radiation data and mean outdoor temperature were provided by the local Agro-Meteorological Observatory Debrecen, Hungary [24].

In order to determine the local meteorological data, the calculation method for the selected days was presented in several articles [17,18].

According to the Hungarian Meteorological Service, those days that have a maximum temperature higher than 25 °C are called "Summer days". If the maximum temperature value is higher or equal to 30 °C, the day is defined as a "Hot day". "Torrid days" have the maximum daily temperature equal to or higher than 35 °C [25].

Between 2009 and 2019, 514 summer days, 247 hot days, and 21 torrid days were registered.

The daily maximum temperature of the selected summer day was 25.3 °C. The chosen hot day had a maximum daily external air temperature of 32.4 °C, while the torrid day had a maximum external air temperature of 35.6 °C.

The local climatic data for the summer, hot and torrid days are presented in Table 1 (and Figures A3–A6).

Table 1. Maximum radiation values and energy yield for hottest days [26–28].

Day	Measure Units	I_{GH}	I_{difH}	I_{difV}	$I_{dirSouth}$	$I_{dirEast}$	$I_{dirWest}$	$I_{dirNorth}$
clear sky day								
Summer day	W/m^2	927	153	76	390	676	673	169
	Wh/m^2 day	8110	1338	669	2295	3091	3136	416
Hot day	W/m^2	884	146	73	377	577	689	166
	Wh/m^2 day	7687	1268	634	2236	3038	3140	372
Torrid day	W/m^2	871	144	72	368	531	777	99
	Wh/m^2 day	7423	1225	612	2231	2643	3148	266
different standards								
MSZ-04-140-2 Hungary	W/m^2	900	188	192	317	561	561	193
	Wh/m^2 day	8156	2049	1969	1752	2695	2862	802
Urban INCD Romania	W/m^2	914	149	75	373	681	527	221
	Wh/m^2 day	8327	1493	747	2210	3636	3233	862
DIN 4710 Germany	W/m^2	861	150	192	407	542	542	84
	Wh/m^2 day	7479	1680	2442	2409	2597	2597	338

The clear sky days are important because the cloudiness is omitted (direct radiation values are higher than for real days, while the diffuse radiation values are lower).

Furthermore, the input meteorological data given by different national standards are included in Table 1.

Standard 04-140-2 (national Hungarian Standard) was previously in use [26]. This standard provides the solar radiation and external air temperature data for heat load calculation until 2012.

In standard [27], local meteorological data are mentioned, related to Oradea city, Romania, which is located near Debrecen, Hungary. The distance between the two cities is 60 km.

Standard [28] contains only statistic days, and real days are not included.

4. Results and Discussion

In the following, the measured and calculated operative temperatures and the determined daily energy need for cooling are presented.

4.1. Measurements of the Operative Temperature

The gathered values of the operative temperature and outdoor dry temperature are illustrated in Figure 4a,b.

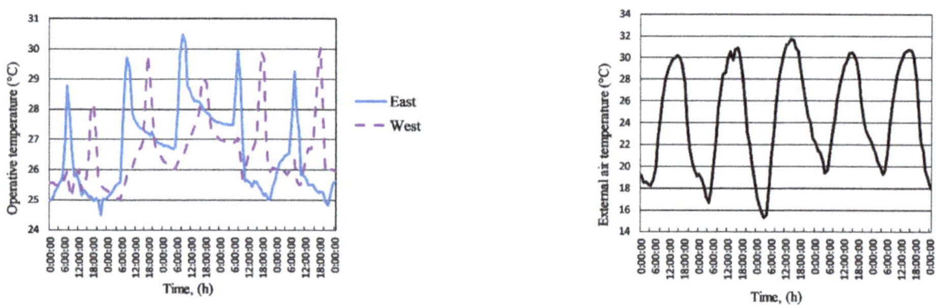

(**a**) operative temperature in offices (**b**) external air temperature (3 August 2018–7 August 2018)

Figure 4. Measured operative and external temperature.

It can be observed that the operative temperature in the offices followed the variation of the external temperature (three subsequent days increase, two subsequent days decrease). Moreover, the maximum values can be similar (second and fourth days) or different, depending on the cloudiness. As in the offices, the working hours starts at 8^{00} a.m. and ends at 17^{00} p.m. the west-oriented offices are advantageous from the cooling point of view. The question is: how much cooling energy can be saved renting a west-oriented office instead of an east-oriented office?

4.2. Calculation of the Operative Temperature

The results of maximum operative temperature for all analyzed cases are presented in Tables 2 and 3.

It can be observed that in Tables 2 and 3, the highest operative temperatures were obtained to east and west orientation of window for all analyzed days and windows to wall ratios.

The hourly operative temperature for south- and north-orientation glazed areas are presented in Figures A7–A14.

Table 2. Maximum operative temperature for heavy construction buildings.

Heavy Construction	50% Window				100% Window			
Day	South	North	East	West	South	North	East	West
	Maximum Operative Temperature (°C)							
Summer day	26.78	24.95	27.38	27.61	29.69	25.89	30.95	31.56
Hot day	27.58	25.71	28.19	28.48	30.56	26.87	31.82	32.62
Torrid day	27.98	26.13	28.27	28.90	31.03	27.13	31.60	33.01

Table 3. Maximum operative temperature for light construction buildings.

Light Construction	50% Window				100% Window			
Day	South	North	East	West	South	North	East	West
	Maximum Operative Temperature (°C)							
Summer day	34.85	26.99	36.43	39.36	45.96	30.59	50.33	55.59
Hot day	37.15	29.93	37.96	42.45	48.26	33.67	50.94	58.95
Torrid day	38.97	31.09	38.60	43.54	50.26	34.42	49.44	59.30

In Figures 5–8, both for the east and west orientation of the glazed area, the hourly operative temperatures are presented.

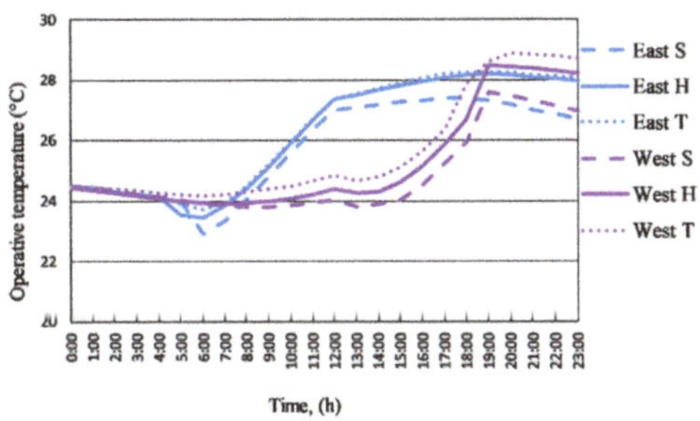

Figure 5. Heavy construction, 50% window (summer, hot, torrid day).

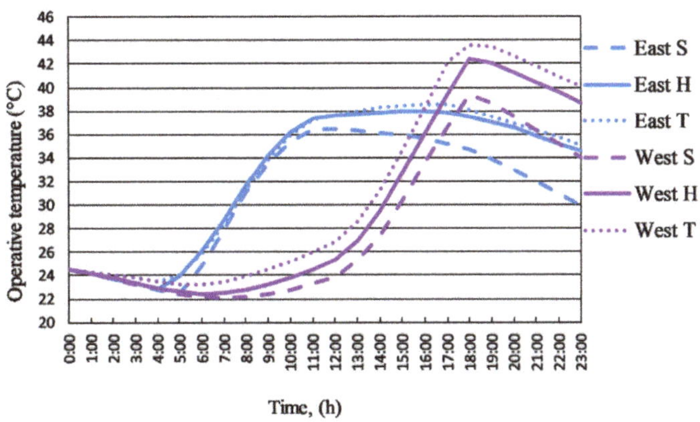

Figure 6. Light construction, 50% window (summer, hot, torrid day).

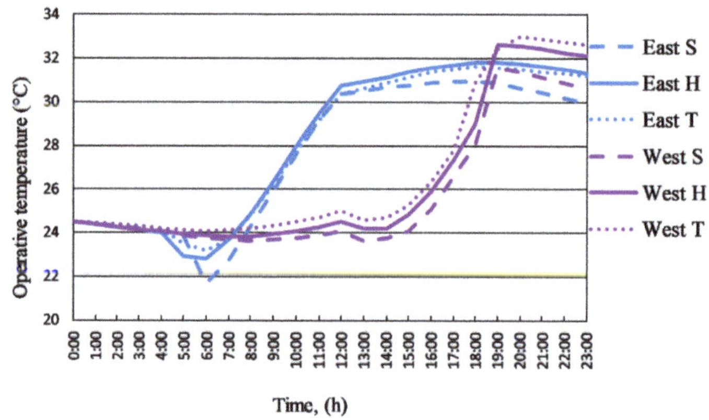

Figure 7. Heavy construction, 100% window (summer, hot, torrid day).

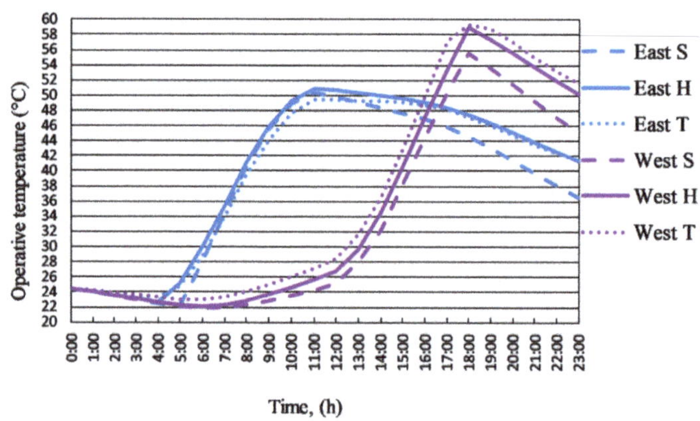

Figure 8. Light construction, 100% window (summer, hot, torrid day).

Assuming the upper limit of comfort operative temperature 26.0 °C, the overheating hours in the summer, hot and torrid days for east and west orientation of window can be determined based on Figures 5–8.

It can be observed that the 26 °C setpoint temperature is exceeded for both orientations in each analyzed day. If there is no cooling system installed, the overheating generates thermal discomfort, which leads to health problems and a decrease in work productivity [29–31]. In the case of offices with air cooling systems, the length of overheating period determines the operation time of the cooling system, while the amplitude of overheating determines the installed cooling capacity.

4.3. Daily Energy Need for Cooling

In the Technical Report CR 1752:1998 Ventilation for buildings—Design criteria for the indoor environment, different building types, and spaces, the design criteria are defined [32].

Single/landscape office operative temperatures in summer and cooling season for different comfort categories:

- A 24.5 ± 1;
- B 24.5 ± 1.5;
- C 24.5 ± 2.5 [32].

In the following, the energy calculations were performed for the heavy and light construction using the clear sky day meteorological data, the daily energy need for cooling was determined. For each building category, the minimum and maximum acceptable operative temperatures were taken into consideration. The results are presented in Tables 4–9. Obviously, with intelligent and properly operated external shading structures, the daily energy need for cooling can be reduced substantially [33].

It can be observed that for all analyzed cases, the maximum value of daily energy demand for cooling was obtained for the east orientation of the glazed area.

Table 4. Daily energy need for cooling of analyzed room. heavy construction, summer day (Wh/day).

Heavy Construction		50% Window				100% Window			
Comfort Category	**Operative Temperature**	**West**	**East**	**West**	**East**	**West**	**East**	**West**	**West**
A	23.5	27,005	19,047	36,211	24,288	38,663	22,176	59,367	34,463
	25.5	7588	111	15,008	6623	20,359	2657	38,445	16,754
B	23	34,599	26,641	43,810	31,882	46,108	29,621	65,500	40,908
	26	5147	0	11,376	5036	17,715	838	34,598	15,058
C	22	49,786	41,828	59,008	47,069	60,999	44,512	80,402	55,799
	27	957	0	4800	2448	12,693	0	27,320	11,874

Table 5. Daily energy need for cooling of analyzed room. Light construction, summer day (Wh/day).

Light Construction		50% Window				100% Window			
Comfort Category	**Operative Temperature**	**South**	**North**	**East**	**West**	**South**	**North**	**East**	**West**
A	23.5	34,115	15,897	51,263	31,905	61,148	24,443	94,710	58,189
	25.5	24,934	6125	40,369	23,826	51,958	14,018	84,105	49,777
B	23	37,770	19,552	54,936	35,561	64,547	27,987	98,395	61,507
	26	23,414	4738	38,223	22,653	50,269	12,265	81,924	48,364
C	22	45,082	26,864	62,281	42,872	71,882	35,322	105,765	68,842
	27	20,390	2314	34,239	20,396	46,892	9099	77,561	45,795

Table 6. Daily energy need for cooling of analyzed room. Heavy construction, hot day (Wh/day).

Heavy Construction		50% Window				100% Window			
Comfort Category	**Operative Temperature**	**South**	**North**	**East**	**West**	**South**	**North**	**East**	**West**
A	23.5	32,218	23,730	41,973	48,066	38,688	27,756	64,523	39,640
	25.5	11,501	2900	19,944	16,497	21,051	7158	43,857	20,844
B	23	39,812	31,324	49,572	59,598	44,548	35,202	71,975	47,085
	26	9060	971	16,290	13,379	18,804	4649	40,010	19,001
C	22	55,000	46,512	64,770	82,663	56,268	50,092	86,877	61,976
	27	4457	0	9421	8107	14,595	1280	32,463	15,541

Table 7. Daily energy need for cooling of analyzed room. Light construction, hot day (Wh/day).

Light Construction		50% Window				100% Window			
Comfort Category	**Operative Temperature**	**South**	**North**	**East**	**West**	**South**	**North**	**East**	**West**
A	23.5	42,149	23,995	60,304	40,598	69,113	32,695	104,538	67,400
	25.5	32,373	13,661	48,711	31,305	59,623	22,214	92,951	58,418
B	23	45,805	27,651	63,976	44,253	72,781	36,363	108,223	71,068
	26	30,720	12,083	46,565	29,927	57,935	20,384	90,586	56,870
C	22	53,116	34,962	71,322	51,565	80,116	43,698	115,593	78,403
	27	27,618	9215	42,273	27,350	54,558	17,063	86,126	53,995

Table 8. Daily energy need for cooling of analyzed room. Heavy construction, torrid day (Wh/day).

Heavy Construction		50% Window				100% Window			
Comfort Category	Operative Temperature	South	North	East	West	South	North	East	West
A	23.5	36,255	28,097	43,568	34,137	48,694	31,763	63,764	44,667
	25.5	14,042	5274	20,607	11,976	27,516	9438	42,073	22,988
B	23	43,849	35,691	51,167	41,730	56,140	39,208	71,215	521,12
	26	11,237	2735	16,898	9979	24,588	6584	38,227	21,144
C	22	59,037	50,879	66,365	56,918	71,030	54,099	86,117	67,003
	27	6295	0	9634	6559	19,193	2167	30,773	17,456

Table 9. Daily energy need for cooling of analyzed room. Light construction, torrid day (Wh/day).

Light Construction		50% Window				100% Window			
Comfort Category	Operative Temperature	South	North	East	West	South	North	East	West
A	23.5	48,662	29,856	61,753	47,665	76,015	38,355	101,151	75,241
	25.5	37,221	18,069	49,415	36,359	64,827	26,573	88,787	64,134
B	23	52,318	33,512	65,426	51,321	79,682	42,022	104,836	78,909
	26	35,473	16,268	47,269	34,707	62,989	24,577	86,453	62,446
C	22	59,629	40,823	72,771	58,632	87,018	49,358	112,206	86,244
	27	32,168	12,997	42,977	31,722	59,487	20,913	82,093	59,211

5. Conclusions

The majority of comfort and energy standards do not contain external meteorological data. In the case of calculation of the operative temperature and energy need for cooling, it is essential to know the values and variation of the external air temperature and solar radiation. Therefore, meteorological data over the past 11 years were analyzed. For this period, 514 summer, 274 hot, and 21 torrid days were identified. In order to determine and investigate the daily cooling energy need, one summer day, one hot day, and one torrid day was chosen.

In the frame of a case study, the operative temperature was measured and analyzed in two similar office rooms, having east and west orientation of glazing, in a 5-day period. The results enlighten that the energy consumption for cooling and the operational costs for west-orientation offices might be advantageous.

With the use of a validated model, the operative temperature was calculated for different building structures and window dimensions. For different comfort categories, the daily energy need for cooling was determined.

The cooling energy demand was calculated in the case of the analyzed days in light and heavy construction buildings, with different sizes and orientations of windows. It was also determined the differences that occur in the case of daily energy demand under summer-hot and torrid days' radiation conditions.

The upper limit of the operative temperature in a B comfort category office is 26.0 °C. In the case of the south orientation of the glazed area, for this operative temperature, assuming 100% of the daily energy need for cooling in a summer day for a heavy structure building with 100% window, it can be observed that for hot days the daily energy need will be 106.1%, while for torrid days 138.8%. In the case of a light structure building, on a summer day is 283.8%, hot day, the daily energy need for cooling is 327%, while for a torrid day, the increase is 355.6%.

Keeping the same reference value for 100% of the daily cooling energy need but decreasing the glazed area to 50%, the daily energy need for cooling decreases to 29.1% (summer day), 51.1% (hot day), and torrid day 63.4% (torrid day) in the case of heavy structure.

For light structure, the daily cooling energy needs increase to light structures 132.2% (summer day), 173.4% (hot day), and 200.2% (torrid day).

In the glazing area north orientation, the daily energy need for cooling is substantially lower in comparison to other glazing orientations.

Analyzing the results obtained for cooling energy need (east and west orientation), it can be observed that for heavy structure, the daily operative temperature for west orientation exceeds with 1.4–2.8 K the values obtained for east orientation, while the daily energy need for cooling can be even 2.3 times higher for east orientation in comparison on the west orientation of the facades.

In the case of light construction, the operative temperature differences can be 4.7–6.7 K, while the daily energy need for cooling can be 1.69 times higher for east orientation in comparison with the west orientation of facades.

The data presented in Tables 4–9 might be useful for building renters and operators. The main point is that torrid days, hot and summer days are the most important because these days will be operating days of the cooling and ventilation systems, giving the energy consumption in a month or for a whole year.

A future research perspective is the analysis of the effect of the use of intelligent external blinds and PCMs on the cooling energy need. Furthermore, the relation between the behavior of occupants and cooling energy use has to be investigated.

Funding: This research was funded by Hungarian Academy of Sciences: BO/00237/18.

Institutional Review Board Statement: Not applicable.

Informed Consent Statement: Not applicable.

Data Availability Statement: Not applicable.

Acknowledgments: This paper was supported by the János Bolyai Research Scholarship of the Hungarian Academy of Sciences. The author would like to express their gratitude to the Agro-Meteorological Observatory, Debrecen, for providing indispensable meteorological data.

Conflicts of Interest: The author declares no conflict of interest.

Nomenclature

H_{ve}	Heat transfer by ventilation (W/K)
θ_{air}	Air temperature node (°C)
θ_{sup}	The supply temperature (°C)
$H_{tr,w}$	Heat transfer by transmission is split into the window segment, (W/K)
$H_{tr,op}$	The thermal mass (W/K)
θ_s	(A mix of θ_{air} and mean radiant temperature) and the node representing the mass of the building zone, θ_m (°C)
C_m	Thermal capacity (J/K)
Φ_{int}	Heat flow rate given by internal heat sources (W)
Φ_{sol}	Solar heat sources (W)
Φ_{HC}	nd Energy need for cooling (W)
A_f	Is the conditioned area (m^2)
$\theta_{air,10}$	The air temperature obtained for a heating power of 10 W/m^2 (°C)
$\theta_{air,0}$	Is the air temperature in free floating conditions (°C)
$\theta_{air,set}$	Air temperature (°C)
S	Summer day
H	Hot day
T	Torrid day
I_{GH}	Is the global solar radiation on horizontal surfaces (W/m^2)
I_{dirH}	Direct solar radiation on horizontal surfaces (W/m^2)
I_{difH}	Diffuse solar radiation on horizontal surfaces (W/m^2)
I_{difV}	Diffuse solar radiation on vertical surfaces (W/m^2)
$I_{dirSouth}$, $I_{dirEast}$, $I_{dirWest}$, $I_{dirNorth}$	Direct solar radiation on vertical surfaces (W/m^2)

Appendix A

Methods for Measurements of the Operative Temperature

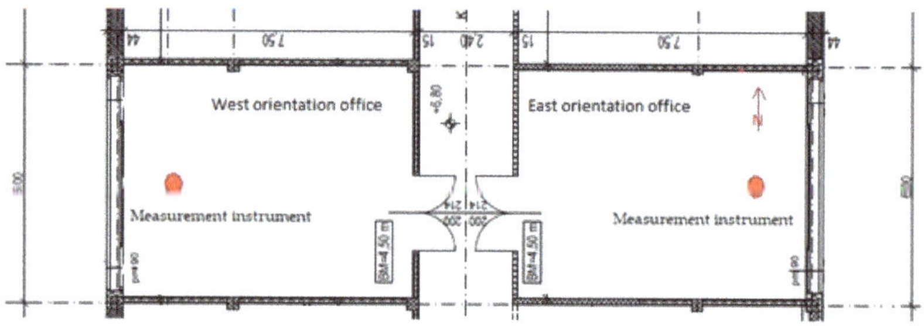

Figure A1. Plan of the offices.

West-orientation façade West-orientation façade East-orientation façade

Figure A2. Façade of the building with windows oriented to east and west.

Input Data for Calculation of the Operative Temperature and daily Energy Need for Cooling

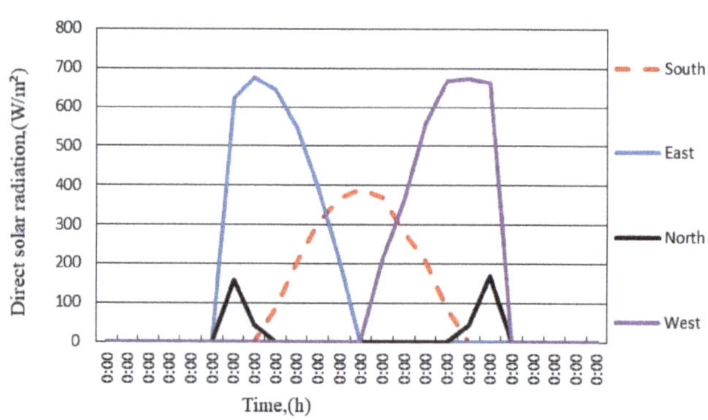

Figure A3. Direct solar radiation values, summer day.

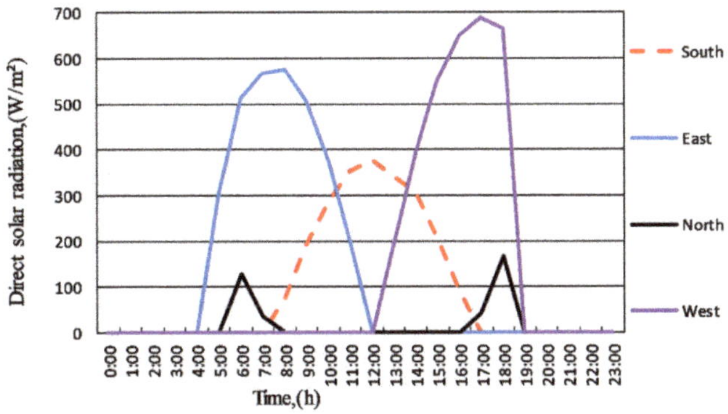

Figure A4. Direct solar radiation values, hot day.

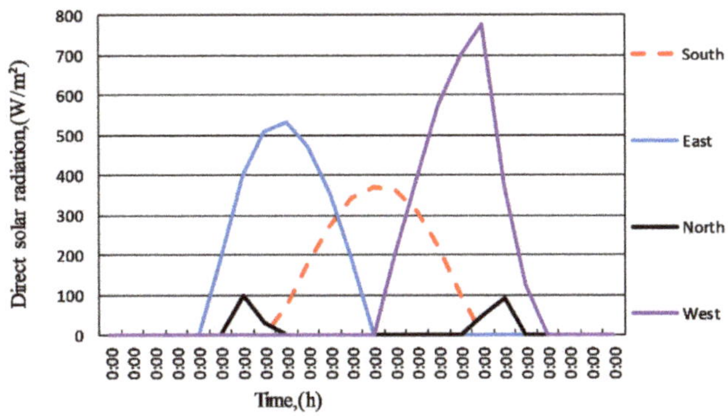

Figure A5. Direct solar radiation values, torrid day.

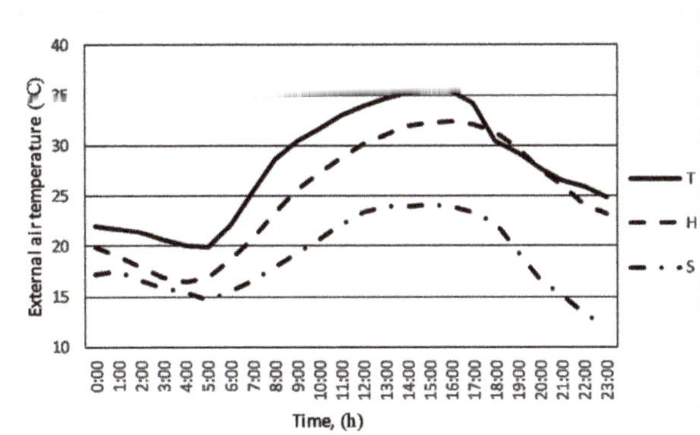

Figure A6. External air temperature, S, H, T days.

Methods for Calculation of the Operative Temperature and Energy Need for Cooling Calculation of the Operative Temperature

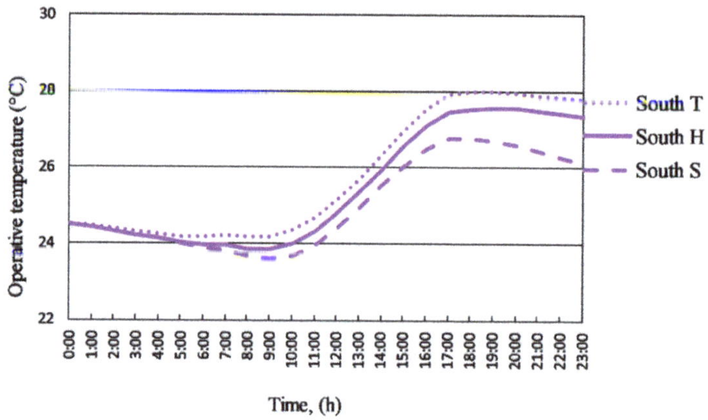

Figure A7. Heavy construction, 50% window (summer, hot, torrid day), south orientation.

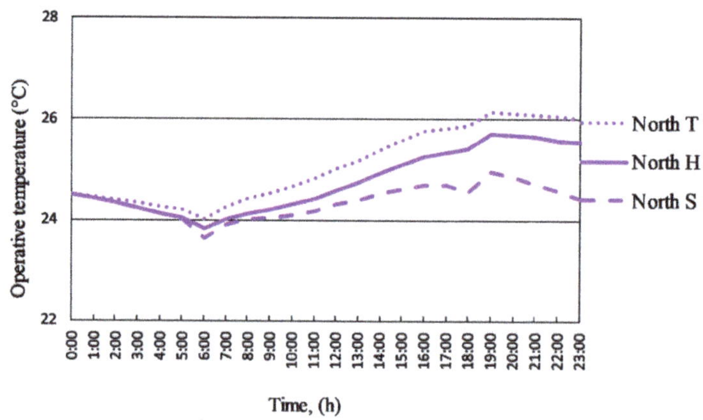

Figure A8. Heavy construction, 50% window (summer, hot, torrid day), north orientation.

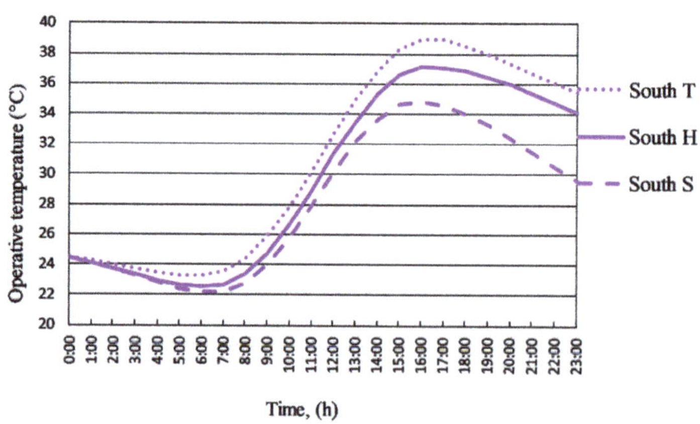

Figure A9. Light construction, 50% window (summer, hot, torrid day), south orientation.

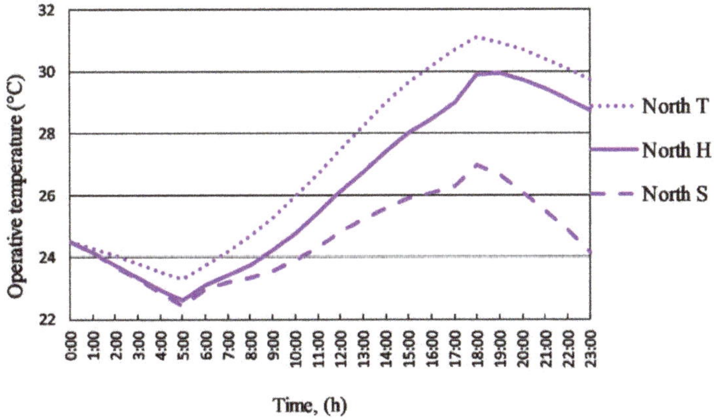

Figure A10. Light construction, 50% window (summer, hot, torrid day), north orientation.

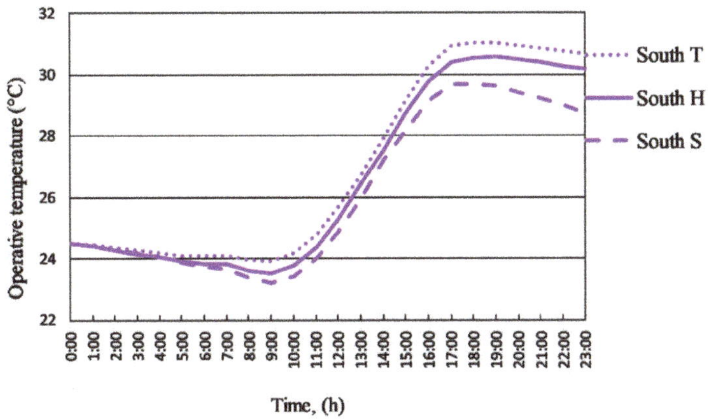

Figure A11. Heavy construction, 100% window (summer, hot, torrid day), south orientation.

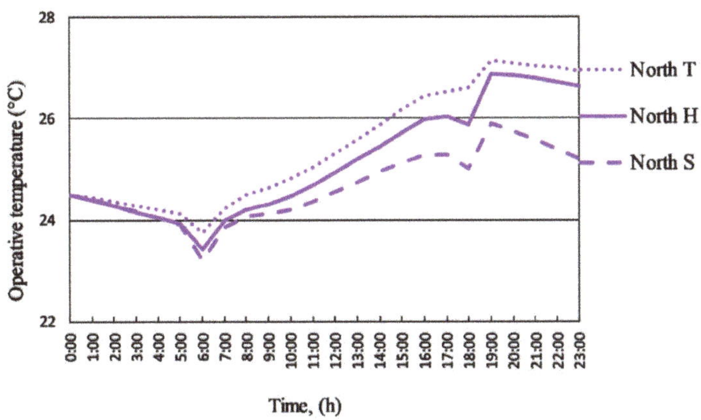

Figure A12. Heavy construction, 100% window (summer, hot, torrid day), north orientation.

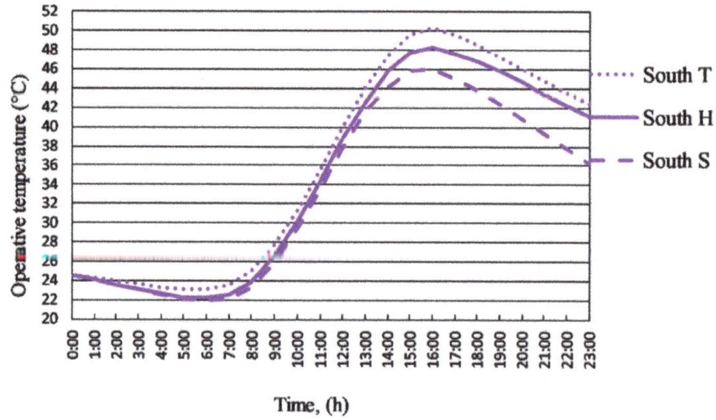

Figure A13. Light construction, 100% window (summer, hot, torrid day), south orientation.

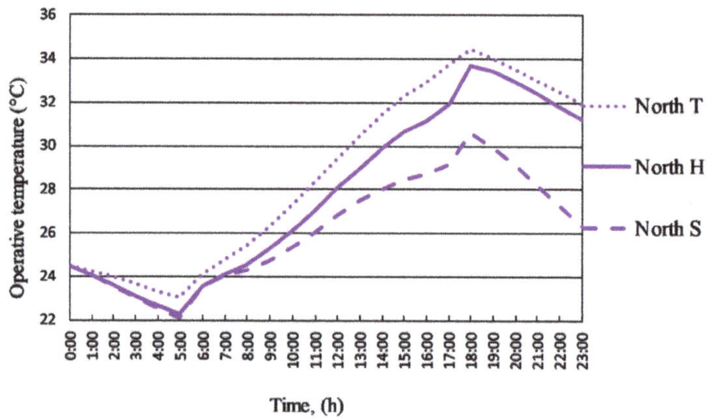

Figure A14. Light construction, 100% window (summer, hot, torrid day), north orientation.

References

1. Szabó, L.G.; Kalmár, F. Parametric Analysis of Buildings' Heat Load Depending on Glazing: Hungarian Case Study. *Energies* **2018**, *11*, 12. [CrossRef]
2. Szabó, L.G.; Kalmár, F. Investigation of energy and exergy performances of radiant cooling systems in buildings: A design approach. *Energy* **2019**, *185*, 449–462. [CrossRef]
3. Manfren, M.; Nastasi, B.; Piana, E.; Tronchin, L. On the link between energy performance of building and thermal comfort: An example. *AIP Conf. Proc.* **2019**, *2123*, 020066. [CrossRef]
4. Majewski, G.; Orman, L.J.; Telejko, M.; Radek, N.; Pietraszek, J.; Dudek, A. Assessment of Thermal Comfort in the Intelligent Buildings in View of Providing High Quality Indoor Environment. *Energies* **2020**, *13*, 1973. [CrossRef]
5. Kalmár, T.; Kalmár, F. Investigation of natural aeration in home offices during the heating season: Case study. *J. Build. Eng.* **2021**, *35*, 102052. [CrossRef]
6. Bullová, I.; Kapolo, P.; Katunský, D. Quantification of Air Change Rate by Selected Methods in a Typical Apartment Building. *Buildings* **2021**, *11*, 174. [CrossRef]
7. Santamouris, M. Cooling the buildings—Past, presents and future. *Energy Build.* **2016**, *128*, 617–638. [CrossRef]
8. Huang, K.-T.; Hwang, R.-L. Future trends of residential building cooling energy and passive adaptation measures to counteract climate change: The case of Taiwan. *Appl. Energy* **2016**, *184*, 1230–1240. [CrossRef]
9. Mousavi Motlagh, S.F.; Sohani, A.; Djavad Saghafi, M.; Sayyaadi, H.; Nastasi, B. The Road to Developing Economically Feasible Plans for Green, Comfortable and Energy Efficient Buildings. *Energies* **2021**, *14*, 636. [CrossRef]

10. European Commission. Commission Staff Working Document Evaluation of Directive 2010/31/EU on the Energy Performance of Buildings. Available online: https://ec.europa.eu/info/sites/default/files/swd-2016-408-final_en_0.pdf (accessed on 15 July 2021).
11. Hungarian Energy and Public Utility Regulatory Authority. Available online: http://www.mekh.hu/evrol-evre-no-a-legkondicionalo-berendezesek-villamosenergia-fogyasztasa (accessed on 19 July 2021).
12. Long, L.; Ye, H. The roles of thermal insulation and heat storage in the energy performance of the wall materials: A simulation study. *Sci. Rep.* **2016**, *6*, 24181. [CrossRef] [PubMed]
13. Bazazzadeh, H.; Pilechiha, P.; Nadolny, A.; Mahdavinejad, M.; Hashemi Safaei, S.S. The Impact Assessment of Climate Change on Building Energy Consumption in Poland. *Energies* **2021**, *14*, 4084. [CrossRef]
14. Reilly, A.; Kinnane, O. The impact of thermal mass on building energy consumption. *Appl. Energy* **2017**, *198*, 108–121. [CrossRef]
15. ISO: 52000-1. *Energy Performance of Buildings—Overarching EPB Assessment—Part 1: General Framework and Procedures*; International Standardization Organization: Geneva, Switzerland, 2017.
16. ISO: EN 13790:2008. *Energy Performance of Buildings—Calculation of Energy Use for Space Heating and Cooling*; European Committee for Standardization: Brussels, Belgium, 2008.
17. Csáky, I.; Kalmar, F. Effects solar asymmetry on building's cooling energy needs. *J. Build. Phys.* **2015**, *1*, 35–54.
18. Csáky, I.; Kalmár, F. Effect of thermal mass, ventilation, and glazing orientation on indoor air temperature in buildings. *J. Build. Phys.* **2015**, *39*, 189–204. [CrossRef]
19. Oh, S.; Song, S. Detailed Analysis of Thermal Comfort and Indoor Air Quality Using Real-Time Multiple Environmental Monitoring Data for a Childcare Center. *Energies* **2021**, *14*, 643. [CrossRef]
20. The Testo Saveris Measurement System. Measurement Data Monitoring with Testo Saveris Professional Edition, Instruction Manual. Available online: https://www.testo.com/hu-HU/saveris-bazis-gsm-modullal/p/0572-0221 (accessed on 9 June 2021).
21. Halverson, R.; Hart, R.; Athalye, W. *ANSI/ASHRAE/IES Standard 90.1—2013 Determination of Energy Savings: Qualitative Analysis*; Pacific Nortwest National Laboratory (PNNL): Richland, WA, USA, 2014.
22. Baumann, M.; Csoknyai, T.; Kalmár, F.; Magyar, Z.; Majoros, A.; Osztroluczky, M.; Szalay, Z. *Building Energetics*; PTE Pollack Mihály Műszaki Kar: Pécs, Hungary, 2009.
23. Guo, H.; Huang, L.; Song, W.; Wang, X.; Wang, H.; Zhao, X. Evaluation of the Summer Overheating Phenomenon in Reinforced Concrete and Cross Laminated Timber Residential Buildings in the Cold and Severe Cold Regions of China. *Energies* **2020**, *13*, 6305. [CrossRef]
24. Csáky, I.; Kalmár, T. Analysis of degree day and energy need for cooling in educational buildings. *Environ. Eng. Manag. J.* **2014**, *13*, 2765–2770. [CrossRef]
25. Hungarian Meteorological Services. Evaluation of Extreme Climate Indices. Available online: http://www.met.hu/en/omsz/tevekenysegek/klimamodellezes/szelsosegek/ (accessed on 18 February 2015).
26. MSZ 04-140-2. Power Engineering. In *Dimensioning Calculus of Buildings and Building Envelope Structures*; Power Engineering: Budapest, Hungary, 1992.
27. Urban INCD Incerc; Ministry of Regional Development and Tourism. *Statistics of Romania Meteorological Data for Air Conditioning Equipment*; Contract nr. 483/2011; Ministry of Regional Development and Tourism: Bucharest, Romania, 2011.
28. *DIN 4710 Statistics on German Meteorological Data for Calculating the Energy Requirements for Heating and Air Conditioning Equipment*; Deutsches Institut für Normung: Berlin, Germany, 2003.
29. Carlucci, S. *Thermal Comfort Assessment of Buildings*; Springer Briefs in Applied Science and Technology Series; Springer: Berlin/Heidelberg, Germany, 2013.
30. Carlucci, S.; Pagliano, L. A review of indices for the long-term evaluation of the general thermal comfort conditions in buildings. *Energy Build.* **2012**, *53*, 194–205. [CrossRef]
31. Lee, V.W.; Steemers, K. Exposure duration in overheating assessments: A retrofit modelling study. *Build. Res. Inf.* **2016**, *45*, 60–82. [CrossRef]
32. *CEN-CR 1752: Ventilation for Buildings. Design Criteria for the Indoor Environment*; CEN/TC 156; European Committee for Standardization: Brussels, Belgium, 1998.
33. Katunský, D.; Lopušniak, M. Impact of shading structure on energy demand and on risk of summer overheating ina low energy building. *Energy Procedia* **2012**, *14*, 1311–1316. [CrossRef]

energies

Article

Transforming a Historic Public Office Building in the Centre of Rome into nZEB: Limits and Potentials

Fabrizio Cumo [1], Fabio Nardecchia [2], Sofia Agostinelli [2] and Flavio Rosa [1,*]

[1] CITERA Interdepartmental Centre, Sapienza University of Rome, 00197 Rome, Italy; fabrizio.cumo@uniroma1.it

[2] Department of Astronautical, Electrical and Energy Engineering (DIAEE), Sapienza University of Rome, 00184 Rome, Italy; fabio.nardecchia@uniroma1.it (F.N.); sofia.agostinelli@uniroma1.it (S.A.)

* Correspondence: flavio.rosa@uniroma1.it; Tel.: +39-6-4991-9172

Abstract: According to the last census of 2019, about two million Italian buildings are more than 100 years old. Building energy retrofitting involves a diverse mix of influencing factors, depending on history, intended use, and construction techniques. This paper aims to assess the energy needs of a historic building by evaluating the variability of climatic conditions and internal loads, as well as the thermal capacity of the building envelope. The energy analysis was conducted using dynamic simulation systems (TRNSYS). The purpose of the study is to provide an analysis of the current energy conditions of the building to identify the main critical issues and suggest the most suitable interventions to be implemented. All the transformations were conducted to meet the nZEB requirements and evaluate technical and economic feasibility, compatibility with architectural and landscape constraints, and large-scale replicability. Specifically, to reach the proposed targets, a 36 kWp PV system was implemented for an area of 210 m^2, in addition to the Air Handling Unit (AHU) already present. The profit index is above the unit, and it yields a time range between three and four years. Therefore, fully respecting the energy performance parameters required by the Italian legislation, the study demonstrated the unattainability of the nZEB class for a listed building.

Keywords: nZEB; historical buildings; TRNSYS; buildings retrofitting; buildings office; economic feasibility

Citation: Cumo, F.; Nardecchia, F.; Agostinelli, S.; Rosa, F. Transforming a Historic Public Office Building in the Centre of Rome into nZEB: Limits and Potentials. *Energies* **2022**, *15*, 697. https://doi.org/10.3390/en15030697

Academic Editors: Andrea Mauri and Francesco Minichiello

Received: 25 November 2021
Accepted: 14 January 2022
Published: 18 January 2022

1. Introduction

The energy efficiency of public buildings in Italy is often a matter of joint solutions to energy issues and the preservation of historical buildings. Maintaining the protection of the architectural heritage and landscape of historical Italian cities while upgrading their energy efficiency status almost always involves complex planning and delicate execution. In public housing, where current intended uses are not those for which buildings were originally designed, it is necessary to extensively alter the building envelope and the technical building system (TBS). The energy efficiency measures (EEMs) for energy retrofitting of historic buildings in cold and hot climates impact the TBS but especially the opaque envelope [1,2]. A list reference was developed by ANSI/ASHRAE Standard 100-2006 as a guide to address commercial and residential occupancies, though much of the content pertains primarily to commercial and institutional buildings [3]. Insulation interventions on the external envelopes of historical buildings that allow improving the transmittance and the consequent dispersions are often impractical due to the protection constraints of the architectural and landscape heritage. The impact of the thermophysical performance of envelopes on the global energy consumption is estimated to be 40–50% [4]. The aims of the construction sector decarbonisation policies can only be achieved by accounting for the impact of the design solutions on public buildings, including those which fall within the historical and architectural heritage [5]. For public buildings, the near-zero energy buildings (nZEB) requirements are not easy to achieve, especially in a restricted area such as the territory of Rome.

Italian public buildings are estimated to be at more than 65,000 units, and approximately 60% consist of buildings constructed prior to the first Italian energy-saving legislation [6]. About 2700 public administration buildings are particularly energy intensive, absorbing 1.2 TWh/year, with a cost of 177 million euros. The energy consumption of historical buildings in the EU is currently estimated at more than 250 kWh/m^2/year [7,8]. The TBS solution is somewhat "outdated"; from 2014, the most used fuel appears to be methane (62%), but diesel is still widely used (22%). Only 34% of the buildings are equipped with a temperature control system for each room, and air-conditioning system is available in only 46% of the buildings.

The European Directive 27/2012 on energy efficiency, implemented in Italy by Legislative Decree 102/2014, found that from 2014 to 2020, at least 3% per year of the indoor floor area of the air-conditioned buildings of Central Public Administration should be renovated, as required by the Energy Redevelopment Program of the Central Public Administration [9]. These buildings, mainly in the historic city centres, were built with traditional and local construction techniques and many have bearing structures of bricks or local stones (tuffs) and mixed iron and brick slabs. In some areas of Rome's historic city centre, there are still slabs in wooden structures. If not replaced as obsolete or retrofitted as part of the redevelopment of the transparent envelope, these buildings have wooden or iron frames and single glass.

The purpose of this paper is to highlight the limitations facing the transformation of a Listed Public Building (LPB) into offices of the nZEB standards while taking into account the interventions on TBS. The unattainability of the Italian legal requirements for nZEB is investigated in light of the latest published research. The analysis of the interactions between the three categories LPB, nZEB, and TBS, using the Banco Napoli building as a case study, points to the potential for improvement in the energy efficiency of Italy's public building of high architectural and cultural value.

The highest energy consumption is found in the residential sector [10]; it is responsible for around 40% of the global primary energy demand [4,11–13], and therefore, it provides an opportunity for saving a significant amount of energy. On this premise, the European energy policy aims to improve energy redevelopment schemes applied to buildings [14].

Mancini et al. [15] suggest that new buildings, as well as historical ones, will have to combine energy-saving requirements with those of comfort both in terms of indoor air quality and thermal comfort. Lo Basso et al. [16], concerning the energy balance of the nZEB, consider the overall sustainability of the urban areas in which they are located to contain the effects of climate change on energy consumption. Nastasi et al. [17] propose solutions to avoid further soil consumption following the installation of new plants powered by renewable energy sources (RES) at the service of nZEB. Mancini et al. [18] consider the effects on electrification and flexibility in nZEB development. The nZEB construction and the transformation of the existing heritage-status buildings are studied by De Santoli et al. [19]. Improvements in the passive performance of the building envelope and the efficiency of the energy systems and the introduction of renewable energy sources in the energy balance of buildings were investigated by Astiaso Garcia et al. [10] and Pennacchia et al. [20].

By 2021, all new buildings, as well as buildings subjected to substantial renovations, will have to meet nearly zero energy needs. The anticipated deadline for new public buildings, including schools, has been 2019 [11,21]. The term near-zero energy buildings (nZEB) is generally intended to indicate a category of buildings with a very high energy performance, characterised by a very low (almost zero) annual energy consumption, almost entirely covered by on-site renewable sources [22–24].

The characteristics of near-zero energy buildings in Italy are established by the Ministerial Decree of 26 June 2015 of the Ministry of Economic Development, "Minimum requirements for buildings" [25]. The construction of nZEB, as reported by Nucara et al. [26], and the transformation of the existing heritage will be carried out through improvements in the passive performance of the building envelope, the TBS efficiency, and the introduction of renewable energy sources for balancing of the buildings energy efficiency, as reported by Rosa et al. [18].

Since buildings are usually connected to external energy supply networks (gas, electricity) and due to the discontinuous nature of renewable sources [27], the nearly zero value concerns the balance between withdrawn energy and self-produced energy, consumed directly or fed into the national grid [28]. That is why city planning becomes crucial for the production and consumption of energy [29].

The challenge is to apply an appropriate retrofitting to historical buildings which, in Central Italy and in particular in climate zone D, as defined by the Italian legislation [30], are subject to strict regulations on renovations [31–34], to generate a significant impact in an area of such great artistic and architectural value as the historical centre of Rome [35,36]. Where installed, the HVAC system is the first component to be refurbished [37] or to be upgraded by innovative technologies [37] or to extreme conditions [38].

Many authors have approached the study of historic buildings and their possible efficiency in the scientific community. Ascione et al. [39] suggest a multicriteria approach for energy efficiency of historical buildings, proposing methodologies for performance analysis, combining different experimental and numerical studies, and applying such methodologies to a historical building in the province of Benevento. Giombini et al. [40] addressed a case study for a historical building in the city of Perugia, indicating a methodology for the analysis of this type of building. This methodology can be useful to address the energy efficiency status of historical buildings. De Bernardis et al. propose a methodology to guide the recovery design, using a "case by case" approach to identify the best intervention method for each context.

Galatioto et al. [41], Martinez-Molina et al. [42], and Webb [5] suggest a state-of-the-art methodology to analyse the possibility of efficiency in historic buildings. In these works, various types of buildings are studied and classified by purposes of use. Ma et al. and other researchers [33,43,44] proposed a systematic approach to selecting and identifying the best retrofit possibilities for existing buildings, providing methodologies and promoting energy conservation and sustainability. This approach requires research to create dedicated databases for benchmarking performance. Jafari et al. [45] proposed a decision-making framework to calculate the economic benefits of energy retrofitting and determine the optimum energy retrofitting strategy.

The impact of the energy improvement measures on historical buildings is studied by Grytli et al. [46], who presented an integrated analysis method that examined different impacts from various energy efficiency measures on a model building. Many other authors have focused on integrating renewable and nonrenewable energy systems in retrofitting historical buildings [47–49]. Few studies tested innovative on-site technologies such as storage devices [50], while the largest scale of planning interventions has been the master planning of entire districts [51]. Careful analysis of noneconomic barriers is fundamental for the feasibility of any intervention [52,53].

Becchio [53] evaluated the economic component for the nZEB design and noted that it is fundamental to consider both the energy and the economic perspective. In the application phase of the study, the work by Ciampi et al. [54] is a valuable reference since the performance and consequent retrofit actions are simulated on an existing historical public building using the TRNSYS simulation software. Meanwhile, De Santoli et al. used a MATLAB/Simulink tool to design the refurbishment scenarios [55] and evaluate the performance of innovative technologies [56]. Synergies among the simulation software are expected to cover all the energy aspects, as reported in Groppi et al. [57].

The study of historical buildings, the possible efficiency improvements, and the achievement of the nZEB class find support in Dalla Mora et al. [58] and Sauchelli et al. [59]. The authors offer specific case studies of the energy efficiency improvements on historic buildings to meet the nZEB requirements and the Italian legislative regulations. Mauri [60] hypothesised different retrofit scenarios to achieve the nZEB objective with the simulation on a historical building in the province of Agrigento. The author conducted a cost analysis to achieve optimal economic solutions and energy savings. The study can contribute to the development of specific policies and the adoption of suitable measures to facilitate optimal design choices for the energy improvement of historical buildings.

A number of studies addressing the transformation of public buildings into nZEB have been carried out, while the interest in buildings bound for use as energy-intensive offices is scarce. This observation also derives from the work of the authors of this study, who have, for years, carried out research on the topic of energy efficiency of the historic building stock [32,47,61–68].

The present paper highlights technical and economic limitations in transforming a listed public building into nZEB. The object is to evaluate the limitations of energy retrofitting aimed at achieving the nZEB objectives based on the Italian regulatory restrictions, focusing on historic office buildings in areas with landscape and architectural constraints. More specifically, the objective is to show that, based on the regulatory requirements of the Ministerial Minimum Decree, it is impossible to carry out external insulation interventions, even on the building's opaque envelope, together with the energy retrofitting of the TBS, resulting in the unattainability of the nZEB status. Dynamic numerical analyses were set up and performed using TRNSYS software based on the data obtained from an energy analysis conducted on the building envelope and energy systems. In Materials and Methods, the characteristics of the case study are illustrated. Section 3.1 describes the numerical model implemented in TRNSYS to evaluate the reachability of the nZEB objectives described in Section 2.3. In Section 4, the results obtained by comparing the outcomes of efficiency improvements in a proposed case study of BancoNapoli are reported. Section 5 contains the diagram of the impacts of the transformations on the Banco Napoli, highlighting the limits and the potentials for its transformation into nZEB.

2. Case Study

Existing regulatory constraints in the municipality of Rome for building redevelopment activities can be classified into two main categories: architectural and landscape. "Cultural assets" and their respective architectural constraints are the assets protected under Part II of the Cultural Heritage Code (Legislative Decree no. 42/2004) [69] and can be individual buildings, small groups of buildings, historic villas, and others. They are assets protected because their history, their forms, or their materials are a testimony of Italian culture, history, and art.

The historic centre of the city of Rome, included in the perimeter of the Aurelian walls, is classified as UNESCO heritage. The constraints of this qualification overlap those present in the Municipality of Rome (RMP). The protection of the buildings in this area falls under the Superintendence of Rome Capital which imposes restrictions on the interventions that can be carried out on the external envelope of the buildings, especially on the opaque components, while for transparent components, there is only the obligation of frames that respect the forms and materials of the existing ones. This allows them to be replaced with the latest generation high-performance window fixtures.

Whenever an intervention of protected assets is carried out, the operations belong to the field of restoration, and the priority aims are those of conservation and transmission as investigated in Jafari et al. [45] and Grytli et al. [46].

The use of historic/listed buildings intended for public use repeats in all its particularity the issue of energy efficiency as a protection tool, not a single regeneration process in contrast with the conservation requirements. The surveys and analyses conducted during the audit were aimed at optimizing consumption related to air-conditioning systems. Therefore, they analyzed the uses and consumption related to air-conditioning systems to evaluate the possibility of reaching the nZEB building targets set by Italian legislation, analyzing the building focusing on the energy generation systems. However, being an asset of considerable historical interest and subject to various types of constraints, the interventions on energy systems of the entire building—with priority to HVAC systems—must be planned in a strategic vision, focusing on the high quality of the building complex, and ensuring maximum efficiency for energy consumption.

The building object of study, "Banco Napoli", belongs to the patrimony of the Italian House of Parliament and is in the historic centre of Rome at an altitude of 20 m above sea level, at a north latitude of 41°53′ and an east longitude of 12°28′. The building is part of the historical city and has a medieval origin, as defined by the technical standards of the RMP. These are specific morphologies of implantation of the "Fabrics of medieval origin" of the Historical City (Figure 1a,b). The technical standards indicate that attention must be paid to the presence and conservation of the remains of ancient buildings, as stated at the beginning of this section.

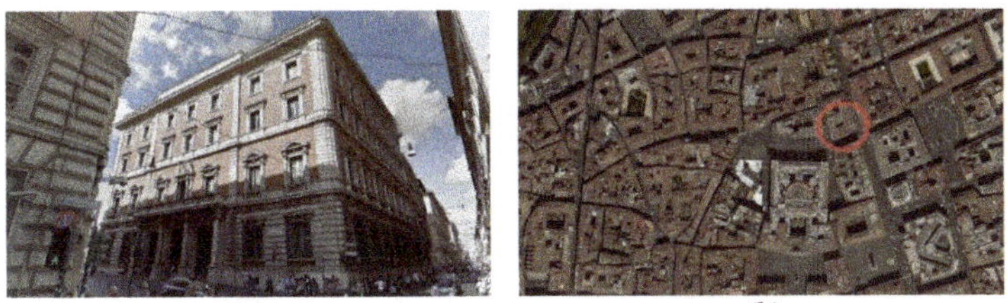

(a) (b)

Figure 1. (**a**) Rome historic centre and case study building detail; (**b**) building view.

In the city of Rome, 35% of the buildings are characterised as historic by age and building techniques and therefore subject to urban planning constraints included in the RMP and QC (quality chart).It is, therefore, necessary to guarantee the preservation of the walls and wall textures that testify to the stratification process over time. The transformation of the TBS of the Banco Napoli building took place over many years with partial interventions on specific areas such as floors or for intended use such as offices or toilets only.

The characteristics of the envelope and TBS of the "Banco Napoli" building are shown below. The raw data collected and classified in this section have been reprocessed and transcribed in Section 3.1 in specific "Types" subsequently implemented in TRNSYS.

2.1. Building Envelope

The current building consists of the union of two building blocks, called A and B-C-D, as shown in Figure 2 below:

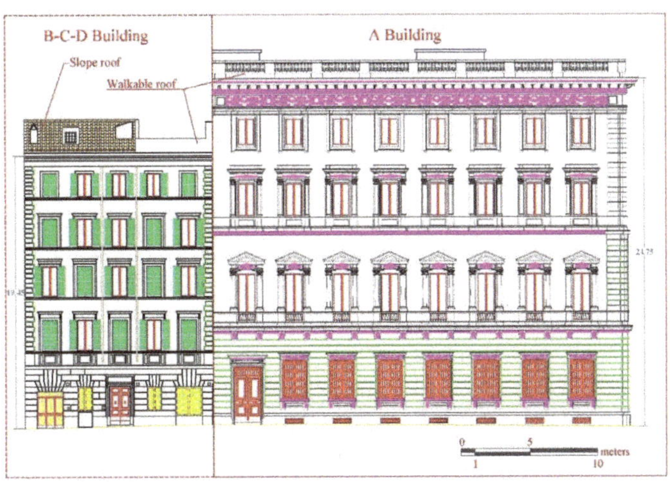

Figure 2. Building overview.

The rooftop type is not thermally insulated and is differentiated into two types: walkable with flooring and nonwalkable (see Figure 2). Regarding the thermal characteristics, a type of vertical wall was identified as made of uninsulated solid masonry with a thickness ranging from 130 cm to 80 cm. The transmittance of the perimeter walls has been assumed to be equal to the average value calculated based on the greater thicknesses on the ground floor and lesser on the top floor, equal to 1.15 W/m^{2}K.

Regarding the external façade, the building is typical of the late nineteenth-century housing in Rome's city centre. A type of floor was identified, and the transmittance is equal to 1.34 W/m^{2}K, extending for a total surface area of approximately 1007 m^2. The thermal transmittance is 1.51 W/m^{2}K. These data were deducted from the audit process, and a check-in reference to the technical report UNI/TR 11552:2014 provides the main thermophysical parameters (thermal transmittance, heat capacity per unit area, and periodic thermal transmittance) of the most used casing opaque components in existing buildings.

The supporting structure is made of bricks with mixed slabs, iron, and bricks. It consists of a basement and four floors above ground. It has total gross and net areas of 6676 m^2 and 4673 m^2, respectively, corresponding to a gross volume of 25,195 m^3 and a net volume of 17,636 m^3.

The load-bearing masonry made of solid bricks has a thickness of 130 cm on the first and second floor, up to 80 cm on the last floor. The internal partitions are made of solid brick for the load-bearing partitions and 10 cm perforated bricks.

The glazed surfaces consist of windows with double-glazed wooden frames and Roman blinds, also in wood.

All the windows of the Banco Napoli building have been redeveloped to improve their energy performance by reducing dispersion.

The windows are made of double glass 4–12–4 wooden frames with thermal break. From an analysis of the transmittance characteristics of the windows, the simple window type has a thermal transmittance of 3.74 W/m^{2}K. All classical window surfaces have solar shutters consisting of Roman blinds. The total area of the glazed elements is 450 m^2.

2.2. Technical Building Systems

The systems that have undergone significant transformations are HVAC and electricity distribution systems. Given the intended use of offices, HVAC systems upgrades are the most widespread and invasive components of the building. The external surfaces of the terraces have been almost entirely occupied, were authorized by the Superintendency offices, with Rooftop HVAC machines.

An energy audit was conducted to determine and classify HVAC by type, power, and areas served. The layouts have been classified into five macro components and the data processed to be able to insert them into the TRNSYS type components:

- Heating systems (thermal and cooling);
- Distribution (heat transfer fluid);
- Settings (Heating systems and terminal units);
- Emission (terminals units);
- Water storage (if any).

The generation plant consists of traditional methane-fueled boilers, electric refrigeration units and 4 Air Handling Units, all located on the terrace of the building.

The air-conditioning of spaces is constituted by primary air systems, fresh air, and fan coil in offices, while in the corridors, toilets, and some common areas, high-temperature radiators are present.

The building's air-conditioned rooms are kept at a uniform temperature through several air-conditioning systems and various centralized and noncentralized technologies: centralized Air Handling Unit (AHU) or Variant Refrigerant Fluid (VRF). The building has been divided into four thermal zones, the AHU BANK (Naples bank building), AHU A, AHU B, and AHU Int, each served by a specific air-conditioning system. The thermal energy generation system consists of four nominal 263 kW methane boilers that provide

DHW and heat the radiator circuit. The fan coil systems and the AHU batteries are powered by two heat pumps and three VRF.

Regarding electricity loads, there are three primary thermal generation systems: two heat pumps that supply the fan coils, the AHU BANK (Naples bank building), the AHU A and the AHU B; four natural gas boilers for the radiator circuit; three Variant Refrigerant Fluid (VRF) with multiple expansion for the AHU in the basement and the air-conditioning of the first floor.

The heating system was built using traditional gas boilers placed on the roofing level of the building, which also produces Domestic Hot Water (DHW) with the integration of solar panels.

There are four traditional boilers (Bongiovanni Bongas 2/14 type) with a capacity of 263.3 kW each, supplied by gas with an efficiency of 90.2%.

The thermal power plant is always in operation as it must guarantee coverage of the demand for DHW production. DHW production is integrated by a system of 13 solar collectors placed on the roof of the building. The total area of the collectors is 26 m^{2,} with a total production of about 4000 kWh per year.

DHW production is centralized in the basement of building A. There are three pumps with a capacity of 250 L/m and a power of 3 HP each. Accumulation is guaranteed by three tanks of 1500 L each.

Inside the building, there are several types of user terminals: the four-pipe fan-coil, radiators, and split units connected to outdoor multiunits serving some offices.

The user terminals, fan-coils, and radiators do not have thermoregulation systems. In particular, the fan-coils have only a manual on/off regulation system, while the radiators do not have a thermostatic valve.

The rooms are heated, as reported in Table 1.

Table 1. Operating hours of heating systems.

Winter/Summer	Conditioning			Heating		
	Mon–Fri	Sat	Sun	Mon–Fri	Sat	Sun
Building A	8–19	Off	Off	7–13 16–19	7–13	Off
Building B-C-D				7–13 16–22	7–13 16–22	7–13 16–22

In the various buildings, there are three different Air Handling Units (AHU) in addition to one located in the underground rooms, whose characteristics are shown in Table 2.

Table 2. AHU position and characteristics.

AHU Position	Technical Characteristics
AHU Bank	Brand: Atisa 17,500 m^3/h
AHU Int/Basement	Brand: Atisa 6000 m^3/h
AHU A	Brand: Atisa 12,500 m^3/h
AHU B	Brand: Atisa 6000 m^3/h

Two heat pumps for the thermal power generation system provide refrigeration with a cooling capacity of 362 kW and a compressor power input of 113 kW and EER of 2.7 each. The thermoregulation of the central unit operates through a single centralized reading from the various circuits served. As a result of the decrease in the return temperature, the refrigeration load delivered by the group will decrease. The distribution subsystem is mainly made up of concealed pipes or interspaces, while the emission has four-pipe fan-coil terminals and metal radiators.

2.3. Lighting Systems and Electric Equipments

To determine the loads due to lighting, the installed power density value (W/m^2) was multiplied by the surface area. A strategy for controlling the ignition of the lighting system within the thermal zone was selected through TRNSYS. For the Banco di Napoli building, the following reference values have been set according to UNI EN 12464-1 Standard: (a) Power density for lighting: 10 W/m^2; (b) Power density for electronic equipment: 0.2 W/m^2.

The building managers have already made interventions to improve efficiency with LED lighting systems only in the offices of the building, with presence sensors and elevators with inverter motors and centralized controls of consumption. In corridors and other spaces, fluorescent lamps are installed.

3. Materials and Methods

The present energy analysis identifies and defines the main technological, air-conditioning, and electrical systems and the thermal zones served by them: each area has been characterized by surface and volume concerning the number of floors in the buildings.

According to a stringent redevelopment constraint, nothing can be altered on the outside of the building to reduce winter heat losses due to the protection regulations of the Italian directives regarding the building analyzed. The solutions were evaluated with consumption dynamics simulations to achieve the aims of the nZEB Italian legislation, acting on the climate systems and maximum energy production from photovoltaic systems installed on the available roof space. Finally, an economic analysis of the interventions is proposed to assess an acceptable economic return.

3.1. Numerical Model

In the following section, the simulation model used to carry out the energy audit for the Banco Napoli building will be presented. The energetic diagnosis is carried out through a dynamic simulation that foresees the use of the TRNSYS software [70].

The TRNSYS is used for dynamic simulations of established phenomena to validate, in this specific application, a new energy management concept, which can range from the evaluation of the behaviour of a DHW production system from solar collectors to the dynamic analysis of the behavior of a "multizone" building. The software gives the user the possibility to easily develop customized components using the most common programming languages (C, C++, PASCAL, FORTRAN, etc.).

The implemented building model needs inputs that provide information regarding the climatic data of the geographical area to which the building belongs and the characteristics of the building's air-conditioning system. The climate data used are provided directly by the software, which contains a library with the climate information of various sites worldwide coming from the Meteonorm database and information on over 1000 sites in 150 different countries. The climate data included in the model relate to Ciampino (the year 2018), as it represents the location (among the available) geographically closest to the Banco Napoli headquarters.

The model was built following the directives imposed by the UNI/-ISO 52016-1 and UNI-ISO 52017-1 standard [71] "Energy performance of buildings—Energy needs for heating and cooling, internal temperatures and sensible and latent heat loads—Part 1: Calculation procedures". The legislation defines a building as "consisting of one or more buildings (building envelopes) or portions of a building, air-conditioned through a single generation system".

The model was calibrated and validated with actual consumption, both for the gas part and for the electricity part. The calibration of the model is presented in Annex A to make the manuscript more readable.

This was followed by identifying the thermal zones to be analyzed during the study. The distribution of the various floors in thermal zones has always been carried out following the directives imposed by the UNI/TS 11300-1 standard that dictates the conditions for the subdivision of the building into sub-areas called "Thermal Zones": "each part of the building, air-conditioned at a certain temperature with the same regulation mode, constitutes a thermal zone". Adhering to this definition, it was possible to divide the whole building into 63 thermal zones.

The breakdown of the building is shown in Figures 3–7, where the spaces for office use (blue), hallways and stairs (yellow), and toilets (green) are highlighted:

Once the distribution of the building was defined, each thermal zone was characterized by determining the indispensable properties for calculating the thermal regime inside the building, such as dimensions of the thermal zone, the definition of walls bordering with other thermal zones or with the external environment, thermal characteristics of opaque and transparent surfaces, the definition of thermal loads inside the thermal zone, hourly frequency of air changes [34], characteristics of the heating system, characteristics of the cooling system, characteristics of the ventilation system, and characteristics of hygrometric air.

Transmittance values of the building envelope were extracted by UNI/TR 11552:2014 Opaque envelope components of buildings—Thermophysical parameters [72]. Table 3 shows the transmittance values of the external envelope.

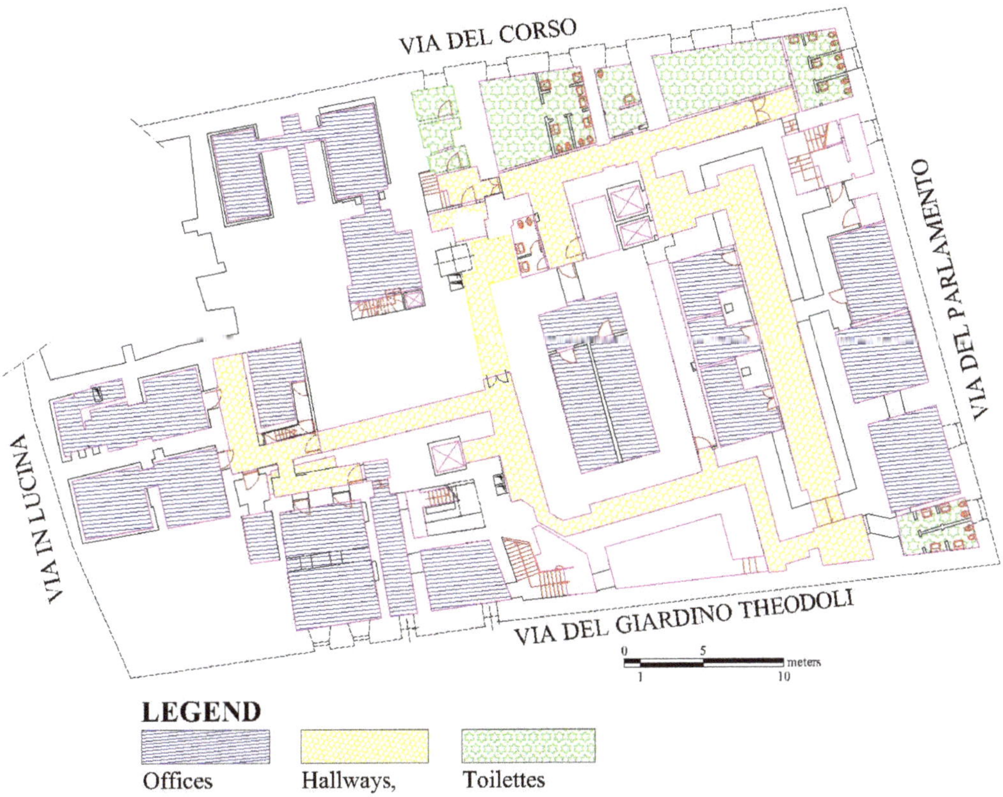

Figure 3. Designated uses on the Int/basement floor.

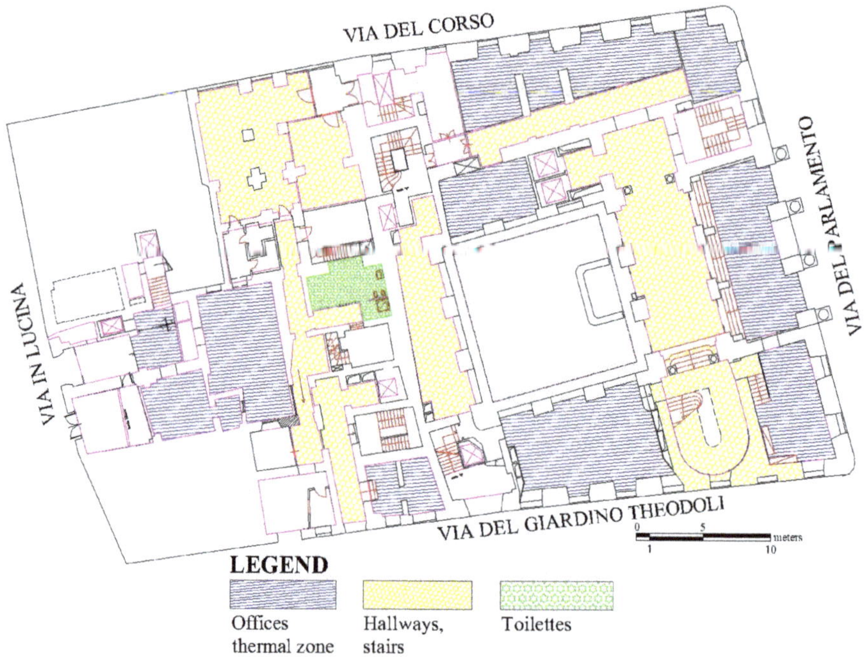

Figure 4. Designated uses on the ground floor.

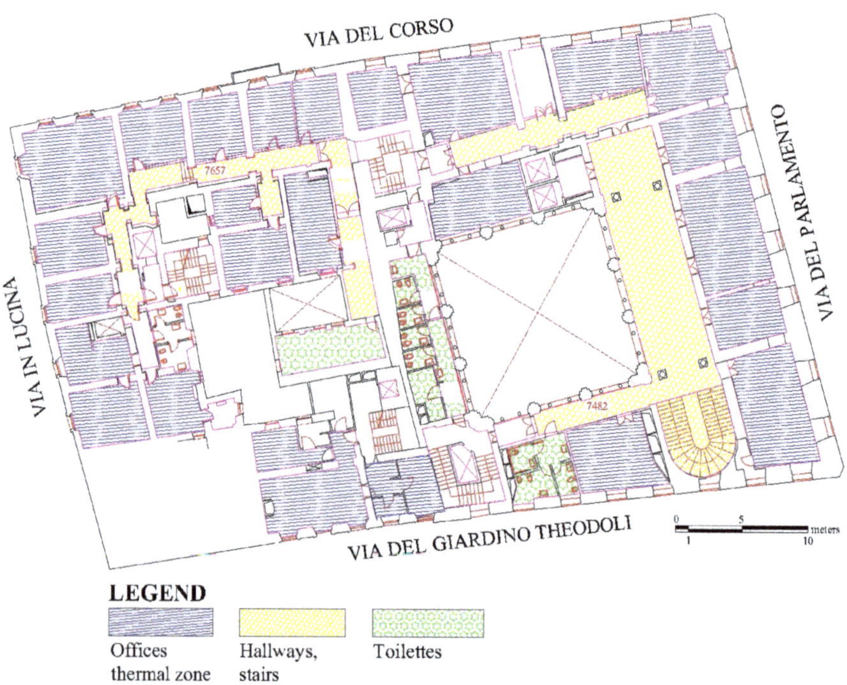

Figure 5. Designated uses on the first floor.

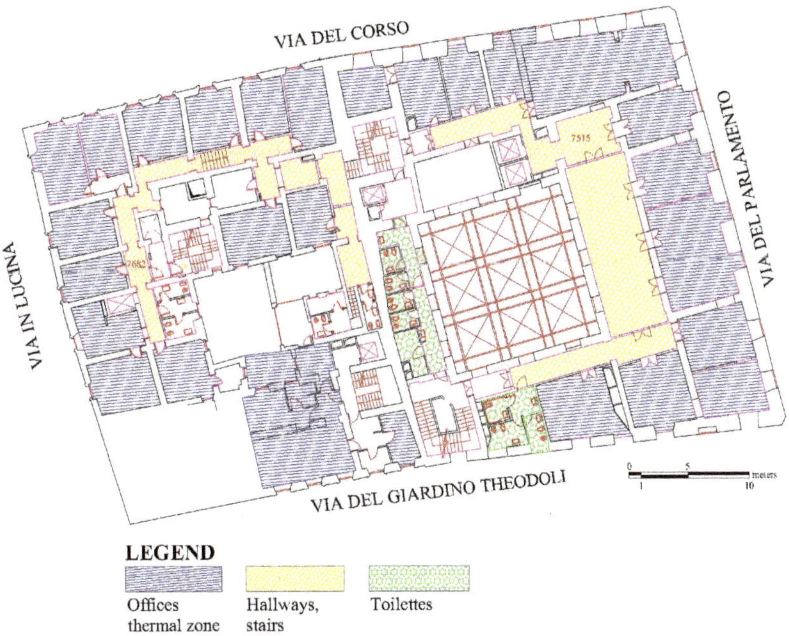

Figure 6. Designated uses on the second floor.

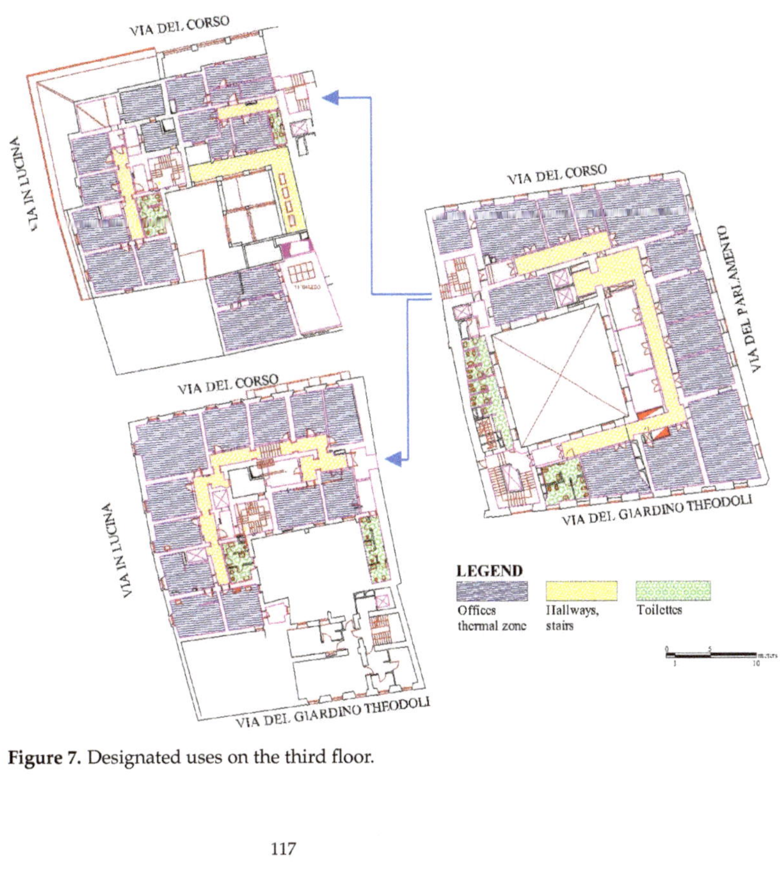

Figure 7. Designated uses on the third floor.

Table 3. Transmittance value of the building envelope.

Type	Transmittance (W/m^{2}K)
Vertical opaque areas	1.15
Inter-floor slab	1.34
Covering slab	1.51
Slab-on-grade floor	1.08
Windows	3.74

The thermal loads for the Banco Napoli building (occupants, lighting equipment) were set using the following values taken from the literature (Fasano et al., ENEA) [70,71] as shown in Table 4.

Table 4. Thermal loads form.

	Power Density
People	150 W/people
Illumination	10 W/m^2
Electronic devices	3.2 W/m^2

To determine the loads due to lighting, the value of installed power density (W/m^2) was multiplied according to the extension of the surface. The software was selected to control the lighting system ignition within the thermal zone. At present, there is an absence of control systems and centralized lighting system management in the Banco Napoli building. For this reason, manual switching was chosen. Currently, lighting is essentially supplied by LED in offices and fluorescent lamps in corridors and other spaces, dissipating about 25% of the absorbed energy by radiation towards the surrounding surfaces. The remaining 50% is dissipated by conduction and by convection. In addition to the lighting thermal loads, other thermal loads deriving from electrical equipment such as computers, printers, coffee dispensers, and elevators were implemented. For each of these, the operating time and the heat output generated based on the occupancy has been defined.

Table 5 are reported the data inserted in TRNSYS Building Model.

Table 5. TRNSYS input model.

Input
Heating set-point temperature: 19 °C
Relative humidity percentage indoor before humidification: 38%
Cooling set-point temperature: 26 °C
Relative humidity percentage indoor before dehumidification: 52%
External air temperature: variable according to outdoor simulation period
Incident solar radiation each cardinal axis varies according to the simulation period

For each thermal zone, it is possible to set the window opening frequency. The average value of fresh air flow rate inside the single room was set to 11 m^3/h according to current legislation [72].

The action carried out by the four AHUs present in the building were implemented by setting the air flows introduced into the environment considering the characteristics of Table 2. The airflow rate is assumed to be constant during operation as currently, the system is not equipped with flow regulation systems. Two separate airflow inputs have been implemented according to the type of distribution organ present inside the environment: delivery hoses AHU = 250 (m^3/h) and diffusers = 150 (m^3/h).

3.2. *nZEB Targets*

The energy performance of a building is determined considering all energy services (heating, cooling, ventilation, DHW, artificial lighting, and the transport of people or things, the latter limited to nonresidential buildings). The calculation of the energy performance can be carried out with different algorithms and methodologies, depending on the available data and the required accuracy. It seems clear that as the complexity of the algorithm increases, the approximation of the results improves. The current legislation (Decree of the Ministry of Economic Development 26 June 2015) [73] provides a methodology for the calculation of the energy performance in buildings, including the use of renewable sources in compliance with national technical standards following the development of EN standards to support Directive 2010/31/EU [22]. In most cases, the calculation of the energy requirement is performed to achieve a standardized assessment of the needs and a certified classification of the building from an energy consumption point of view. It is, therefore, important that the calculation procedures not only have a recognized scientific validity but also meet the regulatory requirements of the sector.

For research purposes of reaching more precise assessments of energy needs, according to the current occupation of the building, it is possible to use more sophisticated simulation methods that provide dynamic modelling.

The regulatory obligations compare the building to a "reference" building, identical in geometry, orientation, territorial location, intended use, and surrounding situation, with different thermal characteristics and pre-established energy parameters. The reference building is therefore considered to be equipped with the same energy production facilities as the actual building, assigning a reference average efficiency for both the utilization subsystems of the reference building and the generation subsystems. Furthermore, the reference building is considered equipped with generation systems for winter heating energy services, summer air-conditioning, and DHW production of the same type as those present in the actual building.

The set of buildings Banco Napoli has a gross area of 6676 m^2 and a gross volume of 25,195 m^3 with a form factor S/V equal to 2.6. The parameters to be respected for the case study were therefore obtained from a table [32] or determined by the calculations of the reference building, modelled with the mandatory limit values as of 1 January 2019 for public buildings. These parameters are summarized in Table 6 below.

Table 6. Threshold limit value.

Parameter		u.m.	Requirements
$\mathbf{H'_T}$	Average overall heat transfer coefficient for transmission per unit of surface dispersant	(W/m^{2}K)	≤0.57
$\mathbf{A_{sol.est}/A_{sup\ utile}}$	Summer equivalent solar area per unit of useful surface	(-)	≤0.040
η_H	Seasonal average efficiency for domestic hot water production	(%)	≥81%
$\mathbf{Ep_H}$	Specific primary energy index for winter heating. (1)	(kWh/m^2)	≤80.35
$\mathbf{Ep_{H.nd}}$	Useful thermal performance index for winter heating	(kWh/m^2)	≤99.19
η_w	Seasonal average efficiency for domestic hot water production	(%)	≥81%
$\mathbf{Ep_w}$	Specific primary energy index for hot water supply. (1)	(kWh/m^2)	≤18.45
$\mathbf{Ep_{w.nd}}$	Useful thermal performance index for hot water supply.	(kWh/m^2)	≤26.35
η_c	Seasonal average efficiency for air cooling systems (including the possible humidity control)	(%)	≥81%
$\mathbf{Ep_c}$	Specific primary energy for summer cooling (including the possible humidity control). (1)	(kWh/m^2)	≤45.22
$\mathbf{Ep_{c.nd}}$	Useful thermal performance index for summer cooling	(kWh/m^2)	≤52.77
$\mathbf{Ep_t}$	Energy performance index of the service for the transport of people and things (lift systems and escalators). (2)	(kWh/m^2)	≤15.46
$\mathbf{Ep_v}$	Energy performance index for ventilation. (1).	(kWh/m^2)	≤22.47
$\mathbf{Ep_L}$	Energy performance index of the service for lighting. (2)	(kWh/m^2)	≤21.15
$\mathbf{Ep_{gl}}$	Index overall energy performance of the building. (1)	(kWh/m^2)	≤165.17
$\mathbf{P_{el.ren}}$	Photovoltaic power plant	(kW)	≥35.20
$\mathbf{CFE_{ren.DHW+CLIM}}$	Requirements for DWH and air-conditioning due to renewable sources	(%)	≥55
$\mathbf{CFE_{ren.DHW}}$	Requirements for DWH due to renewable sources	(%)	≥55

(1) It is expressed in nonrenewable primary energy ("nren" index) or total ("tot" index). (2) This index is not calculated for category E.1 (residential buildings), except for colleges, convents, dormitories, barracks, as well as for category E.1.

The analyzed building represents a common type of building of historical importance, as defined above, on which it is impossible to intervene on the outer shells and living space, focusing all the performance improvement on technological systems only. Furthermore, the Minimum Requirements Decree imposes the highest percentage of energy requirement coverage with renewable sources on near-zero energy buildings. However, it is noted that this coverage relates exclusively to the heating, cooling and production needs of DHW, while the contribution due to lighting is wholly neglected, although it affects the overall value of global energy performance. Other energy services are also neglected, such as ventilation and transport of things or people.

4. Results

This section displays the results related to the dynamic simulations of the actual state of the building under examination and subsequent retrofit interventions. Attention is focused on calculating the parameters that contribute to the building being defined as nZEB.

The proposed feasibility intervention will demonstrate its suitability by addressing an economic analysis covering the time period of 20 years.

4.1. Analysis of the State of the Dynamic Simulations

The following illustrates the results obtained from the dynamic simulations carried out in TRNSYS relating to the current state of the Banco Napoli building-TBS system. The study was conducted by analyzing the consumption resulting from the conditioning of the interior spaces, as shown in Figures 3–7, highlighting the distribution of electricity consumption and for cooling and heating between the various thermal zones. The results obtained were then compared with the consumption recorded by the energy audit. As previously illustrated, the air-conditioning system is a "primary air" system mixed with four-pipe fan coils placed in each room, except for services. Figures 8 and 9 show the hourly thermal energy supplied to the building by fan coils during heating and cooling mode. The HVAC maintenance is managed by a facility management company that receives directly from the Administration of the House of Parliament an indication of when to turn off the boilers and turn on the refrigeration groups to make the reversal of the system. Seasonal transients and conditioning of some areas that require special climatic conditions managed by some local autonomous air-conditioning systems were not considered in the simulation as they had a negligible impact on the final consumption. From the graphs in Figures 8 and 9 (Hours of operation on the abscissa and kWh for the energy generated on the ordinate), it is possible to notice that the duration of the heating period is higher than that of the cooling, but on the other hand, the peak of cooling power (about 280 kW) is well above the peak of thermal power (about 190 kW).

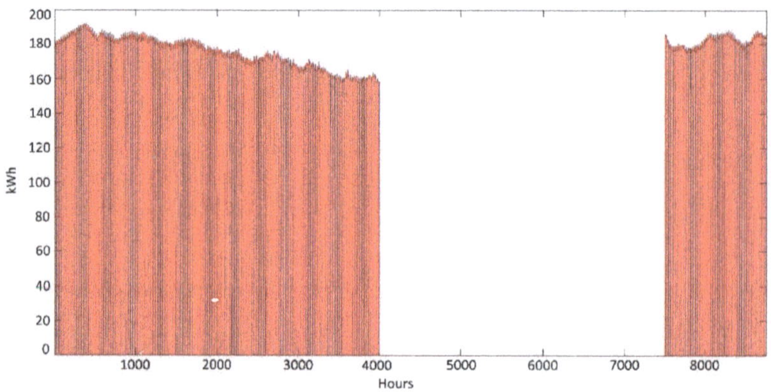

Figure 8. Fan coils hourly thermal energy trend.

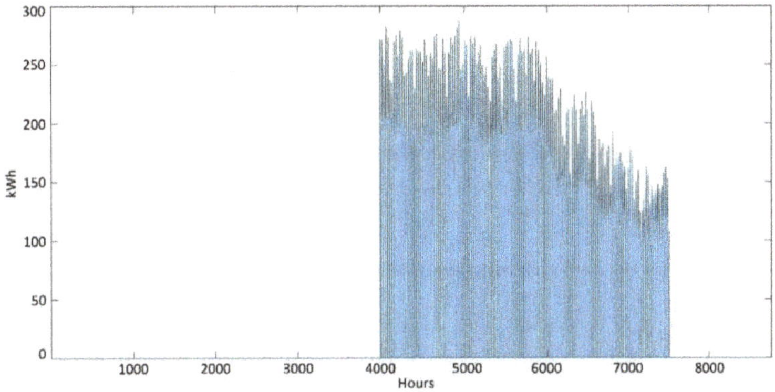

Figure 9. Fan coil hourly cooling energy trend.

The trends obtained in Figures 8 and 9 reflect the consumption obtained in most common office buildings types. It can be clearly seen that the cooling trend decreases after the first half. This corresponds to the period of vacation from work that generally takes place in Italy. The AHUs are left on, but they work at their lowest power. On the other hand, the heating period has an approximately constant trend over time. In the spring season, the plant still works at high speed despite the higher outside temperatures. This is due to the large volume of the building and, therefore, the longer time to heat it than in a standard building.

Figures 10 and 11 show the cooling energy and hourly thermal performances provided by the four AHUs present in the building if all AHUs process their nominal capacity.

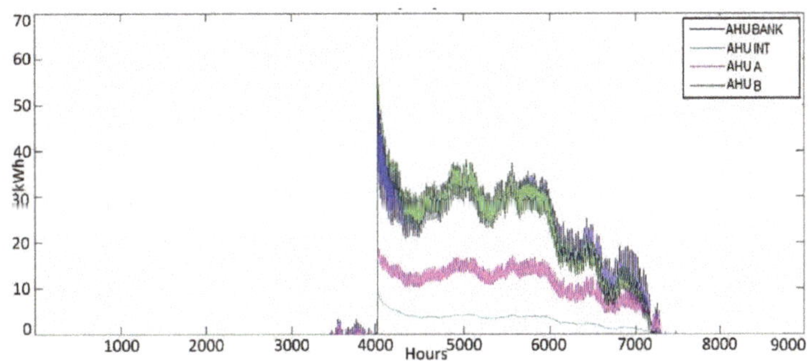

Figure 10. AHU hourly Cooling energy consumptions.

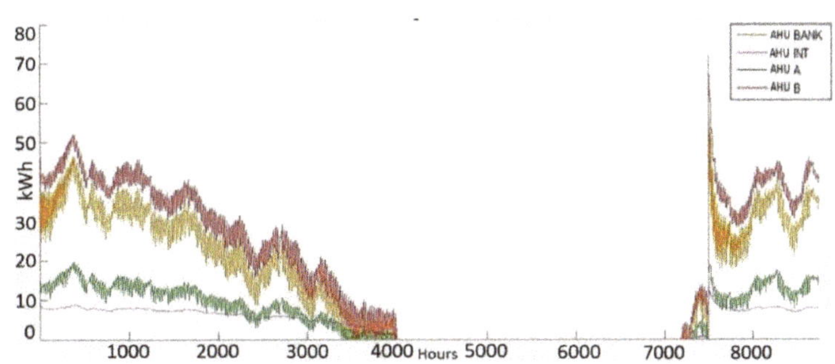

Figure 11. AHU hourly thermal energy consumptions.

It is evident that the AHU Bank provides a quantity of thermal energy in heating and cooling, similar to the AHU B, despite a higher nominal airflow. Thus, the AHU B can easily supply the average needs of the entire structure, but to meet the peak demand, the other AHUs are also used. It is interesting to note the behaviour of the AHU INT; this machine must adopt the minimum variations of the temperature inside the room it serves. Being designed to meet the needs of the basement, considering that the temperature trend of this environment is minimally affected by changes in the external temperature, its operation is not adaptive but almost constant.

The values showing the average energy performance (calculated over a year), the hourly maximum for the AHU and the fan coils are shown in Table 7.

Table 7. Supply hourly thermal/cooling energy values.

Plant	kWh_{avg}/Heating Hour	kWh_{max}/Heating Hour	kWh_{avg}/Cooling Hour	KWh_{max}/Cooling Hour
FAN COILS	52.87	191.87	37.16	287.51
AHU BANK	14.51	59.75	9.09	62.57
AHU B	18.44	72.07	8.88	67.32
AHU A	6.08	19.69	4.16	22.03
AHU INT/Basement	3.98	13.79	1.18	10.89

Once the time thermal profile of the building was defined, it was possible to obtain the annual thermal and cooling energy. Table 8 and Figure 12 show the values of annual thermal and cooling energy divided according to the distribution apparatus. All data are derived from the House of Parliament monitoring data centre.

Table 8. nZEB parameters of the current state.

Parameters		u.m.	Requirements	Current State Data
H'_T	Average overall heat transfer coefficient for transmission per unit of surface dispersant	(W/m^2K)	≤0.53	0.78
$A_{sol.est}/A_{sup\ utile}$	Summer equivalent solar area per unit of useful surface	(-)	≤0.040	0.035
η_H	Seasonal average efficiency for domestic hot water production	(%)	≥81%	85%
Ep_H	Specific primary energy index for winter heating. (1)	(kWh/m^2)	≤80.35	79.10
$Ep_{H.nd}$	Useful thermal performance index for winter heating	(kWh/m^2)	≤99.19	93.06
η_w	Seasonal average efficiency for domestic hot water production	(%)	≥81%	85%
Ep_w	Specific primary energy index for hot water supply. (1)	(kWh/m^2)	≤18.45	17.65
$Ep_{w.nd}$	Useful thermal performance index for hot water supply.	(kWh/m^2)	≤26.35	24.85
η_c	Seasonal average efficiency for air cooling systems (including the possible humidity control)	(%)	≥81%	85%
Ep_c	Specific primary energy for summer cooling (including the possible humidity control). (1)	(kWh/m^2)	≤45.22	43.24
$Ep_{c.nd}$	Useful thermal performance index for summer cooling	(kWh/m^2)	≤52.77	52.34
Ep_t	Energy performance index of the service for the transport of people and things (lift systems and escalators). (2)	(kWh/m^2)	≤5.46	3.58
Ep_v	Energy performance index for ventilation. (1).	(kWh/m^2)	≤22.47	17.45
Ep_L	Energy performance index of the service for lighting. (2)	(kWh/m^2)	≤21.15	18.65
Ep_{gl}	Index overall energy performance of the building. (1)	(kWh/m^2)	≤171.95	168.64
$P_{el.ren}$	Photovoltaic power plant	(kW)	≥35.20	0
$CFE_{ren.DHW+CLIM}$	Requirements for DWH and air-conditioning due to renewable sources	(%)	≥55	71%
$CFE_{ren.DHW}$	Requirements for DWH due to renewable sources	(%)	≥55	60%

(1) It is expressed in nonrenewable primary energy ("nren" index) or total ("tot" index). (2) This index is not calculated for category E.1 (residential buildings), except for colleges, convents, dormitories, barracks, as well as for category E.1.

The winter thermal loads are quite different from the summer season regarding the total consumption of the building (Figure 12). A total thermal energy requirement of around 1140 MWh/year is observed, compared to a cooling energy requirement of about 530 MWh/year.

The winter thermal load is therefore double the summer one. This result reflects the trends seen in Figures 10 and 11, where, as mentioned above, in the end part of the summer period, there is a decrease in the thermal load due to the nonoccupation of the building.

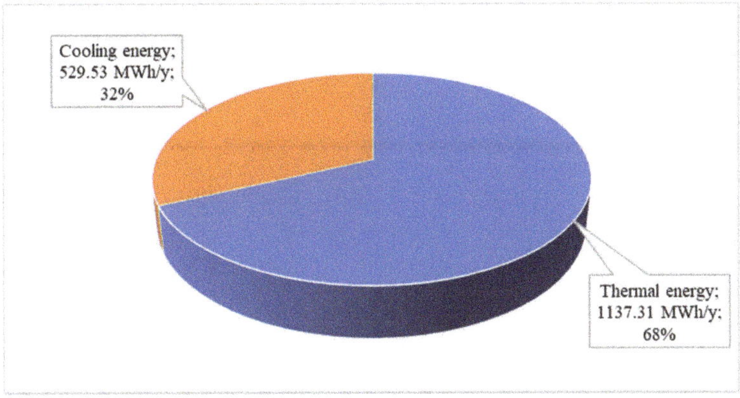

Figure 12. Distributed thermal and cooling energy.

The main part of the energy is supplied to fan coils and radiators (67% for heating and 61% for cooling), and only 35% is absorbed by AHUs as reported in Figure 13.

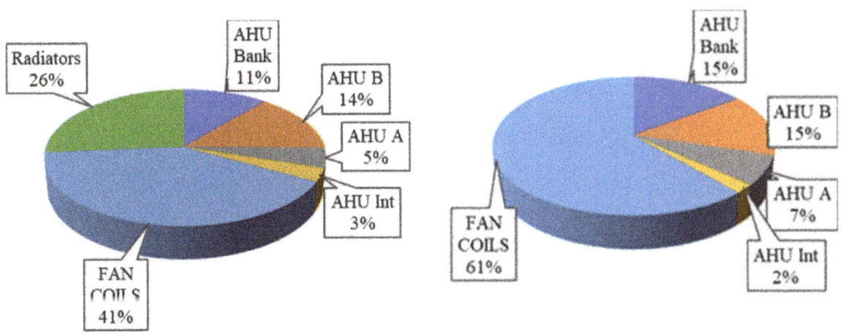

Figure 13. Energy distribution: heating (on the **left**) and cooling (on the **right**).

It is interesting to note that the radiators cover almost the same percentage of all the AHUs together during the winter period. During the summer period, the portion relating to the AHUs always remains the same, and the fan coils cover the portion of energy that was due to the radiators during the winter.

Figure 14 shows the annual demand for electricity from the distribution network, with a breakdown between the two generation systems and the electrical consumption of the VRF and the two heat pumps. The correspondent hourly electric power demand is shown in Figures 15 and 16.

The trend observable in Figures 15 and 16 reflects all the above considerations. The electricity consumption, first at maximum during the coldest period, decreases until it reaches a minimum before the summer period. From here, we notice another peak due to the cooling phase. The electricity consumption does not remain constant during this period but decreases until it reaches a minimum due to the nonoccupation of the building already mentioned above. With the start of the winter period, there is another peak in consumption again.

Table 8 shows the parameters for calculating nZEB requirements referred to the current state. The unmet parameters are in bold.

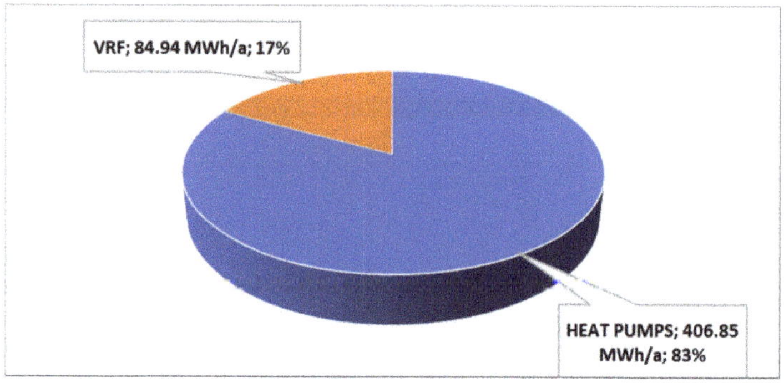

Figure 14. Electric power yearly consumption.

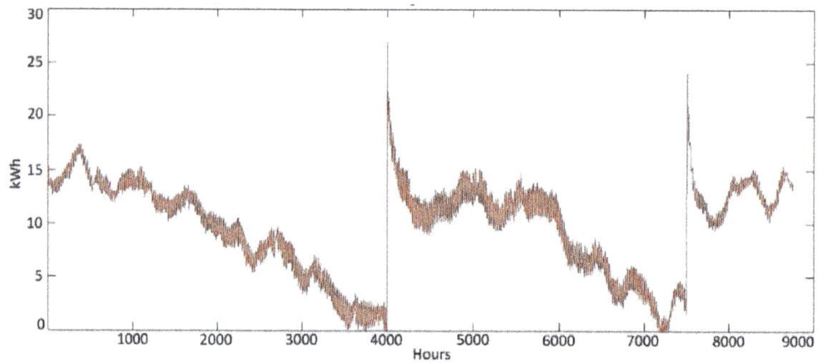

Figure 15. Hourly absorbed electric power VRF.

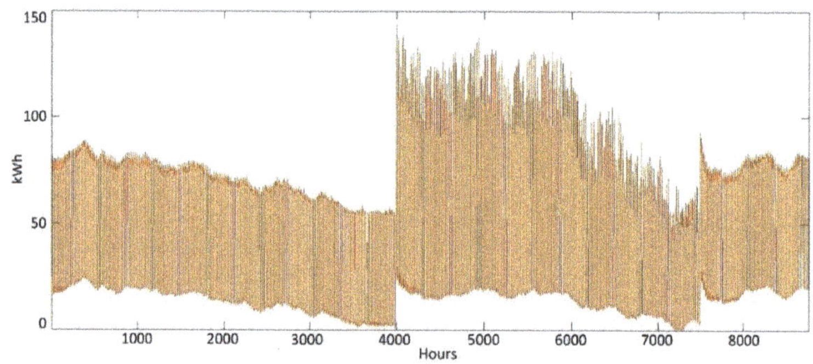

Figure 16. Hourly absorbed electric power heats pumps.

The table above shows how currently, the building does not fit with some of the parameter ranges to be classified as nZEB. The energy performance indexes for the individual energy services in the building are below the threshold values established following the study of the reference building, as reported in the UNI TS 11,300 standard [74]. The presence of heat pumps and VRF grants thermal energy (for DHW and DHW with heating and cooling) from renewable sources above standard thresholds.

As a listed building, it is impossible to consider any kind of intervention that includes an external opaque envelope. That is why it is difficult to reach the threshold value of 0.53 of the H'_T building's global thermal exchange parameters.

The two parameters not respected in the verification concern the performance of the external envelope H'T and Pel Ren.

To achieve the legal requirements on Italian nZEB buildings, an energy efficiency intervention is proposed by installing a photovoltaic system as described in the following Section 4.2.

4.2. Retrofit Efficiency Analysis

The Italian legislation requires the use of a significant increase of energy coming from renewable sources (26 June 2015, "minimum decree" transposition of Directive 2010/31/EU). The proposed intervention aims to reach the threshold value to produce electric power, which in the case study consists of a value of 35.20 kWp, from renewable sources. To reach the goal, it is necessary to install a 36 kWp photovoltaic system according to the available surface of the building roof (about 210 m^2). The intervention, therefore, proposes the installation of about 110 photovoltaic modules of 325 Wp south-east oriented at 45° angle. All the energy produced is supplied to the electricity users of the Banco Napoli set buildings The chart shows the characteristics of the suggested intervention and the relative costs of supply and installation, assuming the cost of a single PV module of €315.56 and the installation cost equal to 20% of the supply cost [75] as reported in Table 9.

Table 9. Photovoltaic system characteristics.

Monocrystalline module power	(Wp)	325
Module efficiency	(%)	20
Modules number	(-)	110
Total power plant	(kWp)	35.75
Inverter number	(-)	1
Auxiliaries efficiency	(%)	85
Electric power consumption preintervention	(MWh/year)	876
Electric power consumption postintervention	(MWh/year)	832
Energy saving	(kWh/year)	43,900
PV total cost	(€)	11,281
Inverter cost	(€)	6000
Additional costs: parallel switchboard, conduits, connections, etc.	(€)	2000
Total supply costs	(€)	19,281
Cost of labour	(€)	3800
Investment	(€)	24,001

An energy analysis of the proposed intervention was conducted. Figure 17 shows energy production during each month of the year.

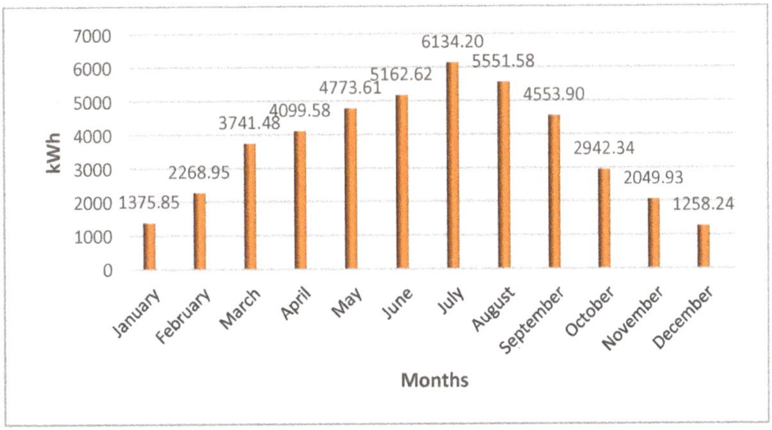

Figure 17. Monthly photovoltaic production.

The annual energy production is about 43.9 MWh. Therefore, by determining the annual production of the proposed PV plant, the economic feasibility of the intervention was analyzed, estimating an intervention lifespan of 20 years, considering an electricity fee of 0.18 €/kWh.

The present net value (NPV) was calculated as a function of different values of the interest rate (i), and for easier reading, the results were merged into a single graph. The following figure shows the NPV according to different values of the interest rate (i) between 0 and 10% and the relative recovery periods of the initial investment as represented in Figure 18.

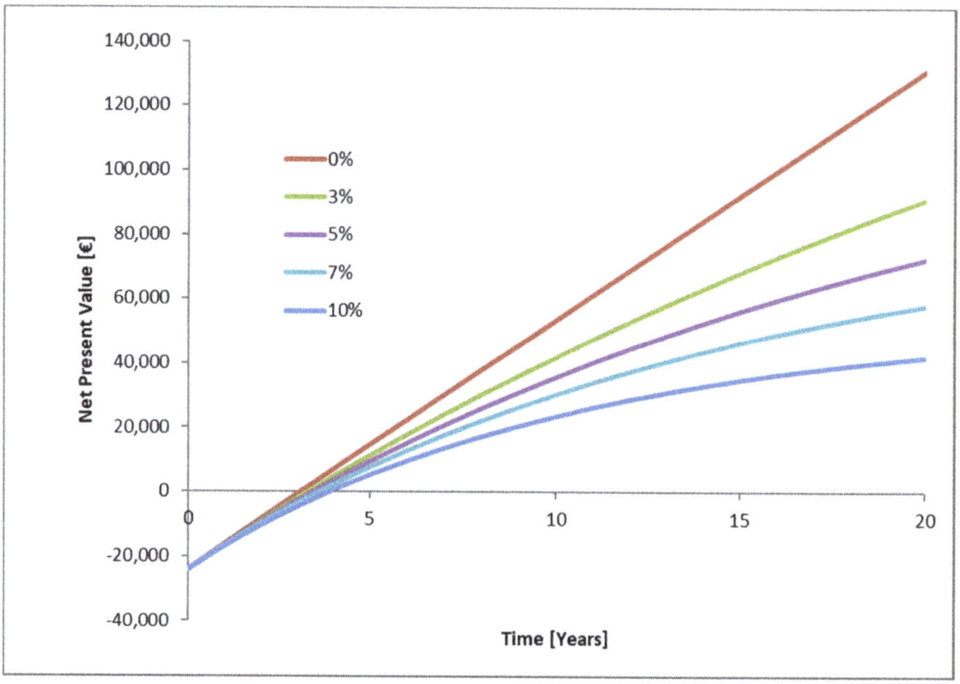

Figure 18. The net present value at fixed rate variation according to with 2020 Regione Lazio price list.

Yields of investments, the NPV and the profit rate based on the interest rate (Profitability Index PI) are shown in Table 10.

Table 10. Economic indicators intervention.

i (%)	In (Years)	NPV (€)	PI (€/€)
0	3	130,791.00	5.44
3	3	91,142.66	3.79
5	3	72,448.51	3.02
7	3	57,988.67	2.42
10	4	41,885.98	1.74

According to the economic analysis, the proposed intervention is economically advantageous, generating a positive NPV. The profit index is always above the unit, and it yields time a range between 3 and 4 years, depending on the interest rate.

In conclusion, this will reduce the network absorbed electric energy consumption by almost 10%. The proposed intervention exceeds the self-production threshold from renewable sources but does not allow, on its own, to reach all the parameters for the classification of the building as nZEB. Therefore, the values of the NZEB parameters of the solution with photovoltaics were calculated to verify the achievement of the expected requirements. These results are shown in Table 11.

Table 11. Italian standard parameters nZEB post retrofit efficiency with 36 kW PV system.

Parameters		u.m.	Requirements	Post Retrofit Data
H'_T	Average overall heat transfer coefficient for transmission per unit of surface dispersant	(W/m^{2}K)	≤0.53	0.78
$A_{sol.est}/A_{sup\ utile}$	Summer equivalent solar area per unit of useful surface	(-)	≤0.040	0.035
η_H	Seasonal average efficiency for domestic hot water production	(%)	≥81%	85%
Ep_H	Specific primary energy index for winter heating. (1)	(kWh/m^2)	≤80.35	79.10
$Ep_{H.nd}$	Useful thermal performance index for winter heating	(kWh/m^2)	≤99.19	93.06
η_w	Seasonal average efficiency for domestic hot water production	(%)	≥81%	85%
Ep_w	Specific primary energy index for hot water supply. (1)	(kWh/m^2)	≤18.45	17.65
$Ep_{w.nd}$	Useful thermal performance index for hot water supply.	(kWh/m^2)	≤26.35	24.85
η_c	Seasonal average efficiency for air cooling systems (including the possible humidity control)	(%)	≥81%	85%
Ep_c	Specific primary energy for summer cooling (including the possible humidity control). (1)	(kWh/m^2)	≤45.22	43.24
$Ep_{c.nd}$	Useful thermal performance index for summer cooling	(kWh/m^2)	≤52.77	52.34
Ep_t	Energy performance index of the service for the transport of people and things (lift systems and escalators). (2)	(kWh/m^2)	≤5.46	3.58
Ep_v	Energy performance index for ventilation. (1).	(kWh/m^2)	≤22.47	17.45
Ep_L	Energy performance index of the service for lighting. (2)	(kWh/m^2)	≤21.15	18.65
Ep_{gl}	Index overall energy performance of the building. (1)	(kWh/m^2)	≤171.95	168.64
$P_{el.ren}$	Photovoltaic power plant	(kW)	≥35.20	36
$CFE_{ren.DHW+CLIM}$	Requirements for DWH and air-conditioning due to renewable sources	(%)	≥55	71%
$CFE_{ren.DHW}$	Requirements for DWH due to renewable sources	(%)	≥55	60%

(1) It is expressed in nonrenewable primary energy ("nren" index) or total ("tot" index). (2) This index is not calculated for category E.1 (residential buildings), except for colleges, convents, dormitories, barracks, as well as for category E.1.

The data in Table 11 show that, even with an energy efficiency intervention with a new 36 kWp photovoltaic system with power exceeding the legal threshold, the nZEB requirements for the Italian legislation are not met as the H'_T parameter is in any case not respected.

5. Discussion

The primary purpose of this paper was to evaluate the transformation of an office LPB into an nZEB in the historic centre of Rome and to highlight its limits and potential. The design and construction path indicated by the European and Italian regulations on restricted buildings, as in the case study presented, does not always allow to reach the regulatory requirements. Installing a photovoltaic system with a higher rated power than the minimum required on the building roof would eventually reach the minimum value of self-produced electric power from the building and respect the established criteria.

Despite this intervention, it is not possible to fit all the nZEB parameters due to the impossibility of acting on the external vertical opaque structure and, therefore, on the parameter H'_T, which is, however, fulfilled for the limit value established for the reference building.

To overcome the problem in some rooms, internal insulation of walls and ceilings has been created using a 10 cm thick natural insulation, completed by a 2 cm finish of interior plaster. This solution, which would have made it possible to satisfy the HT requirement fully, was practically impossible to achieve as it involves a series of logistical and functional problems. The main problem is making the rooms unusable for the occupants for at least 15 days due to the insulating interventions. Moreover, a construction site area inside a functioning building involves a series of delays in the technical works and disturbance to the activities of the employees. Finally, the internal insulation of all the perimeter walls would lead to a reduction from 2% to 5% of the floor area of the building, with evident negative repercussions from the functional and economic point of view.

The results show that, for listed buildings, if the H'_T parameter is not considered, all the parameters to achieve nZEB status can be obtained by applying adequate efficiency measures.

A graphical summary of what has been discussed in this work is the impact diagram in Figure 19.

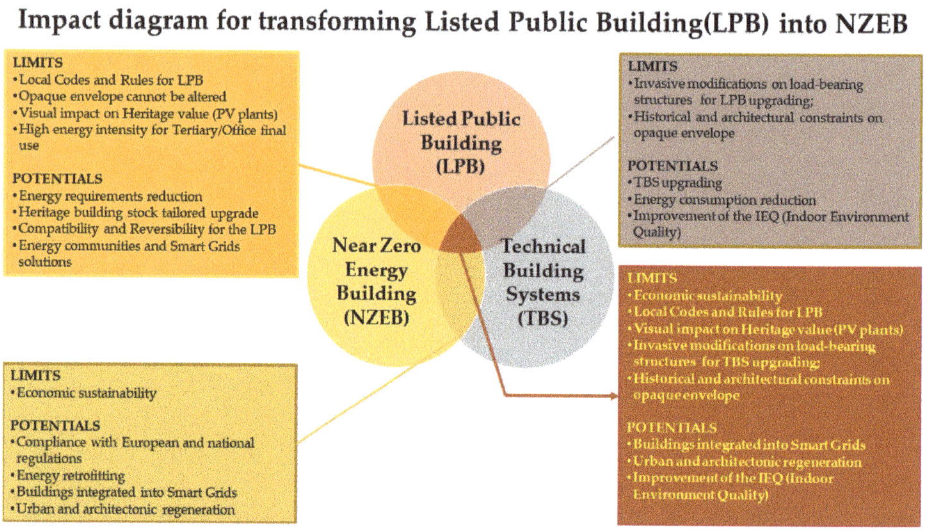

Figure 19. Impact diagram of limits and potentials on transforming LPB in nZEB.

With a logic of overlap of the effects with pairwise comparisons of the three categories (LPB, NZEB, TBS), the significant impacts on a historic building such as that of the case study have been highlighted, from the point of view of the limits and potential for a transformation in NZEB.

6. Conclusions

In this paper, the limits have been highlighted in the transformation of a Listed Public Building to an office destination in nZEB, taking into account the interventions on TBS.

Based on the data obtained from an energy analysis conducted both on the building envelope and energy systems, dynamic numerical analyses were set up and performed using TRNSYS software

This highlighted the potential for improvement through the case study by analysing and presenting the interactions between the three categories LPB-nZEB and TBS.

The Banco Napoli has a building envelope that does not lend itself energetically. Although it should be necessary to carry out the redevelopment of the enclosure to reach the parameters for the nZEB definition, this intervention is not possible because of the architectural and cultural value of the external walls.

Through the suggested intervention of installing a 36 kWp PV plant (higher than the required limit value), integrated with the generation of thermal energy through the existing

VRV systems, the requirements of consumption of thermal energy from renewable sources and consumption electricity from renewable sources are met.

The proposed intervention allows to exceed the self-production threshold from renewable sources but does not allow, on its own, to reach all the parameters for the classification of the building as nZEB. Therefore, it is not possible to reach the nZEB class even if all the parameters shown in Table 7 have been respected.

Hence, the only possible interventions for achieving the nZEB class consists of installing new fixtures with transmittance values in compliance with the new limits established by legislation. Such interventions would attain the energy parameters related to the thermal energy needs for heating but are not compatible with the constraint imposed on the historic building.

The unattainability of the nZEB class for historic buildings is further emphasized by the case of the building chosen in this paper since Banco Napoli has an excellent degree of efficiency not easy to improve.

The cost–benefit analysis showed the suggested intervention as economically advantageous in generating a positive NPV. The profit index is always above the unit, and the returns range between three and four years, depending on the interest rate. The analysis also shows an innovative approach that can guarantee the safeguarding and protection of historic buildings with high energy consumption, such as those intended for offices, is the energy communities model [76,77]. The aggregation of utilities, for example, entire buildings, within an energy community allows for the management of overall incoming and outgoing flows with the availability of energy from renewable sources also produced externally. The balance between imported and exported energy is assessed on the perimeter of the entire energy community or cell [78]. Indeed, what is presented in this paper represents the first phase of ongoing research that will be extended from the single building to cells/energy communities in historical urban contexts. This approach can reduce the impacts on architectural and landscape sensitive historical contexts and buildings.

Author Contributions: Conceptualization: F.R.; methodology: F.R., F.N., and S.A.; software: F.N.; data collection: S.A.; resources: F.R.; writing—review and editing: F.N., F.R., and S.A.; supervision: F.C. All authors have read and agreed to the published version of the manuscript.

Funding: This research received no external funding.

Institutional Review Board Statement: Not applicable.

Informed Consent Statement: Not applicable.

Data Availability Statement: Not applicable.

Acknowledgments: The authors are grateful for the support provided by the administrative and technical officials at the offices of the Italian House of Parliament.

Conflicts of Interest: The authors declare no conflict of interest.

References

1. Ye, Y.; Hinkelman, K.; Lou, Y.; Zuo, W.; Wang, G.; Zhang, J. Evaluating the energy impact potential of energy efficiency measures for retrofit applications: A case study with U.S. medium office buildings. *Build. Simul.* **2021**, *14*, 1377–1393. [CrossRef]
2. Foley, H.C. Challenges and opportunities in engineered retrofits of buildings for improved energy efficiency and habitability. *AIChE J.* **2012**, *58*, 658–667. [CrossRef]
3. Kouma, L. Energy Efficiency Measures to Consider. *ASHRAE Add.* **2011**, *62*, 1–12. Available online: https://xp20.ashrae.org/PCBEA/Files/EEMs-to-Consider-2011-09-15.pdf (accessed on 24 June 2021).
4. Mavromatidis, L.E.; Bykalyuk, A.; Lequay, H. Development of polynomial regression models for composite dynamic envelopes' thermal performance forecasting. *Appl. Energy* **2013**, *104*, 379–391. [CrossRef]
5. Webb, A.L. Energy retrofits in historic and traditional buildings: A review of problems and methods. *Renew. Sustain. Energy Rev.* **2017**, *77*, 748–759. [CrossRef]
6. Italian Parliament. *Ordinary Law of Parliament n. 373/76: Standards for the Reduction of Energy Consumption for Heating in Buildings*; Italian Parliament: Rome, Italy, 1976.

7. Biele, E.; Di Santo, D.; Petraccaro, A.; Tomassetti, G.; Romeo, C. The nZEB Buildings in the Public Administration—Gli Edifici nZEB Nella Pubblica Amministrazione. 2016. Available online: https://www.qualenergia.it/wp-content/uploads/2018/03/NZEB_PA_enea_fire.pdf (accessed on 12 September 2021).
8. Energy Consumption per m2 2. Available online: https://ec.europa.eu/energy/content/energy-consumption-m2-2 en (accessed on 12 November 2021).
9. Guidelines to the Presentation of Projects for the Program for Energy Redevelopment of the Buildings of the Central Public Administration PREPAC (Ministerial Decree of 16 September 2016)—Linee Guida alla Presentazione dei Progetti per il Programma per la Riqualificazione Energetica degli Edifici della Pubblica Amministrazione Centrale PREPAC. 2017. Available online: https://www.efficienzaenergetica.enea.it/pubblicazioni/linee-guida-alla-presentazione-dei-progetti-per-il-programma-per-la-riqualificazione-energetica-degli-edifici-della-pubblica-amministrazione-centrale-prepac.html (accessed on 14 September 2021).
10. De Santoli, L.; Garcia, D.A.; Groppi, D.; Bellia, L.; Palella, B.I.; Riccio, G.; Cuccurullo, G.; d'Ambrosio, F.R.; Stabile, L.; Dell'Isola, M.; et al. A General Approach for Retrofit of Existing Buildings towards NZEB: The Windows Retrofit Effects on Indoor Air Quality and the Use of Low Temperature District Heating. In Proceedings of the 2018 IEEE International Conference on Environment and Electrical Engineering and 2018 IEEE Industrial and Commercial Power Systems Europe (EEEIC/I&CPS Europe), Palermo, Italy, 12–15 June 2018; pp. 1–6. [CrossRef]
11. European Parliament. Directive 2012/27/EU of the European Parliament and of the Council of 25 October 2012 on energy efficiency. *Off. J. Eur. Union Dir.* **2012**, 1–56. [CrossRef]
12. Direttiva 2009/28/CE del Parlamento Europeo e del Consiglio del 23 aprile 2009 sulla Promozione dell'Uso dell'Energia da Fonti Rinnovabili, Recante Modifica e Successiva Abrogazione delle Direttive 2001/77/CE e 2003/30/CE. Available online: http://eur-lex.europa.eu/legal-content/IT/TXT/PDF/?uri=CELEX:32009L0028&from=IT (accessed on 8 February 2018).
13. da Graça Carvalho, M. EU energy and climate change strategy. *Energy* **2012**, *40*, 19–22. [CrossRef]
14. Comprehensive Study of Building Energy Renovation Activities and the Uptake of Nearly Zero-Energy Buildings in the EU | Energy. Available online: https://ec.europa.eu/energy/studies/comprehensive-study-building-energy-renovation-activities-and-uptake-nearly-zero-energy_en?redir=1 (accessed on 7 April 2020).
15. Mancini, F.; Nardecchia, F.; Groppi, D.; Ruperto, F.; Romeo, C. Indoor Environmental Quality Analysis for Optimizing Energy Consumptions Varying Air Ventilation Rates. *Sustainability* **2020**, *12*, 482. [CrossRef]
16. Mancini, F.; Lo Basso, G. How climate change affects the building energy consumptions due to cooling, heating, and electricity demands of Italian residential sector. *Energies* **2020**, *13*, 410. [CrossRef]
17. Nastasi, B.; Di Matteo, U. Solar Energy Technologies in Sustainable Energy Action Plans of Italian Big Cities. *Energy Procedia* **2016**, *101*, 1064–1071. [CrossRef]
18. Mancini, F.; Nastasi, B. Energy Retrofitting Effects on the Energy Flexibility of Dwellings. *Energies* **2019**, *12*, 2788. [CrossRef]
19. Calice, C.; Clemente, C.; de Santoli, L.; Fraticelli, F. Guidelines for the retrofit of the school building stock for sustainable urban regeneration of the city of Rome. *WIT Trans. Ecol. Environ.* **2011**, *155*, 417–428. [CrossRef]
20. Romano, G.; Pennacchia, E.; Agostinelli, S. Historical analysis and refurbishment proposal of the "red schools" in Viterbo. *E3S Web Conf.* **2020**, *197*, 02003. [CrossRef]
21. Directive 2009/28/EC of the European Parliament and of the Council of 23 April 2009 on the Promotion of the Use of Energy from Renewable Sources and Amending and Subsequently Repealing Directives 2001/77/EC and 2003/30/EC. Available online: http://eur-lex.europa.eu/legal-content/EN/TXT/PDF/?uri=CELEX:32009L0028&from=IT (accessed on 8 February 2018).
22. EN; EU. Directive 2010/31/EU of the European Parliament and of the Council of 19 May 2010 on the energy performance of buildings. *Off. J. Eur. Union.* **2010**, 13–35. Available online: https://eur-lex.europa.eu/LexUriServ/LexUriServ.do?uri=OJ:L:2010:153:FULL:EN:PDF (accessed on 14 September 2021). [CrossRef]
23. *CEN ISO/TR 52000-2:2017*; Energy Performance of Buildings—Overarching EPB Assessment—Part 2: Explanation and Justification of ISO 52000-1. CEN: Milan, Italy, 2017.
24. Nearly Zero Energy Buildings Definitions across Europe. Episcope Project, Brussels. 2015. Available online: http://bpie.eu/uploads/lib/document/attachment/128/BPIE_factsheet_nZEB_definitions_across_Europe.pdf (accessed on 27 April 2018).
25. Costanzo, E. *Osservatorio Degli Edifici a Energia Quasi Zero (nZEB) in Italia*; ENEA: Rome, Italy, 2019.
26. Marino, C.; Nucara, A.; Panzera, M.F.; Pietrafesa, M. Towards the nearly zero and the plus energy building: Primary energy balances and economic evaluations. *Therm. Sci. Eng. Prog.* **2019**, *13*, 100400. [CrossRef]
27. Nezhad, M.M.; Neshat, M.; Groppi, D.; Marzialetti, P.; Heydari, A.; Sylaios, G.; Garcia, D.A. A primary offshore wind farm site assessment using reanalysis data: A case study for Samothraki Island. *Renew. Energy* **2021**, *172*, 667–679. [CrossRef]
28. Tronchin, L.; Manfren, M.; Nastasi, B. Energy analytics for supporting built environment decarbonisation. *Energy Procedia* **2019**, *157*, 1486–1493. [CrossRef]
29. Prina, M.G.; Groppi, D.; Nastasi, B.; Garcia, D.A. Bottom-up energy system models applied to sustainable islands, Renew. Sustain. *Energy Rev.* **2021**, *152*, 111625. [CrossRef]
30. Official Journal—Gazzetta Ufficiale Presidential Decree 412/1993, Regolamento Recante Norme per la Progettazione, l'Installazione, l'Esercizio e la Manutenzione degli Impianti Termici degli Edifici ai Fini del Contenimento dei Consum di Energia, Attuazione dell'art. 4, Comma 4, della Legge 9 Gennaio 1991, n. 10. 1993. Available online: https://www.gazzettaufficiale.it/eli/id/1993/10/14/093G0451/sg (accessed on 2 September 2020).

31. de Santoli, L. Guidelines on energy efficiency of cultural heritage. *Energy Build.* **2015**, *86*, 534–540. [CrossRef]
32. Flavio, R.; Elisa, C.; Rosa, F.; Carbonara, E. An analysis on technological plant retrofitting on the masonry behaviour structures of 19th century Traditional Historical Buildings (THB) in Rome. *Energy Procedia* **2017**, *133*, 121–134. [CrossRef]
33. Ma, Z.; Cooper, P.; Daly, D.; Ledo, L. Existing building retrofits: Methodology and state-of-the-art. *Energy Build.* **2012**, *55*, 889–902. [CrossRef]
34. Mazzarella, L. Energy retrofit of historic and existing buildings. The legislative and regulatory point of view. *Energy Build.* **2015**, *95*, 23–31. [CrossRef]
35. Albo, A.; Rosa, F.; Tiberi, M.; Vivio, B. High-efficiency and low-environmental impact systems on a historical building in Rome: An InWall solution. *WIT Trans. Built Environ.* **2014**, *142*, 529–540. [CrossRef]
36. ICOMOS. Guidance on Heritage Impact Assessments for Cultural World Heritage Properties A publication of the International Council on Monuments and Sites. *ICOMOS* **2011**, *66*, 37–39.
37. Nastasi, B. Hydrogen policy, market, and R&D projects. *Sol. Hydrog. Prod. Process. Syst. Technol.* **2019**, 31–44. [CrossRef]
38. Di Matteo, U.; Pezzimenti, P.M.; Garcia, D.A. Methodological proposal for optimal location of emergency operation centers through multi-criteria approach. *Sustainability* **2016**, *8*, 50. [CrossRef]
39. Ascione, F.; Bianco, N.; De Masi, R.F.; De'Rossi, F.; Vanoli, G.P.; Mauro, G.M.; Vanoli, G.P.; De'Rossi, F.; Vanoli, G.P. Energy retrofit of an educational building in the ancient center of Benevento. Feasibility study of energy savings and respect of the historical value. *Energy Build.* **2017**, *95*, 172–183. [CrossRef]
40. Giombini, M.; Pinchi, E.M. Energy functional retrofitting of historic residential buildings: The case study of the historic center of Perugia. *Energy Procedia* **2015**, *82*, 1009–1016. [CrossRef]
41. Galatioto, A.; Ciulla, G.; Ricciu, R. An overview of energy retrofit actions feasibility on Italian historical buildings. *Energy* **2017**, *137*, 991–1000. [CrossRef]
42. Martínez-Molina, A.; Tort-Ausina, I.; Cho, S.; Vivancos, J.-L. Energy efficiency and thermal comfort in historic buildings: A review. *Renew. Sustain. Energy Rev.* **2016**, *61*, 70–85. [CrossRef]
43. Pisello, A.L.; Petrozzi, A.; Castaldo, V.L.; Cotana, F.; Pisello, A.L.; Petrozzi, A.; Castaldo, V.L.; Cotana, F.; Yan, J.; Lee, D.; et al. On an innovative integrated technique for energy refurbishment of historical buildings: Thermal-energy, economic and environmental analysis of a case study. *Appl. Energy* **2016**, *162*, 1313–1322. [CrossRef]
44. Troi, A.; Bastian, Z. *Energy Efficiency Solutions for Historic Buildings: A Handbook*; EURAC Research: Basel, Switzerland, 2015; ISBN 9783038216506.
45. Jafari, A.; Valentin, V. An optimization framework for building energy retrofits decision-making. *Build. Environ.* **2017**, *115*, 118–129. [CrossRef]
46. Grytli, E.; Kværness, L.; Rokseth, L.S.; Ygre, K.F. The impact of energy improvement measures on heritage buildings. *J. Archit. Conserv.* **2012**, *18*, 89–106. [CrossRef]
47. Lo Basso, G.; Rosa, F.; Garcia, D.A.; Cumo, F. Hybrid systems adoption for lowering historic buildings PFEC (primary fossil energy consumption)—A comparative energy analysis. *Renew. Energy* **2018**, *117*, 414–433. [CrossRef]
48. Burattini, C.; Nardecchia, F.; Bisegna, F.; Cellucci, L.; Gugliermetti, F.; De Lieto Vollaro, A.; Salata, F.; Golasi, I. Methodological approach to the energy analysis of unconstrained historical buildings. *Sustainability* **2015**, *7*, 10428–10444. [CrossRef]
49. Astiaso Garcia, D. Can radiant floor heating systems be used in removable glazed enclosed patios meeting thermal comfort standards? *Build. Environ.* **2016**, *106*, 378–388. [CrossRef]
50. Manfren, M.; Nastasi, B.; Tronchin, L.; Groppi, D.; Garcia, D.A. Techno-economic analysis and energy modelling as a key enablers for smart energy services and technologies in buildings. *Renew. Sustain. Energy Rev.* **2021**, *150*, 111490. [CrossRef]
51. Motlagh, S.F.M.; Sohani, A.; Saghafi, M.D.; Sayyaadi, H.; Nastasi, B. The Road to Developing Economically Feasible Plans for Green, Comfortable and Energy Efficient Buildings. *Energies* **2021**, *14*, 636. [CrossRef]
52. PANZEB. Piano d'Azione Nazionale per Incrementare gli Edifici ad Energia Quasi Zero, Roma. 2015. Available online: http://www.sviluppoeconomico.gov.it/images/stories/documenti/PANZEB_13_11_2015.pdf (accessed on 31 October 2017).
53. Barthelmes, V.M.; Becchio, C.; Corgnati, S.P.; Guala, C. Design and construction of an nZEB in Piedmont Region, North Italy. *Energy Procedia* **2015**, *78*, 1925–1930. [CrossRef]
54. Ciampi, G.; Rosato, A.; Scorpio, M.; Sibilio, S. Energy and economic evaluation of retrofit actions on an existing historical building in the south of Italy by using a dynamic simulation software. *Energy Procedia* **2015**, *78*, 741–746. [CrossRef]
55. De Santoli, L.; Basso, G.L.; Garcia, D.A.; Piras, G.; Spiridigliozzi, G. Dynamic simulation model of trans-critical carbon dioxide heat pump application for boosting low temperature distribution networks in dwellings. *Energies* **2019**, *12*, 484. [CrossRef]
56. Manfren, M.; Nastasi, B. Parametric Performance Analysis and Energy Model Calibration Workflow Integration—A Scalable Approach for Buildings. *Energies* **2020**, *13*, 621. [CrossRef]
57. Groppi, D.; Nastasi, B.; Prina, M.G.; Astiaso Garcia, D. The EPLANopt model for Favignana island's energy transition. *Energy Convers. Manag.* **2021**, *241*, 114295. [CrossRef]
58. Mora, T.D.; Cappelletti, F.; Peron, F.; Romagnoni, P.; Bauman, F. Retrofit of an Historical Building toward NZEB. *Energy Procedia* **2015**, *78*, 1359–1364. [CrossRef]
59. Sauchelli, M.; Masera, G.; D'antona, G.; Manzolini, G. ISIS Facchinetti: A nearly zero energy retrofit in Italy. *Energy Procedia* **2014**, *48*, 1326–1335. [CrossRef]

60. Mauri, L. Feasibility Analysis of Retrofit Strategies for the Achievement of NZEB Target on a Historic Building for Tertiary Use. *Energy Procedia* **2016**, *101*, 1127–1134. [CrossRef]
61. Astiaso Garcia, D.; Di Matteo, U.; Cumo, F. Selecting eco-friendly thermal systems for the "Vittoriale Degli Italiani" historic museum building. *Sustainability* **2015**, *7*, 12615–12633. [CrossRef]
62. Astiaso Garcia, D.; Cumo, F.; Tiberi, M.; Sforzini, V.; Piras, G. Cost-Benefit Analysis for Energy Management in Public Buildings: Four Italian Case Studies. *Energies* **2016**, *9*, 522. [CrossRef]
63. Rosa, F.; Cumo, F.; Calcagnini, L.; Vivio, B. Redevelopment of Historic Buildings through the Implementation of Green Roofs: A Study of a Design Methodology. (n.d.). Available online: http://www.academia.edu/2421781/Redevelopment_of_historic_buildings_through_the_implementation_of_green_roofs_a_study_of_a_design_methodology (accessed on 9 September 2013).
64. Groppi, D.; de Santoli, L.; Cumo, F.; Astiaso Garcia, D. A GIS-based model to assess buildings energy consumption and usable solar energy potential in urban areas. *Sustain. Cities Soc.* **2018**, *40*, 546–558. [CrossRef]
65. Roversi, R.; Cumo, F.; D'Angelo, A.; Pennacchia, E.; Piras, G. Feasibility of municipal waste reuse for building envelopes for near zero-energy buildings. *WIT Trans. Ecol. Environ.* **2017**, *224*, 115–125. [CrossRef]
66. Mattoni, B.; Nardecchia, F.; Bisegna, F. Towards the development of a smart district: The application of a holistic planning approach. *Sustain. Cities Soc.* **2019**, *48*, 101570. [CrossRef]
67. Pompei, L.; Nardecchia, F.; Mattoni, B.; Bisegna, F.; Mangione, A. Comparison between two energy dynamic tools: The impact of two different calculation procedures on the achievement of nZEBs requirements. *Build. Simul. Conf. Proc.* **2019**, *6*, 4259–4266. [CrossRef]
68. Romano, G.; Pennacchia, E.; Agostinelli, S. Refurbishment of a house in a historical building: Energy saving, electrification and flexibility. *E3S Web Conf.* **2020**, *197*, 02010. [CrossRef]
69. Ministero per i Beni e le Attività Culturali (MIBACT). Legislative Decree n. 42 of 22 January 2004. Code of Cultural and Landscape Heritage. 2004. Available online: http://www.unesco.org/culture/natlaws/media/pdf/italy/it_cult_landscapeheritge2004_engtof.pdf (accessed on 23 March 2017).
70. University of Wisconsin-Madison. TRNSYS 17. Multizone Building Modeling with Type 56 and TRNBuild. 2010, Volume 6. Available online: http://web.mit.edu/parmstr/Public/Documentation/06-MultizoneBuilding.pdf (accessed on 10 September 2021).
71. *EN ISO 52016-1:2018*; Energy Performance of Buildings—Energy Needs for Heating and Cooling, Internal Temperatures and Sensible and Latent Heat Loads—Part 1: Calculation Procedures. Ente Nazionale Italiano di Unificazione (UNI): Milan Italy, 2018.
72. Opaque Envelope Components of Buildings—Thermo-Physical Parameters UNI/TR 11552: 2014. Available online: http://store.uni.com/catalogo/uni-tr-11552-2014?___store=en&josso_back_to=http%3A%2F%2Fstore.uni.com%2Fjosso-security-check.php&josso_cmd=login_optional&josso_partnerapp_host=store.uni.com&___from_store=it (accessed on 2 September 2020).
73. Decree of 26 June 2015 Concerning New Minimum Requirements and Methodology for Calculating Energy Performance of Buildings. 2015. Available online: https://www.mise.gov.it/index.php/it/normativa/decreti-interministeriali/2032966-decreto-interministeriale-26-giugno-2015-applicazione-delle-metodologie-di-calcolo-delle-prestazioni-energetiche-e-definizione-delle-prescrizioni-e-dei-requisiti-minimi-degli (accessed on 7 February 2020).
74. *UNI/TS 11300-1*; Prestazioni Energetiche degli Edifici-Parte 1: Determinazione del Fabbisogno di Energia Termica dell'Edificio per la Climatizzazione Estiva ed Invernale. Ente Nazionale Italiano di Unificazione (UNI): Milan, Italy, 2014.
75. Regione Lazio-Price List-Public works-Tariffa dei Prezzi (Lavori Pubblici). 2016. Available online: http://www.regione.lazio.it/rl_infrastrutture/?vw=contenutiDettaglio&cat=1&id=122 (accessed on 20 March 2019).
76. Council Directive 2011/70/Euratom of 19 July 2011 Establishing a Community Framework for the Responsible and Safe Management of Spent Fuel and Radioactive Waste. 2011. Available online: https://eur-lex.europa.eu/legal-content/EN/TXT/?uri=celex:32012L0027 (accessed on 23 January 2021).
77. Development of Energy Communities and the Strategic Implications for the Country and Its Economy—The European House Ambrosetti, (n.d.). Available online: https://www.ambrosetti.eu/en/whats-hot-en/cities-and-territories/development-of-energy-communities/ (accessed on 24 June 2021).
78. Cumo, F.; Astiaso Garcia, D.; Calcagnini, L.; Rosa, F.; Sferra, A.S. Urban policies and sustainable energy management. *Sustain. Cities Soc.* **2012**, *4*, 29–34. [CrossRef]

energies

MDPI

Article

Renewable Energy System Controlled by Open-Source Tools and Digital Twin Model: Zero Energy Port Area in Italy

Sofia Agostinelli [1], Fabrizio Cumo [2,*], Meysam Majidi Nezhad [1], Giuseppe Orsini [1] and Giuseppe Piras [1]

[1] Department of Astronautics, Electrical and Energy Engineering (DIAEE), Sapienza University of Rome, 00184 Roma, Italy; sofia.agostinelli@uniroma1.it (S.A.); meysam.majidinezhad@uniroma1.it (M.M.N.); giuseppe.orsini@uniroma1.it (G.O.); giuseppe.piras@uniroma1.it (G.P.)

[2] CITERA Research Centre, Sapienza University of Rome, 00197 Rome, Italy

* Correspondence: fabrizio.cumo@uniroma1.it

Abstract: The present paper deals with an infrastructure digitization policy to optimize maintenance processes and energy efficiency to transform port areas to ZED (Zero Energy District). The Lazio Region started the process for all its ports in 2020. The Anzio port started and developed as a pilot project as it is a particularly representative sample for the Mediterranean Sea reality due to its geomorphological conformation. The study aimed to develop energy-saving procedures and strategies and integrate production systems from Renewable Energy Systems (RESs) for sustainable mobility. In the article, these strategies are described in detail and energy analysis is carried out, starting from the current state and demonstrating the potential energy self-sufficiency of the infrastructure. Finally, the investigation's potential utilizing a Digital Twin (DT) of the area is highlighted. Furthermore, the BIM (Building Information Modeling) and GIS (Geographic Information System) combining possibility to maximize the energy efficiency measures beneficial impact are discussed.

Keywords: Renewable Energy Systems (RESs); Zero Energy District (ZED); Digital Twin (DT); Building Information Modelling (BIM); Geographic Information System (GIS); Revit software's

Citation: Agostinelli, S.; Cumo, F.; Nezhad, M.M.; Orsini, G.; Piras, G. Renewable Energy System Controlled by Open-Source Tools and Digital Twin Model: Zero Energy Port Area in Italy. *Energies* **2022**, *15*, 1817. https://doi.org/10.3390/en15051817

Academic Editors: Andrea Mauri, Benedetto Nastasi and Gerardo Maria Mauro

Received: 10 January 2022
Accepted: 25 February 2022
Published: 1 March 2022

Publisher's Note: MDPI stays neutral with regard to jurisdictional claims in published maps and institutional affiliations.

1. Introduction

Increasing energy demand due to human society's population growth has led to rising energy prices [1], pollution and Greenhouse Gas (GHG) emissions. In this case, energy costs can be a significant overhead for ports [2]. Reducing GHG emissions and air pollution directly contributes to ports sustainability and green landscape [3]. Energy efficiency in ports is mainly related to providing the same services with less energy consumption and Renewable Energy Sources (RESs) and environmentally friendly [4]. Energy efficiency has a critical role for ports to reduce energy consumption and provide environmentally friendly services. The weather conditions that can influence port policymakers, sustainability and adaptation strategies are vital to helping create green ports [5]. Since many of the ports are located near large cities, they play an essential role in air pollution [6–8]. Ports, especially container ports, have three functional areas: quayside, yard, and landside [9,10]. Reducing GHG and pollutant emissions directly results from energy efficiency, equipment electrification, alternative fuels [11] and RESs. Along with increasing operational efficiency, these aspects can form a large part of the ports concept in the next generation [12].

There is a strong relationship between port operational efficiency and port energy efficiency. The increased operational efficiency of sources reduces energy consumption and significantly increases energy efficiency in ports [13]. Energy consumption in ports can be either electricity or fossil fuel. In recent years, practical steps have been taken to electrify equipment using electricity generated in ports through RESs, including the increasing advances in electricity generation, storage, distribution, conversion, and consumption technologies [14]. Furthermore, those technologies used in ports can significantly increase energy efficiency [15]. The significant increase in renewable energy technology, accompanied by the control technologies development and the converter installation, has led to the development efforts to develop energy in ports considering the existing potential [16].

Today, new technologies can increase energy efficiency and reduce GHG emissions in ports as solutions. On the other hand, using boats and ships with electricity can prevent severe marine pollution caused by oil spills. These solutions include the electricity use as RESs for independent vehicles, energy storage devices, cooling technologies and clean fuels such as cold-ironing [17], equipment [18,19], reefer containers [19], technologies in lighting. This technological improvement can dramatically guarantee energy efficiency using the Light-Emitting Diode (LED) lamps instead of high-pressure sodium lamps in port storage facilities, management buildings, high lighting towers in the wind space terminal [20]. For example, the Netherlands Delta terminal uses LED bulbs to save 922 MWh of annual power consumption, equivalent to € 300,000 [21]. In addition to using LED technology, focusing on lighting levels and designing armatures in ports can help save energy.

Renewable energies have been evaluated and identified as clean sources, such as tidal [22] and wave energy [23], geothermal energy [24], wind [25], and solar energy [26,27] are available due to the geographical location areas for ports. Many studies are addressing port energy management such as the ports of Singapore [28], Hamburg [29], Rotterdam in the Netherlands [30], Antwerp [31], Istanbul [32], Lübeck [8], Vancouver [33], La Spezia [34]. Sadek et al. [35] focused on RESs to replace fossil fuels of the Mediterranean ports. The offshore wind turbines and fuel cell units have been used as two examples of energy sources in ports. Their research shows that the combined system of wind turbines and fuel cells is the best choice for the unit cost of electricity generation with 0.101 and 0.107 of Alexandria port. Furthermore, they state that using fuel cells [36] and offshore wind turbines [37] as a green power concept will reduce "CO_2", "NOx" and "CO" emissions per year. Finally, they point to using a combination of renewable energy and green energy supply in the port of Alexandria, possibly reducing 22.31% of annual electricity costs.

In this regard, maritime transport is under increasing pressure to reduce the harmful effects of climate and the environment. Maritime transport can be responsible for 10% to 15% of the annual emissions of sulfur (SOx) and nitrogen oxides (NOx) and also approximately 3% of the global carbon dioxide (CO_2) emissions [38]. In 2018, the International Maritime Organization (IMO) member states set a 50% reduction in greenhouse gas emissions by 2050 compared to 2008, referred to as the "Paris Agreement on Shipping". Achieving this goal in 2050 requires the different sectors' strenuous installation and development of new technologies and other political measures by governments to adopt the maritime sector to zero emissions [39]. Hence, five critical factors in reducing port pollution have been identified by the World Port Climate Declaration at the international level: (1) reduce CO_2 from ships that sail the long and deep sea to ports, (2) evaluate how CO_2 is reduced from port operations, (3) evaluate how CO_2 is reduced from inland shipping, (4) study how to use RESs as an alternative way, and (5) further develop methods for calculating CO_2 in ports [34].

The Digital Twin (DT) system can be proposed essentially based on integration of softwares that has already been used, such as Building Information Modelling (BIM) and Geographic Information System (GIS). In addition, sensors aimed at the database implementing [40] and the functioning Artificial Intelligence (AI) systems optimizing [41] can be inserted in a sharing platform and powered in real-time by a series of the Internet of Things (IoT). The simulations can be carried out using specific, compatible tools that will allow the use of the DT in multiple fields of study, from architecture to engineering and economics [42]. Implementing a systemic, digital approach applied to industrial areas and urban systems produces diversified digital city models [43] based on the scale of analysis [44]. The DT methodology presents the traditional urban basis map's evolution and progressive technological transformation and is developed, managed [45], and constantly monitored in three dimensions through models based on intelligent geo-databases [46].

In buildings, energy-related parameters can be linked to the DT building portion of the neighborhood to monitor energy consumption, related costs and optimization [47]. Opening up renewed scenarios such as intelligent and evolutionary cities, and these monitoring possibilities can find application in the management of individual infrastructures or more extensive scale of entire commercial and industrial areas. Creating an information model based on objects, specific properties and attributes to develop an accurate DT model [48], means configuring a tool for analysing and evaluating possible scenarios supporting the decision-making process. Information models are powered by a constant flow of data generated [49] to update the DT model in real-time of the main digital model, i.e., sensors, cameras and smart grids [50]. The DT model can progressively collect a large number of data using sensors installed inside the buildings and port infrastructures, returning a virtual mirror of reality at any time. In particular, the DT thus collects operational and environmental information in the process components [51], later processed with analytical techniques and algorithmic simulations.

This study aims to develop energy-saving and increase energy efficiency methods and strategies using the RESs integration production systems in the energy ports. Firstly, energy-saving strategies are described in energy analysis detail [52]. Furthermore, the research potential of DT has been through the integration of BIM and GIS software. Secondly, several open-source (online and free access) platform-tools have been used to evaluate the Anzio port's wind and solar energy potential. Finally, these open-source platform tools are used to discover, extract, and process RESs data mapping, assessment and modelling to understand better the port of Anzio with very high time resolution data.

This paper is organized as follows: Section 2 explains the materials and methods, Section 3 describes the Anzio port as a case study, the results are provided in Section 4. Section 5 presents research discussion and Section 6 presents conclusions.

2. Materials and Methods

The DT includes a physical model, a virtual model and a connection between the physical and virtual models. In particular, several online open-platform data such as "Renewables. Ninja", "RSE Wind Atlas" platforms and MERRA-2 reanalysis data were used as input data of three software types such as, BIM, GIS and Revit software were used for analysis and integrating the model.

2.1. Efficient Strategies Development and DT Model

Firstly, the IoT data is collected through sensors and actuators sharing information to the DT in the cloud. Furthermore, the DT simulates its operation based on the information collected and uses these simulations either as a benchmark for comparison with the actual performance or to modify the operation/setting of the duplicated physical object.

The realisation of the DT model related to the product, system, organisation or activity process investigated. In the case of DT in the urban context, the idea of "Smart City" is the focus of this study. The model can be created using BIM software, such as Graphisoft ArchiCAD and Autodesk Revit (Figure 1).

Furthermore, the efficient strategies developed through the identification of individual technologies able to reduce consumption (e.g., lighthouse towers equipped with Light-Emitting Diode (LED) lighting) (Figure 2) and through energy diagnoses on buildings performances using dedicated software such as MC4 and TerMus. Autodesk's simulation engine, Green Building Studio (GBS), has been utilized; it enables the energy analysis functions in Revit and Insight, the web interface for interacting with the results produced by GBS.

As an example, the use of the electric multi-scale digital BIM and GIS model allows the detailed analysis of energy consumption, both through a punctual computation of all local loads such as lighting fixtures and electric recharging columns for boats and through the calculation of the actual consumption of public buildings [51] in the area (harbour master's office, ticket office, etc.).

The inclusion of appropriate shared parameters in the BIM model has allowed a detailed calculation [53], estimate and description of the energy consumption associated with the various facilities in the port area. For example, the consumption of the lighting terminals, which is about 67% of the total electrical consumption (see Table 1), is shown below.

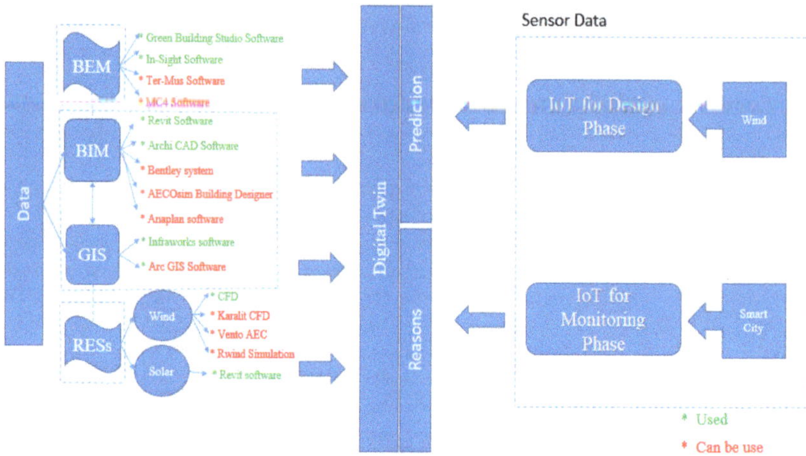

Figure 1. The main stages of the method developed in the port of Anzio. The software's used in this research is shown in green and the software's that can be used is shown in red.

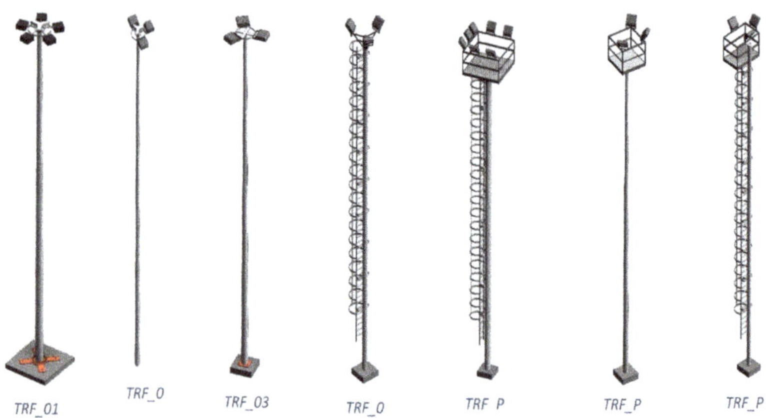

Figure 2. The lighting structures under studies.

The consequent efficiency improvement hypothesis by replacing the old energy-consuming floodlights of the light towers LED lighting structures is shown in Figure 2. The total load, following the efficiency measures, is therefore equal to:

$$C_l + C_{cr} \tag{1}$$

With C_l the consumption of the new lighting system and C_{cr} that of the charging stations. Therefore, after the interventions, the overall energy saved is equal to:

$$E_{AO} - E_{PO} = MWh\ per\ Year \tag{2}$$

where E_{AO} and E_{PO} represent the pre- and post-construction consumption respectively.

Table 1. Actual electrical load lighting system of the Anzio port.

Existing Phase—Consumption Estimate—Lighting										
Data Sheet	**Description**	**Family and Type**	**Quantity (n)**	**Spotlight (n)**	**Spotlight Power (W)**	**Type Power (W)**	**Total Power (W)**	**Hours of Use (h)**	**Days of Use (n)**	**Annual Consumption (Wh)**
STF-A_004.1	Streetlight	LMP_01: H_4m	3	1	60	60	180	10	365	657,000
STF-A_004.2	Streetlight	LMP_02: H_6m	18	1	80	80	1440	10	365	5,256,000
STF-A_004.3	Streetlight	LMP_03: H_6m - 3_Proiettori	4	3	100	300	1200	10	365	4,380,000
STF-A_004.4	Streetlight	LMP_04: H_3m	2	1	80	80	160	10	365	584,000
STF-A_004.5	Streetlight	LMP_05: H_3m	12	1	100	100	1200	10	365	4,380,000
STF-A_004.6	Streetlight	LMP_06: H_6m	4	1	80	80	320	10	365	1,168,000
STF-A_004.7	Streetlight	LMP_07: H_6m - 1_Proiettore	3	1	100	100	300	10	365	1,095,000
STF-A_004.8	Streetlight	LMP_07: H_6m - 2_Proiettori	1	2	100	200	200	10	365	730,000
STF-A_005.1	Light tower with platform	TRF_P_01: H_12m - 7_Proiettori	1	7	850	5950	5950	10	365	21,717,500
STF-A_005.2	Light tower with platform	TRF_P_02: H_12m - 3_Proiettori	1	3	850	2550	2550	10	365	9,307,500
STF-A_005.3	Light tower with platform	TRF_P_03: H_12m - 3_Proiettori	1	3	850	2550	2550	10	365	9,307,500
STF-A_006.1	Light tower	TRF_01: H_12m - 3_Proiettori	7	3	850	2550	17,850	10	365	65,152,500
STF-A_006.2	Light tower	TRF_01: H_12m - 5_Proiettori	6	5	850	4250	25,500	10	365	93,075,000
STF-A_006.3	Light tower	TRF_02: H_12m - 2_Proiettori	1	2	850	1700	1700	10	365	6,205,000
STF-A_006.4	Light tower	TRF_02: H_12m - 3_Proiettori	1	3	850	2550	2550	10	365	9,307,500
STF-A_006.5	Light tower	TRF_03: H_12m - 3_Proiettori	2	3	850	2550	5100	10	365	18,615,000
STF-A_006.6	Light tower	TRF_04: H_12m - 3_Proiettori	1	3	850	2550	2550	10	365	9,307,500
STF-A_007.1	Signal light	FRR: H_4m	1	1	100	100	100	10	365	365,000
STF-A_007.2	Signal light	FRV: H_4m	1	1	100	100	100	10	365	365,000
							71,500			260,975,000
										260,975 kWh

2.2. Wind Energy Potential Assessment

In this context, MERRA-2 reanalysis data has been used to wind speed potential mapping and time series analysis of the areas of Anzio port over 41 years between 1980 to 2021. The wind speed time-series analysis and mapping can help in decision-making about the RESs in the Anzio port areas.

Furthermore, the numerical analysis, the MATLAB software "Curve Fitting Toolbox" tool used to obtain the value of the annual energy produced. With the turbine control, it is possible to produce:

$$N_t \times E_t\ (kWh)\ per\ year \tag{3}$$

With N_t the number of devices and E_t the energy produced in a year by each of them.

2.3. Solar Energy Potential Assessment

MERRA-2 reanalysis data has been used for solar irradiation potential analysis of the areas of Anzio port over 41 years between 1980 to 2021. The solar energy sources time-series analysis and mapping can help decision-making and better understand the RESs in the Anzio port areas.

In this step, two areas for installing solar panels have been investigated, (i) Photovoltaic (PV) asphalt, (ii) the parking PV shelter.

Finally, the total area of the PV plant was calculated:

$$S = \frac{P_p}{\eta} = m^2 \tag{4}$$

where P_p is the peak power of the system and η the average yield of the modules. With the same energy produced with solar asphalt:

$$P_p = \frac{E_{as}}{E_{1kw(p)}} \approx kWp \tag{5}$$

With E_{as} the electricity produced overall by the solar asphalt and $E_{1kWp(pens)}$ produced by a 1 kWp system mounted on the shelters. Therefore, the number of shelters can be easily calculated:

$$\frac{P_p}{P_{pp}} = PV \tag{6}$$

Since each shelter covers an area of 50 m^2, the surface occupied by all the infrastructure is equal to:

$$S_p \times N_p = m^2 \tag{7}$$

Corresponding exactly to that of the parking area identified.

2.4. Energy Produced Balance

The total energy produced by the two RESs, it means wind and solar is equal to:

$$E_f + E_e = MWh\ per\ year \tag{8}$$

The goal of transforming the area into a ZED has been achieved. The optimized annual energy requirement of the port area is fully covered by the on-site production of wind and photovoltaic systems.

$$E_{FER} + C = +2.8\ MWh\ per\ year \tag{9}$$

With E_{FER} the energy produced from RESs and C the consumption of the port.

2.5. CO_2 Emissions Avoided

The energy upgrading of the port area can significantly contribute to the reduction of CO_2 emissions and the reduction of energy absorption from the national electricity grid.

The calculation of avoided CO_2 emissions consists of the electricity generated from RESs by the average annual CO_2 emission factor associated with the electricity grid.

The table with the emission factors, taken from the "Joint Research Center", shows the quantity in tons of CO_2 emitted per MWh of energy produced by some energy carriers. Therefore, first of all, it is possible to calculate the CO_2 not emitted as a result of the area's consumption reduction interventions:

$$C_{a-p} \times F_{re} = t \ Per \ CO_2 \quad (10)$$

where C_{a-p} is the difference in pre- and post-construction consumption, and F_{re} is the considered emission factor. On the other hand, the share of CO_2 emissions not emitted into the environment following the installation of RESs, is equal to:

$$E_{FER} \times F_{re} = t \ Per \ CO_2 \quad (11)$$

With E_{re}, the energy produced by RESs. Therefore, the CO_2 not released into the environment compared to the starting situation is:

$$CO_{2e} + CO_{2FER} = t \ Per \ CO_2 \quad (12)$$

To this value, an additional quantity of CO_2 not emitted into the environment should be added, i.e., that of the boats that would use the recharging service through the columns installed along the quays of the port (whose CO_2 emission factors from the electricity network are in any case lower than those of conventional fuels).

Unfortunately, the estimation of the "carbon footprint" applied to boats parked on the quay or when approaching/leaving the port is a rather complex operation since the data relating to the turnout of boats in the harbour are not available. Furthermore, the environmental benefit of the measures adopted should be extended to the lack of emissions of pollutants such as $PM_{2.5}$ and NOx from boats.

3. Case Study

Ports can be considered one of the most well-known places where human activities and environmental issues are indirect. Many important ports are focused on local and regional development and have been able to preserve local traditions for a long time because they are directly related to the development of their immediate city. Ports are therefore publicly owned in most parts of the world, although they have been privatized in the operational sector, which is usually required due to the nature, size, and long-term prospects of the investments required [54]. Hence, energy management plans in ports are highly dependent on energy management strategies in the nearby city [24]. Therefore, ports are directly dependent on the national electricity grid as an energy source to meet the needs of domestic electricity and ships stationed [35].

Anzio (41°26′52.61″ N–12°37′44.59″ E) is a city with 43.43 km^2 (16.77 sq mi) and commune on the coast of the Lazio region of Italy, about 51 km (32 mi) south of Rome, Lazio (Latium) region, and located on a peninsula jutting into the Tyrrhenian Sea [55] (Figure 3). Anzio is a fishing port and a departure point well known for its seaside harbour setting for ferries and hydroplanes to the Pontine islands of Ponza, Palmarola, and Ventotene [56].

Figure 3. Anzio port.

4. Results

In this section, the research results are presented with a new integrated method using digital and RES models. These results offer practical strategies for evaluating wind and solar energy sources in ports to study the ZED.

4.1. Integration of DT Models and LEDs

The main objective of the DT framework is proposed to support decision-making using integrated multi-scale digital data sources, BIM and GIS information containers for simulation purposes about the implementation of strategies improving energy performance in the entire port area.

Future developments will integrate multi-scale digital simulations into real-time data. The digital models are structured to be interconnected to a cloud platform to acquire valuable data from the models and IoT sensors, configuring the effective DT.

Through Revit software, a series of distinct digital models were created by discipline (architectural, structural and MEP), each representing a specific "layer" within the overall digital model of the port area (Figure 4). Each of these models was then populated with three-dimensional families/objects, relative to the specific discipline under investigation, used as information containers.

Using the Industry Foundation Classes (IFC) file format allows exchanging information through a standard, open and non-proprietary format. As a result, it is possible to exploit all the BIM functionalities, generating an energy model of the building/plant system. Consequently, it is possible to analyse the actual state of different energy loads thanks to a detailed analysis of heating, hot water and cooling requirements in natural conditions, identifying and adjusting the most critical parts in the system's annual energy balance [57]. Once the efficiency of the entire port has been achieved, potential areas for the insertion of renewable energy production technologies are identified [58].

Similarly, with the interoperability between BIM and GIS, operated through Autodesk's Infra works software, the DT information is enriched with geospatial information describing the urban environment. This systems cooperation creates a reliable model where geographic information and design data are integrated to understand different asset interactions with the surrounding city.

Figure 4. BIM model of the Anzio port area.

Among possible applications, GIS information can be exploited in a BIM process to improve energy savings. In this sense, GIS informs BIM that by exploiting data such as building heights and footprints, it is possible to identify areas with high energy loads or those with the highest priority for energy retrofitting [59].

The inclusion of appropriate shared parameters in the BIM model has allowed a detailed calculation, estimate and description of the energy consumption associated with the various facilities in the port area. The lighting terminals' consumption is about 67% of the total electrical consumption. The consequent efficiency improvement hypothesis is shown by replacing the old energy-consuming floodlights of the light towers with LED lighting structures. The results can be summarized as a reduction of about 65% in energy consumption for lighting than the current state as reported (Table 2).

In addition, some charging devices are installed in the Anzio port area for private and public boats (Figure 5). These devices are located in different places and divided into double charging stations and simple interlocked sockets.

Figure 5. BIM model of the charging stations.

Table 2. Optimized electrical load of the port lightning system.

Project Phase—Consumption Estimate—Lighting										
Data Sheet	**Description**	**Family and Type**	**Quantity (n)**	**Spotlight (n)**	**Spotlight Power (W)**	**Type Power (W)**	**Total Power (W)**	**Hours of Use (h)**	**Days of Use (n)**	**Annual Consumption (Wh)**
STF-A_004.1	Streetlight	LMP_01: H_4m	3	1	60	60	180	10	365	657,000
STF-A_004.2	Streetlight	LMP_02: H_6m	18	1	80	80	1440	10	365	5,256,000
STF-A_004.3	Streetlight	LMP_03: H_6m - 3_Proiettori	4	3	100	300	1200	10	365	4,380,000
STF-A_004.4	Streetlight	LMP_04: H_3m	2	1	80	80	160	10	365	584,000
STF-A_004.5	Streetlight	LMP_05: H_3m	12	1	100	100	1200	10	365	4,380,000
STF-A_004.6	Streetlight	LMP_06: H_6m	4	1	80	80	320	10	365	1,168,000
STF-A_004.7	Streetlight	LMP_07: H_6m - 1_Proiettore	3	1	100	100	300	10	365	1,095,000
STF-A_004.8	Streetlight	LMP_07: H_6m - 2_Proiettori	1	2	100	200	200	10	365	730,000
STF-A_005.1	Light tower with platform	TRF_P_01: H_12m - 7_Proiettori	1	7	250	1750	1750	10	365	6,387,500
STF-A_005.2	Light tower with platform	TRF_P_02: H_12m - 3_Proiettori	1	3	250	750	750	10	365	2,737,500
STF-A_005.3	Light tower with platform	TRF_P_03: H_12m - 3_Proiettori	1	3	250	750	750	10	365	2,737,500
STF-A_006.1	Light tower	TRF_01: H_12m - 3_Proiettori	7	3	250	750	5250	10	365	19,162,500
STF-A_006.2	Light tower	TRF_01: H_12m - 5_Proiettori	6	5	250	1250	7500	10	365	27,375,000
STF-A_006.3	Light tower	TRF_02: H_12m - 2_Proiettori	1	2	250	500	500	10	365	1,825,000
STF-A_006.4	Light tower	TRF_02: H_12m - 3_Proiettori	1	3	250	750	750	10	365	2,737,500
STF-A_006.5	Light tower	TRF_03: H_12m - 3_Proiettori	2	3	250	750	1500	10	365	5,475,000
STF-A_006.6	Light tower	TRF_04: H_12m - 3_Proiettori	1	3	250	750	750	10	365	2,737,500
STF-A_007.1	Signal light	FRR: H_4m	1	1	100	100	100	10	365	365,000
STF-A_007.2	Signal light	FRV: H_4m	1	1	100	100	100	10	365	365,000
							24,700			90,155,000
									kWh	90,155

In Table 3 is reported the overall electrical consumption of all the devices.

Table 3. Electrical load for charging systems.

Consumption Estimate—Electrical Device							
Data Sheet	**Description**	**Family and Type**	**Quantity (n)**	**Power (W)**	**Hours of Use (h)**	**Days of Use (n)**	**Annual Consumption (Wh)**
STF-A_003.1	Charging station	CLL_01_QMC200B: GW68832W	2	4	10	90	7200
STF-A_003.2	Charging station	CLL, 02: 4P	5	4	10	90	18,000
STF-A_003.3	interlocked socket	PRI: 2P, 01	10	4	10	90	36,000
STF-A_003.4	interlocked socket	PRI: 2P, 02	15	4	10	90	54,000
STF-A_003.5	interlocked socket	PRI: 3P	2	4	10	90	7200
STF-A_003.6	interlocked socket	PRI: 4P	2	4	10	90	7200
							129,600

The total amount of optimizing electrical loads of the Anzio port area is 90,155 + 129,600 = 219,755 kWh for a year. This is the target of implementing the RESs local grid production in the same place to reach a ZED (Figure 6).

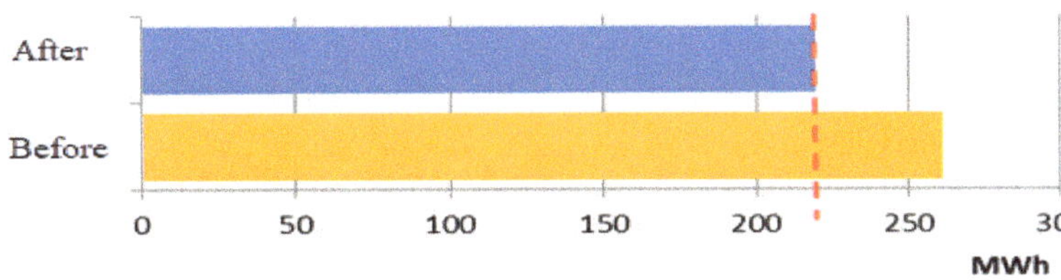

Figure 6. Electricity consumption of Anzio.

4.2. Wind Energy Potential Assessment

More than 40 years of monthly data from the MERRA-2 reanalysis dataset have been used to understand better the wind speed potential and mapping of port areas and understand the wind speed (Figures 7 and 8) in the case of micro wind turbines' installation.

Secondly, two sites have been located as ideal for the placement of the turbines [60]. The locations match the piers at the South and North ends of the harbour (Figure 9) for a total amount of fifteen turbines.

Ten wind turbines will be located next to the previous breakwater points, and the remaining five will follow the second breakwater lines.

The micro wind turbine DS3000 model (ETNEO Italia) was chosen to be installed at Anzio Port. DS3000 model is a 3 kW vertical axis micro-generator, equipped with a Savonius rotor mounted on the central axis of the turbine, valid for the starting of the rotation with low winds, and three Darrieus blades to increase the production with medium/strong winds. The Savonius blades, oriented on the four cardinal points to capture the wind from any direction, do not require the orientation of the rotor (Figure 10).

The annual average wind speed was calculated using the "Renewables. ninja" platform, which provides an hourly average wind speed [37]. For both Region of Interest (ROI), the obtained value at the height of 10 m is around 4.75 m/s (data confirmed by the RSE Wind Atlas platforms and MERRA-2 reanalysis dataset) [60–62]. Regarding calculating the turbine's annual energy production, a reference was made to the data estimated by the supplier ETNEO Italia, as shown in Table 4.

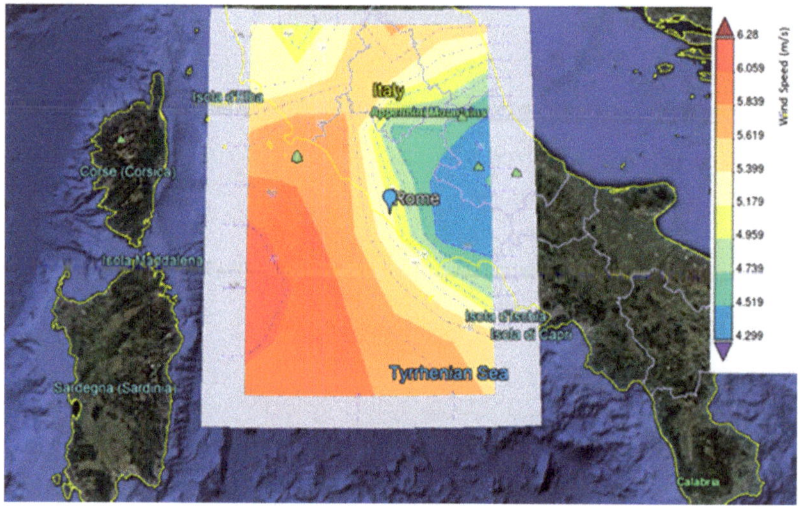

Figure 7. Surface wind speed (m/s) in the Rome City and Anzio port showed a blue point for 1980 to 2021.

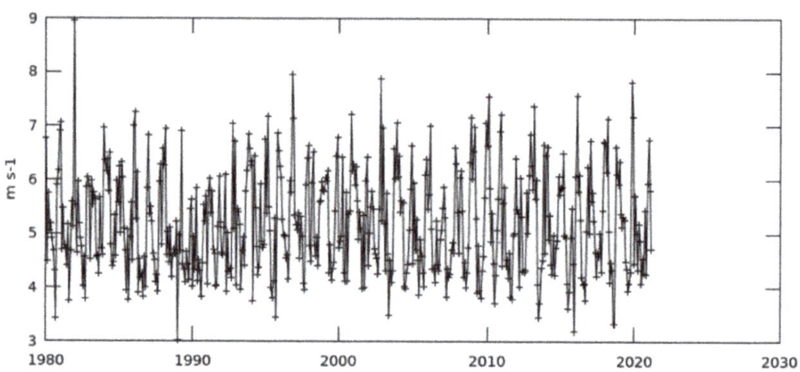

Figure 8. Showed a monthly time-series analysis between 1980 to 2021 for the Anzio port.

With the power curve in MATLAB the energy value produced by the calculating turbine for a vehicle speed of 4.75 m/s is approximately 2420 kWh. Fifteen micro wind turbines are located in external port areas, as shown in Figure 9 as an example. Therefore, the obtained value is approximately 2420 kWh per year. Therefore, it is possible to produce 36,300 kWh per year with fifteen turbines located on the outer side of the pier (Figure 9). Specifically, a spline has been selected as the appropriate interpolation function, as shown in the curve in Figure 11.

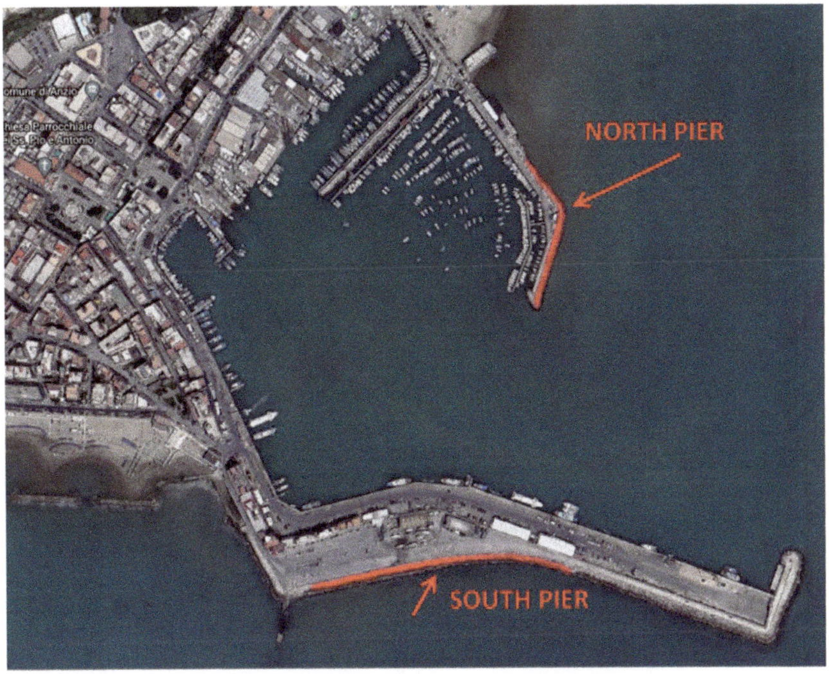

Figure 9. The micro-wind turbine installation locations.

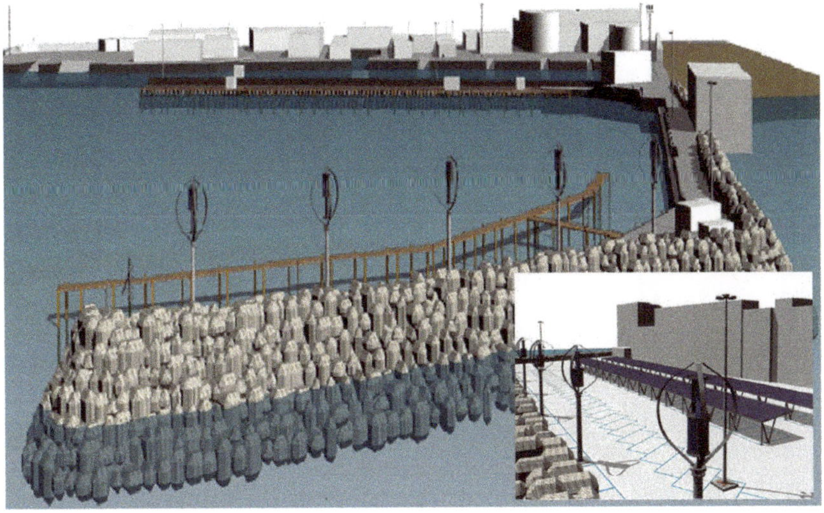

Figure 10. Turbines BIM Model.

Table 4. Turbine's Annual Energy Production (AEP).

Average Wind Speed (m/s)	AEP (kWh)
5	2.851
5.5	3.819
6	4.877
6.5	5.975
7	7.061
7.5	8.088
8	8.945

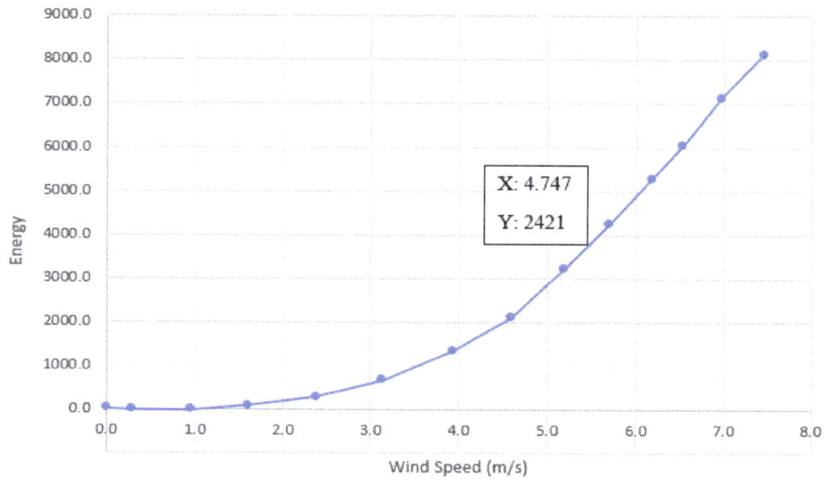

Figure 11. Turbine production data.

4.3. Solar Energy Assessment (PV System)

More than 40 years of monthly data from the MERRA-2 reanalysis dataset have been used to map PV solar irradiation analysis mapping of port areas with exciting potential to evaluate the solar irradiation potential (Figures 12 and 13) for PV installation.

Secondly, electric power production from a solar source should be performed by positioning a dedicated PV panel directly on the ground in port areas without significant cars. The technology of solar roads rapidly increases its penetration, especially on infrastructures and installation characterized by wide spaces without buildings and people presence [63–65].

For installing the photovoltaic modules, the area at the end of the southern pier was chosen. It's a large area, characterized by good exposure and without shading (there are no buildings nearby), as shown in Figure 14, with indicated in the red zone with the arrow. Let's consider the latter solution: the parking area identified is shown in the figure. It is an area of about 800 m^2, with parking spaces oriented at 30° to the south.

Photovoltaic canopies are car park covers that have the dual purpose of covering parked vehicles and generating clean energy. They are mainly made of galvanized steel, a highly resistant material supporting the photovoltaic module system. The photovoltaic system for electricity production is placed on the sloping roof, which can be connected to the grid or even to the charging columns of parked electric cars (Figures 14 and 15). These systems protect cars from prolonged exposure to the sun, transforming a potentially harmful agent for the car into a source of clean energy.

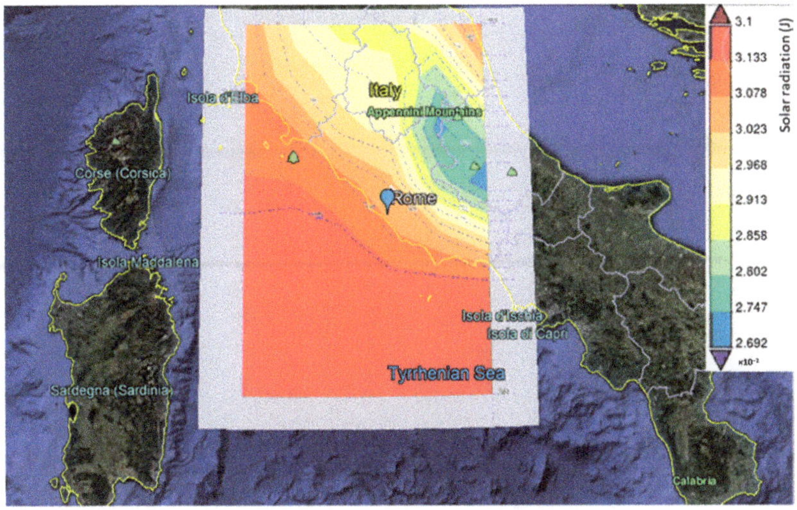

Figure 12. Solar irradiation in the Rome City and Anzio port showed a blue point from 1980 to 2020.

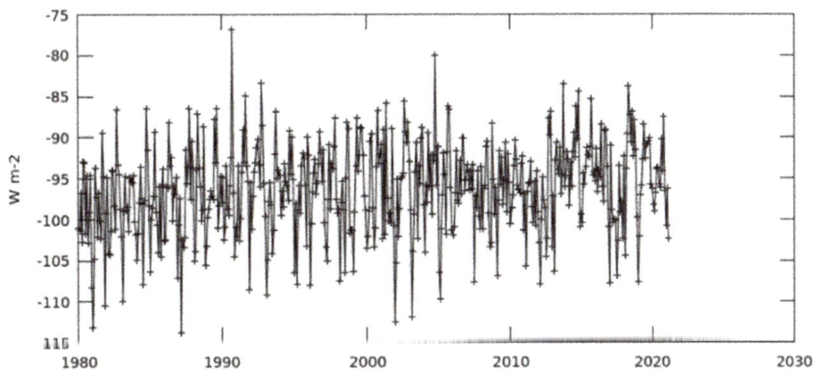

Figure 13. Showed a monthly time-series analysis between 2019 to 2020 for the Anzio port.

The structure also offers shelter from precipitation, hail and snow. From the data collected in the literature, the modules used for solar asphalt have average yields of around 9%. The performance of this technology is lower than standard modules, but it is still worth studying its application in the port also in the perspective of future technological advances. To cover the remaining energy requirement, the peak power required to be available is approximately 150 kWp.

The orientation and inclination have been set to 0°, and also the "integrated in the building" option has been chosen as the mounting position. From the calculation performed with PVGIS, the energy produced is equal to 186,254.63 kWh per year. The results were obtained from the simulation for the 150 kWp plant.

To estimate the energy production of the PV shelters, the azimuth angle of the car parks has been inserted as the "tilt" angle of the wall unit's design. The energy produced per kW is much greater than in the previous case (1334.22 kWh/kW compared to 1241.69 kWh/kW of the solar asphalt).

A single shelter with P_{pp} = 9k the PV system peak power. Taking into account the rounding up, the shelters produce a total of 192.13 MWh per year, with a total peak power of 144 kWp.

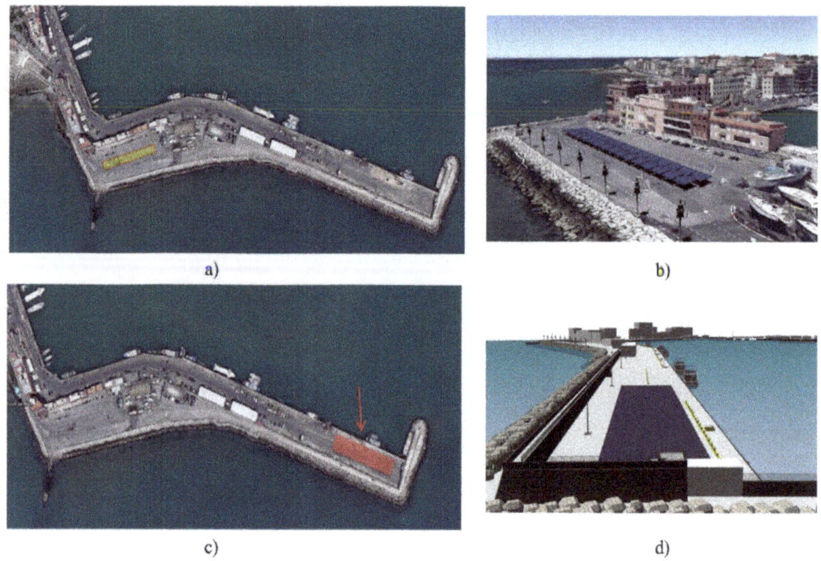

Figure 14. (**a**,**b**) Photovoltaic (PV) solar asphalt area, (**c**,**d**) PV asphalts.

Figure 15. Photovoltaic shelters.

4.4. *Energy Produced Balance*

As shown in the pie chart, the energy contribution of wind turbines is significantly lower than that of solar asphalt (Figure 16 and Table 5). The goal of transforming the area into a ZED has been achieved. The optimized annual energy requirement of the port area is fully covered by the on-site production of wind and photovoltaic systems.

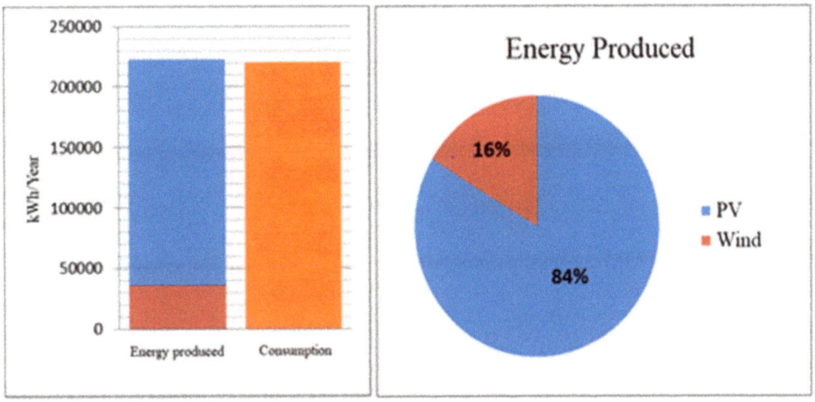

Figure 16. Comparison of consumption - energy produced by RES (**Left**), Subdivision of energy produced by RES (**Right**).

Table 5. Annual production of energy from RESs.

Annual Energy Production of FER-Anzio		
FER	**Technology**	**kwh/year**
PV	Asphalt PV	186,254.63
Wind	wind turbine	36,300
		222,554.63

4.5. CO_2 Emissions

The energy upgrading of the port area can significantly contribute to the reduction of CO_2 emissions and the reduction of energy absorption from the national electricity grid. The calculation of avoided CO_2 emissions consists of the electricity generated from RESs by the average annual CO_2 emission factor associated with the electricity grid.

The table with the emission factors, taken from the guidelines of the "Joint Research Center", shows the quantity in tons of CO_2 emitted per MWh of energy produced by some energy carriers (Table 6):

Table 6. Emission factors.

CO_2 Emission	Amount
Gas	0.202
Diesel	0.267
Electric tariffs	0.276

An additional quantity of CO_2 not emitted into the environment should be added, i.e., that of the boats that would use the recharging service through the columns installed along the quays of the port (whose CO_2 emission factors from the electricity network are in any case lower than those of conventional fuels) (Figure 17).

Unfortunately, the estimation of the "carbon footprint" applied to boats parked on the quay or when approaching/leaving the port is a rather complex operation since the data relating to the turnout of boats in the harbour are not available. Nevertheless, the measures adopted should be the environmental benefit to the lack of pollutants such as $PM_{2.5}$ and NOx from boats.

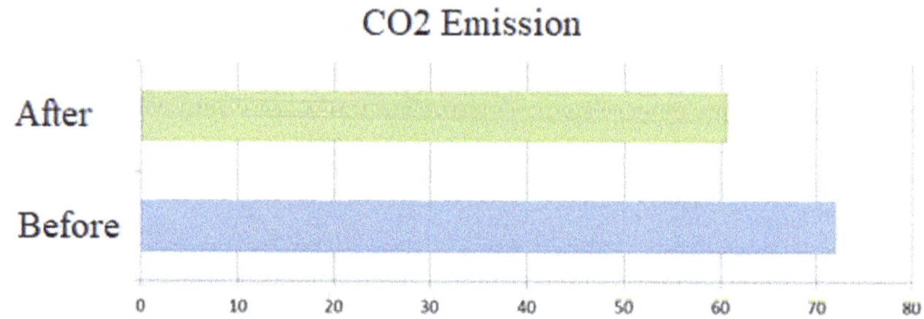

Figure 17. CO_2 emissions before and after construction.

4.6. Economic Evaluation

In this paragraph, we will limit ourselves to making an account of the investment costs relating to installing wind and PV systems. As already explained above, there are no reliable references in the literature concerning the costs of solar asphalt technology. Therefore, for the economic analysis, reference will be made to the installation costs of the photovoltaic shelters. For the latter, the 9 kW model from Kit. Solutions were chosen from the quotes consulted online. The cost of each cover amounts to € 19,650. Regarding the micro-wind plant, the price list was provided to us by the supplier company Etneo Italia Srl. From the quotation of the bidding company (Table 7), it appears:

Table 7. Cost of turbine components.

RES	Cost
DS3000 Turbine	9600 €/cad
On grid turbine	2350 €/cad
Huawei inverter 4 kW	1250 €/cad
Total	13,200 €/cad

The calculation of expenses is shown in the Table 8 below:

Table 8. RESs cost.

FER	P_n [kW_p]	Amount [n]	P_p plant [kW_p]	Energy Produced [kWh/Year]	Energy Produced Plant [kWh/Year]	Cost [€]	Total Cost [€]
PV	9	16	144	12,008	192,128	19,650	314,400
Turbine	3	15	45	2420	36,300	13,200	198,000

The overall cost of the infrastructure installed is, therefore, equal to 512,400 euros. This expense must be added to that relating to installation, which we assume equal to 45% of the cost of the devices (Table 9):

Table 9. Total investment.

	Cost [€]	FPO [€]	Total Investment [€]
PV Shelters	314,400	141,480	455,880
Turbine	198,000	89,100	287,100
			742,980

For micro-wind turbines, there are state incentives that concern the production of energy and not the installation of the system and are provided by the Energy Services

Manager (GSE). This means that the return on investment is more significant the more energy is produced (if the wind turbine is located in a windy area). The two methods of incentive proposed are the all-inclusive rate and the "exchange on-site". All-inclusive tariff involves a gain of EUR0.30 for each kWh fed into the national grid and has a duration of 15 years, after which one can access the free market or the so-called "dedicated collection". The "exchange on-site" incentive is more suitable for small plants.

In this case, the gain amounts to EUR0.20 per kWh without time constraints. The calculation mechanism consists of the difference between the energy produced and the energy consumed: if the result is negative, it is charged to the bill. If positive, there is an energy "credit" scalable on future billing. In addition to the GSE incentives just outlined, for the installation of RES plants, such as wind power, the 2020 Budget Law allows a tax deduction of 65% (micro-generators) up to a maximum expenditure of EUR100 thousand. Photovoltaic systems on canopies are always a convenient solution, even without the incentives of the energy bill. In fact, by exploiting the "Exchange on-site" and sizing the system concerning actual consumption, thus aiming for self-consumption, it is possible to reduce the electricity bill significantly.

5. Discussion

Energy efficiency is one of the hot research topics that advances in existing technology directly affect future research prospects [66]. In this case, innovative approaches, economic analysis [67], operations optimization, the new technological advances effects [68], and management analysis will be the most important for future ports research. Economic and environmental analysis for automated and electric ports are also necessary and inevitable [69]. Therefore, integrating independent and electrical equipment with energy storage devices makes smart meters more potent and straightforward, making the significant range possible for further analysis [70]. Next-generation ports will use automation, electricity and intelligent energy management systems. To this aim, the role of independent or electric vehicles in the smart grid is unavoidable, which should be further discussed for future port operations.

Energy management can be considered one of the management tools prominent features to move ports towards greater sustainability with the environment and reduce greenhouse gases. That is why energy management in ports is more focused on Environmental Management Systems (EMS). For example, the European Maritime Ports Organization (ESPO) supports and encourages European ports to develop environmental management programs [3], but EMS is only approved by half of the European ports. Therefore, interaction with shareholders can be considered basic and essential to implement EMS in ports better [33].

The ports' main challenges in the EMS development can be identified in the following cases; (1) balancing economic and environmental goals in ports and (2) sharing knowledge and joint production and successful experiences in national and international ports [71]. In this regard, several ports define specific programs for energy management, which can be considered with the protection policies of the European Union, including projects for the renewable energy use to reduce air and environmental pollution [27].

Managing sustainable energy development using RESs is an emerging issue for ports [12]. In this regard, a conceptual framework for energy management systems, similar to the model in their construction, can be very effective [72]. Unfortunately, there are no studies analyzing barriers to energy efficiency in ports. These barriers to energy efficiency in ports include most technical, economic and regulatory aspects. There are also barriers to the supply of clean fuels and other technologies [1], so barrier analysis is invaluable to industry and academia. In the port industry literature, most technical reports explain the RESs use.

Researchers can focus on intelligent grid analysis and evaluate smart grids operational and environmental performance using simulation tools. Balancing energy demand and supply in the smart grid is a complex task [73]. Furthermore, more research in the initial step should have high data quality to perform a successful analysis. Assessing and evaluating aspects of monitoring has benefits for ports, (i) assessing the level of environmental pollution in ports, (ii) improving the environmental and energy management impacts, and (iii) helping to reform port strategies and policies, (v) the possibility of evaluating energy consumption studies and environmental risk management in ports to estimate and determine practical goals and measures in the future [74,75].

In this regard, to support decision-makers and port managers in selecting and implementing sustainable technologies, the need for actual empirical research and accurate implementation of the tools selection, technologies and operations [76]. Hence, there is a need to transfer more practical information from the scientific community to decision-makers to help port decision-makers. Furthermore, this knowledge transfer to port decision-makers who have more experience in implementation decisions will help them make the final decision [77].

Furthermore, conducting studies that act as a guide by considering the necessary standards can be very useful in developing a strategy to reduce GHG emissions in ports [78]. These studies can be used from a technical point of view, availability and cost analysis of alternative energy technology study with new development, evaluation, and implementation that increase regulatory regulations in ports [35]. The proposed DT framework aims to configure a digital integrated and multi-scale database for simulation purposes. It is intended to be integrated into real-time data from sensors and improve data management.

6. Conclusions

The project proposed presents an Anzio port digital transformation process of the Lazio Region, starting from their infrastructural centre. The implications of this transformation directly concern the environmental, economic and social spheres, setting the port area as the epicentre and extending to the city. The port has potential for public buildings, water sports schools, boat workshops and association headquarters, not the subject of this paper, which can be incorporated into the DT. Due to their geometric characteristics and location, a maximum of two-level buildings is easily transformable into ZEB buildings. A further implementation of the DT and extending the harbour representation can improve its environmental and economic management.

The data can be entered into a BIM and GIS environment within a sharing platform predictive scenario derived, and the urban fabric resilience with the creation and use of appropriate tools can be estimated. This allows planning in a well conscious way, respecting environmental sustainability and interventions aimed at economic, commercial and social activities. Critical from a design and then construction point of view is the cost estimate accuracy of interventions in such a critical area. Together with the costs, it is fundamental to estimate and plan the risks related to the execution of works, which will change the social impact during their implementation.

The "Port of Anzio" DT implementation digitalization area makes it possible to start from the digital and ecological transformation epicentres and spread throughout the territory. The exchange flows studying with the surrounding territories, linked to transport by sea, land, road, and rail. It would also replace electricity production's economic and environmental costs for public lighting and electricity supply to moored boats by switching from carbon to renewable energy sources.

With repercussions in all fields, this digital transformation will open new scenarios and higher efficiency in managing public finances, especially in the green deal area. There will be substantial energy and environmental benefits already in the construction phase of the works. In the planning and design phase, it will be possible to accurately assess various implications related to the implementation of the work and its territorial, environmental, and social context. The other important factor is the possibility of minimising the project's costs by simulating different scenarios. This means monetary, environmental, and social costs, not least those due to the construction times uncertainty. This factor scares off potential private, national, and international investors. That can be predicted by adding more future factors such as horizontal transport infrastructure, roads and railways; the parameters will significantly increase the analytical data volume under study. Therefore, the data analysis with current methods will be very time consuming and complex. So, given the data diversity, such studies require a rigorous data management method and analysis to define a digital structure integration approach

Various ports appear to have attempted to install intelligent energy management systems. However, it should be borne in mind that the systems effectiveness in different ports is directly related to the port's geographical location. This relationship varies according to the RESs availability and their type for efficiency in ports. On the other hand, various studies show that evaluating new measures and technologies with a high potential for intelligent energy management systems concerning sustainable and long-term goals is very effective.

In this regard, it can be said that the lack of cooperation between stakeholders and decision-makers is the main reason for stopping various projects in ports. In addition, despite the increase in academic studies on port sustainability, the exploitation of real-world research results in port infrastructure has not yet been well implemented. The main reason for this can be the lack of case studies with regional diversity in small ports. The central gap that should be considered in future studies is the energy efficiency report of a technology or techniques studied in ports. This may indicate the need to combine existing and new measures and technologies to promote the ZEDs concept design in leading studies.

Author Contributions: Conceptualization, S.A.; methodology, S.A., F.C., M.M.N., G.O. and G.P.; formal analysis, S.A., F.C. and M.M.N.; investigation, M.M.N.; resources, M.M.N., G.O. and G.P.; data curation, S.A., F.C. and M.M.N.; writing—original draft preparation, S.A., F.C. and M.M.N.; writing—review and editing, M.M.N.; visualization, G.O. and G.P.; supervision, F.C. and G.P.; project administration, F.C. and G.P. All authors have read and agreed to the published version of the manuscript.

Funding: This research has been carried out within the project "An integrated multidisciplinary and multiscale digital approach fostering the decarbonisation of port areas" protocol RG12117A8A106E46, funded by Sapienza university of Rome.

Institutional Review Board Statement: Not applicable.

Informed Consent Statement: Not applicable.

Data Availability Statement: Not applicable.

Acknowledgments: The authors wish to thank Francesco Nardi for his important contribution to the research team. The authors also would like to thank online platforms for providing the open access data.

Conflicts of Interest: The authors declare no conflict of interest.

Abbreviations

BIM	Building Information Modelling
GIS	Geographic Information System
RESs	Renewable Energy Systems
CO_2	Carbon Dioxide
MWh	Mega Watt Hour
GBS	Green Building Studio
ROI	Region of Interest
EC	European Community
IoT	Internet of Things
DT	Digital Twin
ZED	Zero Energy District
DTM	Digital Twin Model
PVs	Photovoltaics
LED	Light-Emitting Diode

CFD	Computational Fluid Dynamics
IFC	International Foundation Class
AI	Artificial Intelligence
GHG	Greenhouse Gas
MERRA-2	Modern-Era Retrospective analysis for Research and Applications, Version 2
GSE	Energy Services Manager

References

1. Barbarelli, S.; Amelio, M.; Florio, G.; Scornaienchi, N.M. Procedure Selecting Pumps Running as Turbines in Micro Hydro Plants. *Energy Procedia* **2017**, *126*, 549–556. [CrossRef]
2. Acciaro, M. Environmental sustainability in seaports: A framework for successful innovation. *Marit. Policy Manag.* **2014**, *41*, 480–500. [CrossRef]
3. Lam, J.S.L.; Notteboom, T. The Greening of Ports: A Comparison of Port Management Tools Used by Leading Ports in Asia and Europe. *Transp. Rev.* **2014**, *34*, 169–189. [CrossRef]
4. Nastasi, B.; Mazzoni, S.; Groppi, D.; Romagnoli, A.; Astiaso Garcia, D. Optimized integration of Hydrogen technologies in Island energy systems. *Renew. Energy* **2021**, *174*, 850–864. [CrossRef]
5. Groppi, D.; Nastasi, B.; Prina, M.G.; Astiaso Garcia, D. The EPLANopt model for Favignana island's energy transition. *Energy Convers. Manag.* **2021**, *241*, 114295. [CrossRef]
6. Yap, W.Y.; Lam, J.S.L. 80 million-twenty-foot-equivalent-unit container port? Sustainability issues in port and coastal development. *Ocean Coast. Manag.* **2013**, *71*, 13–25. [CrossRef]
7. Lam, J.S.L. Designing a sustainable maritime supply chain: A hybrid QFD-ANP approach. *Transp. Res. Part E Logist. Transp. Rev.* **2015**, *78*, 70–81. [CrossRef]
8. Gibbs, D.; Rigot-Muller, P.; Mangan, J.; Lalwani, C. The role of sea ports in end-to-end maritime transport chain emissions. *Energy Policy* **2014**, *64*, 337–348. [CrossRef]
9. Steenken, D.; Voß, S.; Stahlbock, R. Container terminal operation and operations research—A classification and literature review. *OR Spectr.* **2004**, *26*, 3–49. [CrossRef]
10. Bierwirth, C.; Meisel, F. A follow-up survey of berth allocation and quay crane scheduling problems in container terminals. *Eur. J. Oper. Res.* **2015**, *244*, 675–689. [CrossRef]
11. Nastasi, B.; Mazzoni, S.; Groppi, D.; Romagnoli, A.; Astiaso Garcia, D. Solar power-to-gas application to an island energy system. *Renew. Energy* **2021**, *164*, 1005–1016. [CrossRef]
12. Woo, J.K.; Moon, D.S.H.; Lam, J.S.L. The impact of environmental policy on ports and the associated economic opportunities. *Transp. Res. Part A Policy Pract.* **2018**, *110*, 234–242. [CrossRef]
13. Wilmsmeier, G.; Spengler, T. Energy consumption and container terminal efficiency. *FAL Bull.* **2016**, *350*, 10. Available online: https://www.cepal.org/en/publications/list/date/2016?search_fulltext=container+terminal (accessed on 28 January 2021).
14. Parise, G.; Parise, L.; Malerba, A.; Pepe, F.M.; Honorati, A.; Ben Chavdarian, P. Comprehensive Peak-Shaving Solutions for Port Cranes. *IEEE Trans. Ind. Appl.* **2017**, *53*, 1799–1806. [CrossRef]
15. Lamagna, M.; Nastasi, B.; Groppi, D.; Rozain, C.; Manfren, M.; Astiaso Garcia, D. Techno-economic assessment of reversible Solid Oxide Cell integration to renewable energy systems at building and district scale. *Energy Convers. Manag.* **2021**, *235*, 113993. [CrossRef]
16. Ricci, A.; Janssen, W.D.; van Wijhe, H.J.; Blocken, B. CFD simulation of wind forces on ships in ports: Case study for the Rotterdam Cruise Terminal. *J. Wind Eng. Ind. Aerodyn.* **2020**, *205*, 104315. [CrossRef]
17. Zis, T.; North, R.J.; Angeloudis, P.; Ochieng, W.Y.; Bell, M.G.H. Evaluation of cold ironing and speed reduction policies to reduce ship emissions near and at ports. *Marit. Econ. Logist.* **2014**, *16*, 371–398. [CrossRef]
18. Wen, B.; Jin, Q.; Huang, H.; Tandon, P.; Zhu, Y. Life cycle assessment of Quayside Crane: A case study in China. *J. Clean. Prod.* **2017**, *148*, 1–11. [CrossRef]
19. Carlo, H.J.; Vis, I.F.A.; Roodbergen, K.J. Transport operations in container terminals: Literature overview, trends, research directions and classification scheme. *Eur. J. Oper. Res.* **2014**, *236*, 1–13. [CrossRef]
20. Claudius, C.; Hardt, J. LED Technology for Container Terminals. pp. 1–7. Available online: http://europa.eu/legislation_summaries/energy/energy_efficiency/l27064_de.htm (accessed on 28 January 2021).
21. Van Duin, J.H.R.; Geerlings, H.; Froese, J.; Negenborn, R.R. Towards a method for benchmarking energy consumption at terminals: In search of performance improvement in yard lighting. *Int. J. Transp. Dev. Integr.* **2017**, *1*, 212–224. [CrossRef]
22. Tang, H.S.; Qu, K.; Chen, G.Q.; Kraatz, S.; Aboobaker, N.; Jiang, C.B. Potential sites for tidal power generation: A thorough search at coast of New Jersey, USA. *Renew. Sustain. Energy Rev.* **2014**, *39*, 412–425. [CrossRef]
23. Neshat, M. Layout optimisation of offshore wave energy converters using a novel multi-swarm cooperative algorithm with backtracking strategy: A case study from coasts of Australia. *Energy* **2022**, *239*, 122463. [CrossRef]
24. Acciaro, M.; Ghiara, H.; Cusano, M.I. Energy management in seaports: A new role for port authorities. *Energy Policy* **2014**, *71*, 4–12. [CrossRef]
25. Filom, S.; Radfar, S.; Panahi, R.; Amini, E.; Neshat, M. Exploring wind energy potential as a driver of sustainable development in the southern coasts of iran: The importance of wind speed statistical distribution model. *Sustainability* **2021**, *13*, 7702. [CrossRef]

26. Heydari, A.; Astiaso, D.; Keynia, F.; Bisegna, F.; De Santoli, L. A novel composite neural network based method for wind and solar power forecasting in microgrids Group Method of Data Handling. *Appl. Energy* **2019**, *251*, 113353. [CrossRef]
27. Boile, M.; Theofanis, S.; Sdoukopoulos, E.; Plytas, N. Developing a port energy management plan: Issues, challenges, and prospects. *Transp. Res. Rec.* **2016**, *2549*, 19–28. [CrossRef]
28. Song, S.; Poh, K.L. Solar PV leasing in Singapore: Enhancing return on investments with options. *IOP Conf. Ser. Earth Environ. Sci.* **2017**, *67*, 012020. [CrossRef]
29. Hamburg Port Authority. Hamburg Is Staying on Course—The Port Development Plan to 2025. Published by: Free and Hanseatic City of Hamburg—State Ministry of Economic Affairs, Transport and Innovation Hamburg Port Authority. 2012, pp. 1–98. Available online: http://www.hamburg-port-authority.de (accessed on 28 January 2021).
30. Port of Rotterdam. 2021. Available online: https://www.portofrotterdam.com/en/doing-business/setting-up/existing-industry/energy-industry/sustainable-energy (accessed on 15 April 2021).
31. Port of Antwerp. 2021. Available online: https://businessinantwerp.eu/news/port-antwerp-working-today-energy-tomorrow (accessed on 15 April 2021).
32. Schipper, C.A.; Vreugdenhil, H.; de Jong, M.P.C. A sustainability assessment of ports and port-city plans: Comparing ambitions with achievements. *Transp. Res. Part D Transp. Environ.* **2017**, *57*, 84–111. [CrossRef]
33. Poulsen, R.-T.; Ponte, B.S.; Sornn-Friese, H. Environmental upgrading in global value chains: The potential and limitations of ports in the greening of maritime transport. *Geoforum* **2018**, *89*, 83–95. [CrossRef]
34. Fenton, P. The role of port cities and transnational municipal networks in efforts to reduce greenhouse gas emissions on land and at sea from shipping—An assessment of the World Ports Climate Initiative. *Mar. Policy* **2017**, *75*, 271–277. [CrossRef]
35. Sadek, I.; Elgohary, M. Assessment of renewable energy supply for green ports with a case study. *Environ. Sci. Pollut. Res.* **2020**, *27*, 5547–5558. [CrossRef]
36. Caglayan, D.G.; Ryberg, D.S.; Heinrichs, H.; Linßen, J.; Stolten, D.; Robinius, M. The techno-economic potential of offshore wind energy with optimized future turbine designs in Europe. *Appl. Energy* **2019**, *255*, 113794. [CrossRef]
37. Alavirad, S.; Mohammadi, S.; Golombok, M.; Haans, K. Interconnection and generation from a North Sea power hub—A linear electricity model. *Int. J. Electr. Power Energy Syst.* **2021**, *133*, 107132. [CrossRef]
38. Lindstad, H.E.; Eskeland, G.S. Environmental regulations in shipping: Policies leaning towards globalization of scrubbers deserve scrutiny. *Transp. Res. Part D Transp. Environ.* **2016**, *47*, 67–76. [CrossRef]
39. Bjerkan, K.Y.; Seter, H. Reviewing tools and technologies for sustainable ports: Does research enable decision making in ports? *Transp. Res. Part D Transp. Environ.* **2019**, *72*, 243–260. [CrossRef]
40. Talamo, C.; Paganin, G.; Atta, N.; Rota, F. Epistemic uncertainty, risk management and information: The role of the detailed design. *Techne* **2019**, *18*, 164–173. [CrossRef]
41. Fuldauer, E. Smarter Cities are Born with Digital Twins. 2019. Available online: https://tomorrow.city/a/smarter-cities-are-born-with-digital-twins (accessed on 10 April 2021).
42. Atta, N.; Dalla Valle, A.; Campioli, A.; Chiaroni, D.; Talamo, C. Construction technologies for sustainable affordable housing within fragile contexts: Proposal of a decision support tool. *Sustainability* **2021**, *13*, 5928. [CrossRef]
43. Talamo, C.; Pinto, M.R.; Viola, S.; Atta, N. Smart cities and enabling technologies: Influences on urban Facility Management services. *IOP Conf. Ser. Earth Environ. Sci.* **2019**, *296*, 012047. [CrossRef]
44. Centre for Digital Built Britain. 2020. Available online: https://www.cdbb.cam.ac.uk/what-we-do (accessed on 11 April 2021).
45. Talamo, C.; Atta, N.; Martani, C.; Paganin, G. L'integrazione delle infrastrutture urbane fisiche e digitali: Il ruolo dei "big Data". *Techne* **2016**, *11*, 217–225. [CrossRef]
46. Mancini, F.; Nastasi, B. Solar energy data analytics: PV deployment and land use. *Energies* **2020**, *13*, 417. [CrossRef]
47. Manfren, M.; Nastasi, B.; Tronchin, L.; Groppi, D.; Garcia, D.A. Techno-economic analysis and energy modelling as a key enablers for smart energy services and technologies in buildings. *Renew. Sustain. Energy Rev.* **2021**, *150*, 111490. [CrossRef]
48. Mohammadi, N.; Taylor, J.E. Smart city digital twins. In Proceedings of the 2017 IEEE Symposium Series on Computational Intelligence (SSCI), Honolulu, HI, USA, 27 November–1 December 2017; pp. 1–5. [CrossRef]
49. Paganin, G.; Talamo, C.; Atta, N. Knowledge management and resilience of urban and territorial systems. *Techne* **2018**, *15*, 124–133. [CrossRef]
50. Manfren, M.; Nastasi, B.; Groppi, D.; Astiaso Garcia, D. Open data and energy analytics—An analysis of essential information for energy system planning, design and operation. *Energy* **2020**, *213*, 118803. [CrossRef]
51. Manfren, M. Parametric energy performance analysis and monitoring of buildings—HEART project platform case study. *Sustain. Cities Soc.* **2020**, *61*, 102296. [CrossRef]
52. Mohammadi, S.; de Vries, B.; Schaefer, W. *A Comprehensive Review of Existing Urban Energy Models in the Built Environment BT—Planning Support Systems for Sustainable Urban Development*; Geertman, S., Toppen, F., Stillwell, J., Eds.; Springer: Berlin/Heidelberg, Germany, 2013; pp. 249–265.
53. Aste, N.; Buzzetti, M.; Caputo, P.; Manfren, M. Local energy efficiency programs: A monitoring methodology for heating systems. *Sustain. Cities Soc.* **2014**, *13*, 69–77. [CrossRef]
54. Satir, T.; Doğan-Sağlamtimur, N. The protection of marine aquatic life: Green Port (EcoPort) model inspired by Green Port concept in selected ports from Turkey, Europe and the USA. *Period. Eng. Nat. Sci.* **2018**, *6*, 120–129. [CrossRef]
55. Available online: https://www.britannica.com/place/Anzio (accessed on 28 January 2021).

56. Available online: https://en.wikipedia.org/wiki/Anzio (accessed on 28 January 2021).
57. Mohammadi, S.; Soleymani, S.; Mozafari, B. Scenario-based stochastic operation management of MicroGrid including Wind, Photovoltaic, Micro-Turbine, Fuel Cell and Energy Storage Devices. *Int. J. Electr. Power Energy Syst.* **2014**, *54*, 525–535. [CrossRef]
58. Hautala, K.; Järvenpää, M.E.; Pulkkinen, P. Digitalization transforms the construction sector throughout asset's life-cycle from design to operation and maintenance. *Stahlbau* **2017**, *86*, 340–345. Available online: https://onlinelibrary.wiley.com/doi/epdf/1 (accessed on 28 January 2021). [CrossRef]
59. Safamehr, H.; Rahimi-Kian, A. A cost-efficient and reliable energy management of a micro-grid using intelligent demand-response program. *Energy* **2015**, *91*, 283–293. [CrossRef]
60. Orlando, A.; Pagnini, L.; Repetto, M.P. Structural response and fatigue assessment of a small vertical axis wind turbine under stationary and non-stationary excitation. *Renew. Energy* **2021**, *170*, 251–266. [CrossRef]
61. Syawitri, T.P.; Yao, Y.F.; Chandra, B.; Yao, J. Comparison study of URANS and hybrid RANS-LES models on predicting vertical axis wind turbine performance at low, medium and high tip speed ratio ranges. *Renew. Energy* **2021**, *168*, 247–269. [CrossRef]
62. Manatbayev, R.; Baizhuma, Z.; Bolegenova, S.; Georgiev, A. Numerical simulations on static Vertical Axis Wind Turbine blade icing. *Renew. Energy* **2021**, *170*, 997–1007. [CrossRef]
63. Mirzanamadi, R.; Hagentoft, C.E.; Johansson, P. Coupling a Hydronic Heating Pavement to a Horizontal Ground Heat Exchanger for harvesting solar energy and heating road surfaces. *Renew. Energy* **2020**, *147*, 447–463. [CrossRef]
64. Yesner, C.; Jasim, A.; Wang, H.; Basily, B.; Maher, A.; Safari, A. Energy harvesting and evaluation of a novel piezoelectric bridge transducer. *Sens. Actuators A Phys.* **2019**, *285*, 348–354. [CrossRef]
65. Mirzanamadi, R.; Hagentoft, C.E.; Johansson, P. Numerical investigation of harvesting solar energy and anti-icing road surfaces using a hydronic heating pavement and borehole thermal energy storage. *Energies* **2018**, *11*, 3443. [CrossRef]
66. Lee, D.; Cheng, C.C. Energy savings by energy management systems: A review. *Renew. Sustain. Energy Rev.* **2016**, *56*, 760–777. [CrossRef]
67. Lam, J.S.L.; Ko, M.J.; Sim, J.R.; Tee, Y. Feasibility of implementing energy management system in ports. In Proceedings of the 2017 IEEE International Conference on Industrial Engineering and Engineering Management (IEEM), Singapore, 10–13 December 2017; pp. 1621–1625. [CrossRef]
68. Barbarelli, S.; Florio, G.; Lo Zupone, G.; Scornaienchi, N.M. First techno-economic evaluation of array configuration of self-balancing tidal kinetic turbines. *Renew. Energy* **2018**, *129*, 183–200. [CrossRef]
69. Nakada, M. A Comparative Study on Two Types of Automated Container Terminal Systems. *Comput. Appl. Shipp. Shipbuild.* **1980**, *7*, 217–222.
70. Al-Alawi, B.M.; Bradley, T.H. Review of hybrid, plug-in hybrid, and electric vehicle market modeling Studies. *Renew. Sustain. Energy Rev.* **2013**, *21*, 190–203. [CrossRef]
71. Puente-Rodríguez, D.; van Slobbe, E.; Al, I.A.C.; Lindenbergh, D.E.D. Knowledge co-production in practice: Enabling environmental management systems for ports through participatory research in the Dutch Wadden Sea. *Environ. Sci. Policy* **2016**, *55*, 456–466. [CrossRef]
72. May, G.; Stahl, B.; Taisch, M.; Kiritsis, D. Energy management in manufacturing: From literature review to a conceptual framework. *J. Clean. Prod.* **2017**, *167*, 1464–1489. [CrossRef]
73. Coronado Mondragon, A.E.; Coronado, E.S.; Coronado Mondragon, C.E. Defining a convergence network platform framework for smart grid and intelligent transport systems. *Energy* **2015**, *89*, 402–409. [CrossRef]
74. Gonzalez Aregall, M.; Bergqvist, R.; Monios, J. A global review of the hinterland dimension of green port strategies. *Transp. Res. Part D Transp. Environ.* **2018**, *59*, 23–34. [CrossRef]
75. Kang, D.; Kim, S. Conceptual model development of sustainability practices: The case of port operations for collaboration and governance. *Sustainability* **2017**, *9*, 2333. [CrossRef]
76. Di Vaio, A.; Varriale, L. Management innovation for environmental sustainability in seaports: Managerial accounting instruments and training for competitive green ports beyond the regulations. *Sustainability* **2018**, *10*, 783. [CrossRef]
77. Ng, A.K.Y. Port Decision Maker Perceptions on the Effectiveness of Climate Adaptation Actions. *Coast. Manag.* **2018**, *46*, 148–175. [CrossRef]
78. Nastasi, B. Renewable Hydrogen Potential for Low-carbon Retrofit of the Building Stocks. *Energy Procedia* **2015**, *82*, 944–949. [CrossRef]

energies

MDPI

Article

Investigation on Performance of Various Power Control Strategies with Bifilar Coil for Induction Surface Melting Application

Alagarsamy Sureshkumar [1], Ramachandiran Gunabalan [2], Pradeep Vishnuram [1], Sridhar Ramsamy [1] and Benedetto Nastasi [3,*]

1 Department of Electrical and Electronics Engineering, SRM Institute of Science and Technology, Chennai 603 203, India; sureshkumarask2@gmail.com (A.S.); pradeep.kannan03@gmail.com (P.V.); sridharr@srmist.edu.in (S.R.)
2 School of Electrical Engineering, VIT Chennai, Chennai 600 127, India; gunabalan1979@gmail.com
3 Department of Planning, Design and Technology of Architecture, Sapienza University of Rome, 00196 Rome, Italy
* Correspondence: benedetto.nastasi@outlook.com

Abstract: In recent years, induction heating applications assisted by electronic power control have been very appealing. For melting applications, induction heating is widely used as it seems to be appropriate and provides higher efficiency, zero pollutants, non-contamination of material, etc. in comparison with conventional heating. The conventional variable frequency control scheme is not sufficient for melting applications because of its high switching loss, low efficiency, and lower heat rate. A superlative control technique is required to control the output power smoothly, for a high heating rate with minimum power loss, and to lower the number of components. In this paper, a capacitorless self-resonating bifilar coil is proposed for induction surface melting applications. The performance of the system in terms of modular losses, heat rate, and efficiency is analyzed for various power methods such as pulse duty cycle control, phase shift control, pulse density modulation control, and asymmetric duty cycle control. An experimental validation is performed for the 1 kW prototype, and the heating rate, efficiency, and modular losses are calculated. The control technique is digitally validated using a PIC16F877A microcontroller with 30 kHz switching frequency. The temperature distribution is analyzed using a FLIR thermal imager. Among the tested methods, pulse density modulation-based control provides smooth and varied power control from 0% to 100% with minimum modular losses. The efficiency of the system is 89% at a rated output power and is greater than 85% for pulse density modulation control with a fast heating rate.

Keywords: asymmetric duty cycle control; bifilar coil; pulse duty cycle control; induction heating; metal melting; phase shift control; pulse density modulation; series resonant inverter; variable frequency control

Citation: Sureshkumar, A.; Gunabalan, R.; Vishnuram, P.; Ramsamy, S.; Nastasi, B. Investigation on Performance of Various Power Control Strategies with Bifilar Coil for Induction Surface Melting Application. *Energies* **2022**, *15*, 3301. https://doi.org/10.3390/en15090001

Academic Editor: Abu-Siada Ahmed

Received: 23 March 2022
Accepted: 27 April 2022
Published: 30 April 2022

Publisher's Note: MDPI stays neutral with regard to jurisdictional claims in published maps and institutional affiliations.

1. Introduction

Induction heating (IH) is commonly used in residential, industrial, and medical applications because it has features such as being hygienic; using non-contact heating; providing protection; and having higher competence, higher power density, and specific power control characteristics [1]. The electromagnetic induction principle is used to heat the object directly via the eddy current developed on the load due to I^2R loss [2]. The heat depth of penetration of the load depends on the material's skin depth [3] and the magnitude of the heat depends directly on the frequency of the supply. Therefore, for any real-time applications, the selection of an appropriate operating frequency affects the heating capacity. The switching frequency generally varies from 5 kHz to 1000 kHz, with a load handling capacity of 500 kW [4].

In the recent past, high-frequency resonant inverters with minimum switching losses have enticed researchers to study IH applications. The inverter switches are operated either in zero voltage switching (ZVS) or zero current switching (ZCS) with higher power density, higher efficiency, and light weight [5,6]. Based on the arrangement of the resonant capacitor with respect to the IH load, the inverters are termed as either series-resonant or parallel-resonant inverters. The series-resonant inverter with the voltage source inverter offers better overall performance than the parallel-resonant inverter [7]. Generally, a voltage-source full-bridge series-resonant inverter is preferred for medium-power applications, ranging from 1 kW to 5 kW [8].

Single-switch resonant inverters are used in IH systems where the power requirement is less than 2 kW. Shenkman et al. used a single-switch parallel-resonant AC–AC converter for IH applications [9]. High-frequency ac (HFAC) supply was generated with the single switch connected in series with the supply voltage. The source power factor was improved by an LC filter. The variable frequency power control scheme increased the switching losses. This problem was overcome by using a multi-cycle modulation control technique with a fixed frequency [10]. Two bidirectional semi-conductor SiC junction gate field effect transistor devices were used to convert 50 Hz AC to HFAC. Both switches were operated in soft switching mode to reduce the switching losses.

A half-bridge (HB) series-resonant inverter (SRI) is preferred for applications in which the power ratings range from 2 kW to 4 kW. Ahmed et al. developed a boost HB inverter with an LC branch at the source side [11]. This branch reduces the DC and high-frequency components on the input side, resulting in a higher efficiency. An additional LC network increases the system time constant. Forest et al. developed a multi-winding induction coil-fed IH system to handle irregular shapes of the load [12]. In this system, square and rectangular coils were used and power control was carried out by adjusting the inverter switching frequency. Lucia et al. developed a control algorithm for estimating the load parameters under dynamic variations [13]. The inverter switches were operated with a switching frequency from 20 to 100 kHz, and a control scheme was implemented using a field-programmable gate array (FPGA) controller. Load inductance and resistance were measured for various switching frequencies. Lucia et al. developed a multiple coil-based induction cooking system to deliver power to many loads [14]. The output power was controlled using the PDM technique with minimum switching losses. Lucia et al. developed a multi-inverter operated in discontinuous conduction mode for better efficiency during light load conditions [15].

Power control plays an important role in IH systems. For metal melting applications, mainly pulse width modulation (PWM), pulse density modulation (PDM), asymmetric duty cycle control (ADC), phase shift control (PSC), and variable frequency (VF) control are employed to meet the requirements. A simple DC link PWM control using a thyristor-based rectifier with an input inductor and a DC link capacitor were used to regulate the output voltage [16]. The bulky DC link capacitor increases the overall size of the module. In general, the VF control was used to control the output power for a fixed load [17]. However, ZVS is not feasible if the inverter is operated below the resonant frequency. For constant load operation, pulse width modulation is preferred to regulate the output power based on duty cycle control. This results in a smaller range of soft switching [18]. The efficiency of a PSC-fed inverter was found to be high at a duty cycle greater than 90% [19]. In addition, PDM and AVC control techniques provide better soft-switching ranges. PDM achieves greater efficiency, a wide range of set power control, and soft switching among the above control techniques [20].

In wireless power transfer applications, the self-resonating coil is often used, and its preference depends on the dimensions, frequency, and choice of applications [21]. The virtual capacitance of the coil results in either a series or a parallel resonance depending on the coil structure. In parallel resonance, the magnitude of the impedance seen from the source side is higher, which results in less current being drawn from the supply. Thus, the magnetic field stored in the coil reduces, which results in less heat generation. This problem is overcome by a current source inverter (CSI) [22], but it draws more reactive power on the secondary side of the coil. These issues are mitigated in series resonance-based bifilar coil [23]. Hence, for real-time heating applications, an RLC series resonance-based bifilar coil is preferred [24]. A capacitorless IH system is proposed in [25], with an RLC series resonance-based bifilar coil.

The main purpose of this work is to perform power control with various modulation techniques such as duty cycle control, VF, PDM, AD,C and PSC to enhance the performance of the bifilar coil-based IH system developed in [25]. Additionally, well-suited modulation techniques are needed for IH systems with less switching loss, a wide range of power control, and a fast heat rate. The developed system decreases the number of resonating components, which reduces the total system size, and uses a suitable control technique for smooth variation with less switching losses.

The rest of this paper is organized as follows: Section 2 describes the circuit description of a bifilar coil-based IH system. Various power control strategies are summarized in Section 3. An experimental validation of various control techniques is performed in Section 4. A summary of the discussions on the experimental results is provided in Section 5. The conclusions of this paper are presented in Section 6.

2. Circuit Description

The circuit of a full-bridge series-resonant inverter is shown in Figure 1. The equivalent DC source is represented in the circuit diagram. Four IGBTs (S_1, S_2, S_3, S_4) with anti-parallel diodes are used to convert the pure DC into high-frequency AC (HFAC). The output of the system is connected to a bifilar coil-fed IH load, where the bifilar coil consists of the equivalent resistance (R_{eq}) and inductance (L_{eq}) [26,27].

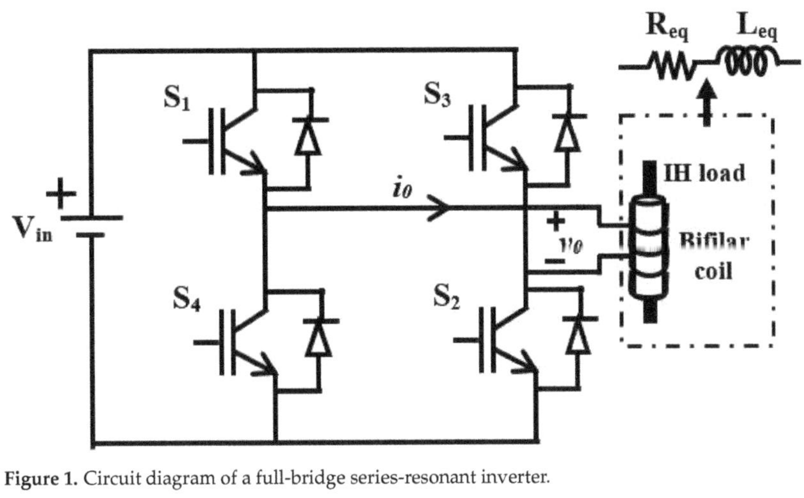

Figure 1. Circuit diagram of a full-bridge series-resonant inverter.

3. Power Control Strategies

In an induction-based melting applications, various temperature profiles are to be considered. As it is required to meet the aforesaid profile, various control techniques are needed to investigate the performance of the system.

3.1. Duty Cycle Control

Duty cycle control is the technique most commonly used to regulate the output power by controlling the duty cycle (D). In this method, the output power is varied by adjusting the on time of the duty cycle. The pulse width is varied in a symmetrical manner in both half cycles to avoid zero sequence components in the output voltage (V_0). During the off period of the IGBTs, the load current freewheels through anti-diodes. Hence, the output power of the inverter can be varied from 0% to 100% of the rated power by controlling the duty cycle. Typical waveforms of duty cycle control are shown in Figure 2. The resonance condition is not maintained when duty cycle is varied, which results in a lower efficiency. The expression that relates the output power with duty cycle is given by [26]

$$P_0 = \frac{16DV_{in}^2 \cos^2 \Phi}{\pi^2 R_{eq}} \tag{1}$$

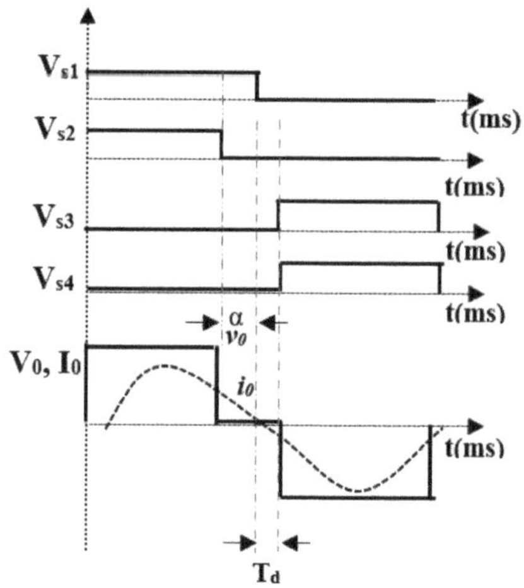

Figure 2. Typical waveforms of duty cycle control.

3.2. Variable Frequency Control

The output power supplied to the work coil can be varied by tuning the switching pulses of the inverter. The frequency of the inverter is decided based on the natural resonance frequency of the resonant tank, which yields maximum output power. If the operating point is moved away from the resonant frequency, the impedance of the system varies, which in turn varies the output power, increases the switching loss, and decreases the efficiency. The output power expression governing the variable frequency is given by [17]

$$P_0 = \frac{1}{1 + (\frac{1-f_n}{2\xi f_n})^2} P_{max} \tag{2}$$

$$\xi = \frac{R_{eq}}{2}\sqrt{\frac{L_{eq}}{C_r}} \tag{3}$$

where ξ is the damping coefficient and f_n is the normalized switching frequency.

3.3. Phase Shift Control

Typical waveforms of the phase shift control are shown in Figure 3. In this control scheme, the power variation is achieved by shifting the position of the switching pulses between adjacent inverter legs. As the phase angle of switching pulses is shifted from 0° to 180°, the power can be controlled from 0% to 100% of the rated power. The expressions governing the PSC are given by [17]

$$i_0 = \frac{4V_{in}^2 \cos^2(\frac{\phi}{2})}{\pi R_{eq}} \tag{4}$$

$$P_0 = \frac{8V_{in}^2 \cos^2(\frac{\phi}{2})}{\pi^2 R_{eq}} \quad (5)$$

For a lower value of ϕ (for 0°), P_0 is maximum as cos 0° = 1, whereas for an increase in ϕ, the output power decreases. In this method, I_0 decreases for an increase in ϕ, which results in a square-shaped source current. This causes a lower input power factor.

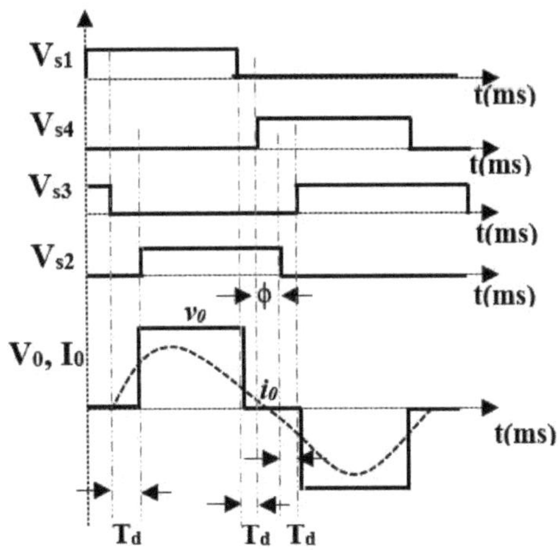

Figure 3. Typical waveforms of phase shift control.

3.4. Asymmetric Duty Cycle Control

For soft switching and power control, asymmetrical duty cycle control is used [28,29]. In this control scheme, the output voltage waveform is made asymmetric, which results in a large dead band for zero crossing of the current. Additionally, the output power is controlled by varying the rms value of the output voltage. Typical waveforms of asymmetric duty cycle control are shown in Figure 4. Because of the asymmetrical voltage waveform, even harmonics will be produced. The expression governing the AVC is given by [29]

$$V_0 = \frac{4V_{in} \cos(\frac{\alpha}{2})}{\pi} \quad (6)$$

$$P_0 = \frac{8V_{in}^2 \cos^2(\frac{\alpha}{2})}{\pi^2 R_{eq}} \quad (7)$$

3.5. Pulse Density Modulation Control

Pulse density modulation is one of the most commonly used techniques for controlling the output power without changing the switching frequency. A constant resonance frequency is maintained throughout the entire switching operation. The density of the switching pulses is varied in order to control the output power by contrasting the high-frequency switching pulses with the low-frequency pulses in compliance with the load specifications. The typical waveforms of pulse density modulation control are shown in Figure 5. The output power expression is given by [30]

$$D_{PDM} = \frac{T_{D,on}}{T_{DPDM}} \quad (8)$$

$$P_0 = D_{PDM} \frac{8V_{in}^2 \cos^2 \Phi}{\pi^2 R_{eq}} \quad (9)$$

where D_{PDM} is the duty cycle of the PDM signal.

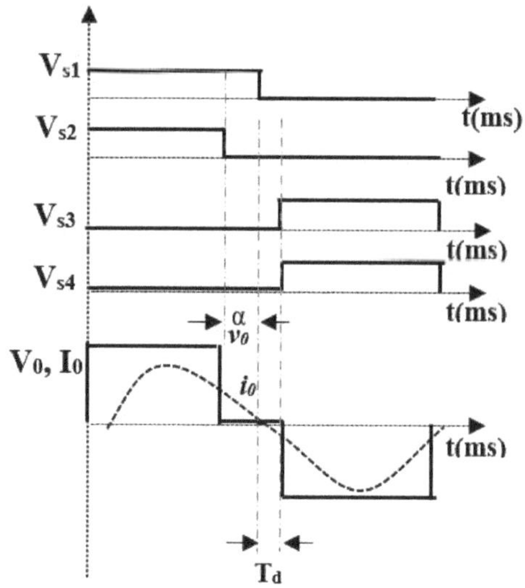

Figure 4. Typical waveforms of asymmetric duty cycle control.

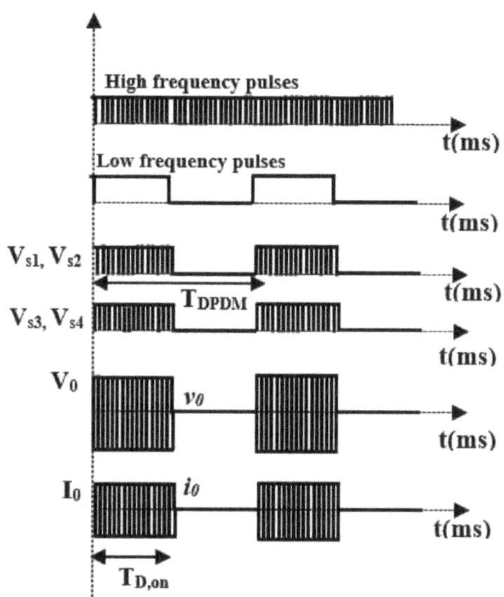

Figure 5. Typical waveforms of pulse density modulation control.

4. Experimental Results

The experimental validation of a bifilar coil-based IH system was performed using an FB inverter with four H20R1203 IGBT power switches. Various control techniques were implemented using PIC16F877A micro controller to obtain HFAC at the load terminal. The TLP250 driver IC amplifies the switching pulses, the SIGLENT CP4060 current probe is used to measure the current, and the digital oscilloscope MDO3024 was used to record the waveform. The advanced FLIR E75 24° thermal imager was used to record the thermal distribution of the load.

The work coil is made with 60 turns of 17 AWG single-layer enamelled-copper wire. The work-piece consists of a metal bar with 25 cm height and 2.5 cm radius. A thermal insulator was placed between the work coil and the work-piece. The design specifications are listed in Table 1. Figure 6 illustrates the experimental setup for the bifilar coil-based IH system. The experiment was performed to analyze the performance of the IH system in terms of efficiency, power control, and heat transfer rate (16.5 °C/min) for various control techniques.

Table 1. Specifications of the bifilar coil configuration.

Parameters	Values
V_{in}	80 V
L_{eq}	0.02 mH
R_{eq}	10 Ω
f_r	29.5 kHz
f_{sw}	30 kHz
P_{max}	1000 W

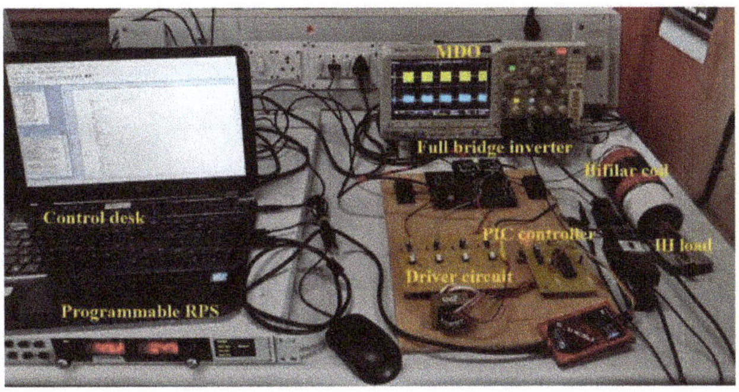

Figure 6. Test setup of the bifilar coil-based IH system.

4.1. Pulse Duty Cycle Control

The experimental results of the IH system with pulse duty cycle control are shown in Figure 7. The switching pulses and output voltage (V_0) and current (I_0) waveforms for 50% pulse duty cycle are shown in Figure 7a,b, respectively. The output voltage and current waveforms for 10% pulse duty cycle is shown in Figure 7c. The pulse width variation reduces V_0 and I_0. At 50% pulse duty cycle and 30 kHz switching frequency, the rms value of the output voltage is 70.71 V and the current is 7.07 A. The output power P_0 is $7.07^2 \times 10 = 500$ W for 10 Ω equivalent resistance. The input power is calculated as $P_{in} = V_{in(dc)} * I_{avg(dc)} = 80 \times 6.97 = 558$ W. The efficiency for 500 W is 89.6%, and for other pulse duty cycle, it is given in Table 2. As it draws more current from the supply, for lower values of duty cycle, the efficiency of the system decreases. The switch (S_1) voltage and current waveforms for 50% pulse duty cycle are shown in Figure 7d. It is inferred that soft switching exists during the on and off periods, which reduces the switching losses. The variation in the temperature with respect to time for 500 W output power is illustrated in Figure 7e. A temperature rise is noted for various duty cycle with the time interval of 5 min, and the heat rate is calculated. The heat rate for various duty cycles is shown in Figure 7f. It is noted that the heat increases and attains 82.5 °C at 50% duty cycle for a period of 5 min. The thermal image for various pulse duty cycle is illustrated in Figure 7g. It is inferred that the heat increases with an increase in pulse duty cycle. The pulse duty ratio with 50% denotes a 50% positive half cycle and a 50% negative half cycle for the 180° conduction mode in which the corresponding load voltage duty ratio is 100%.

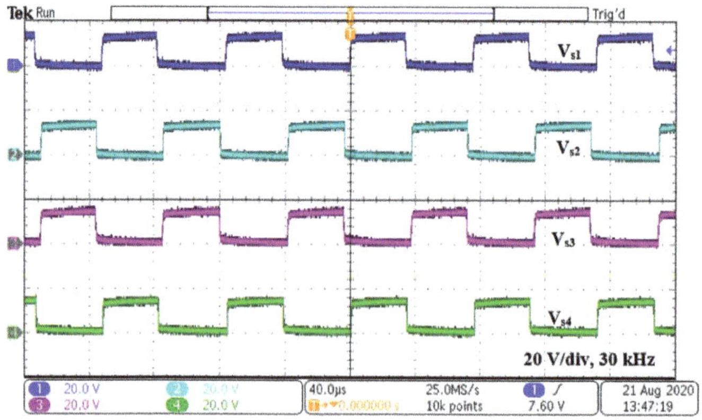

(**a**)

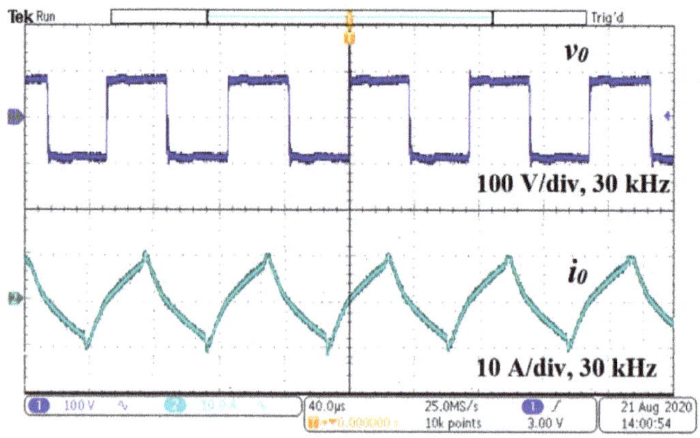

(**b**)

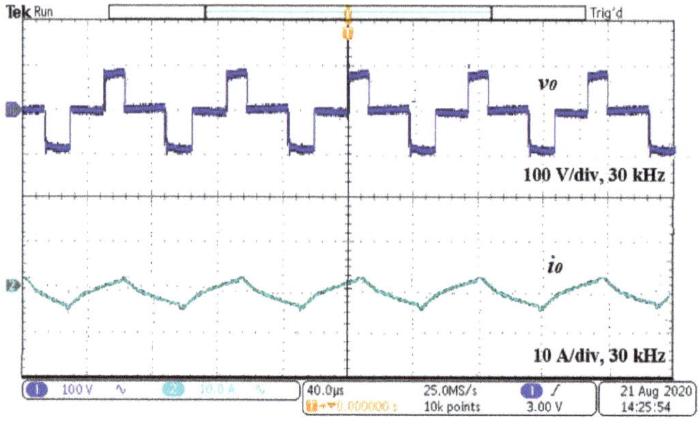

(**c**)

Figure 7. *Cont.*

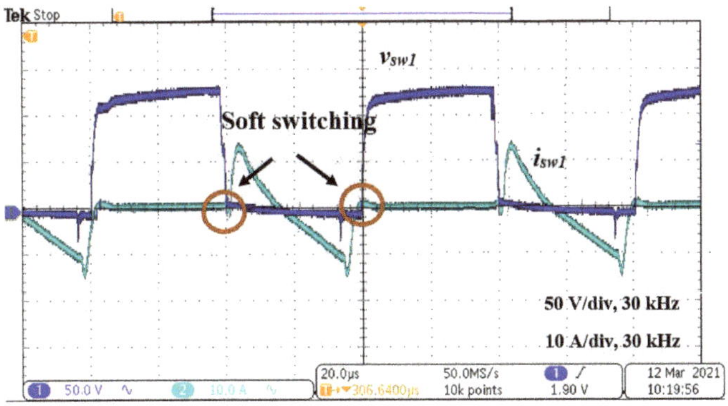

(d)

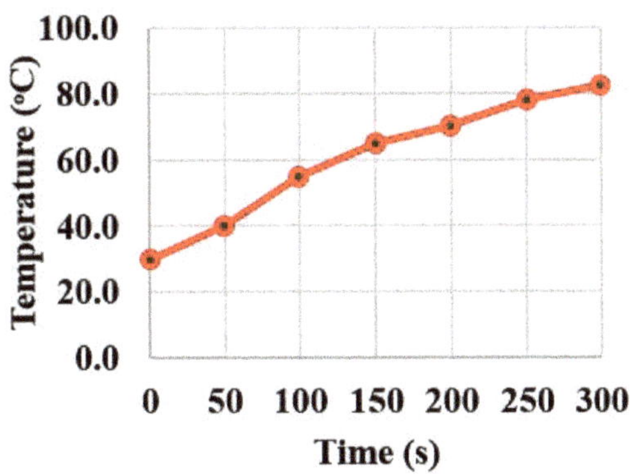

(e)

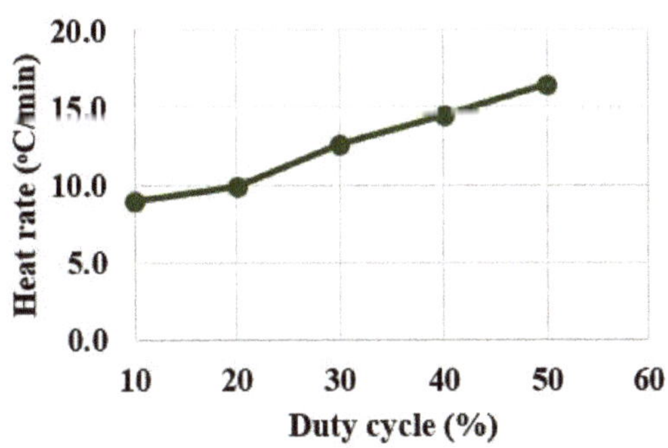

(f)

Figure 7. *Cont.*

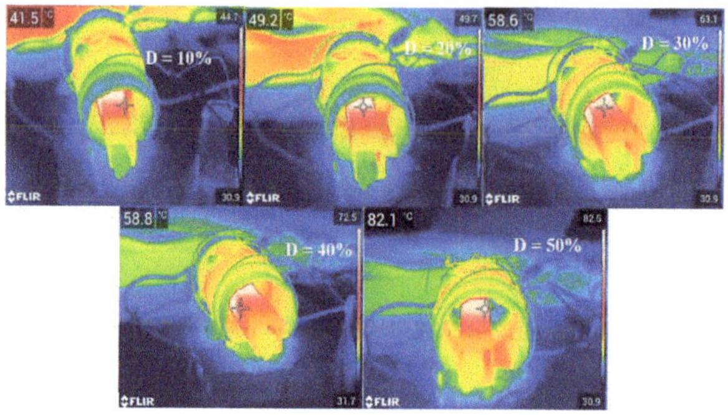

(g)

Figure 7. Pulse duty cycle control with 30 kHz frequency. (**a**) Switching pulses for 50% pulse duty cycle. (**b**) Output voltage and current waveforms for 50% pulse duty cycle. (**c**) Output voltage and current waveforms for 10% pulse duty cycle. (**d**) Switch voltage and current waveforms. (**e**) Temperature variation. (**f**) Heat rate for various pulse duty cycles. (**g**) FLIR thermal image for various pulse duty cycles at t = 5 min.

Table 2. Efficiency of pulse duty cycle control.

S. No	Pulse Duty Cycle (%)	Input Voltage (V)	Input Current (A)	Input Power (W)	RMS Value of Output Voltage (V)	RMS Value of Output Current (A)	Output Power (W)	Efficiency (%)
1	10	80	1.79	143	31.59	3.16	99.8	69.8
2	20	80	3.42	273.6	44.7	4.47	199.8	73
3	30	80	4.87	389.6	54.77	5.48	300	77
4	40	80	6.1	488	63.24	6.32	399.9	81.9
5	50	80	6.97	558	70.71	7.07	500	89.6

4.2. Variable Frequency Control

The experimental results of variable frequency control are shown in Figure 8. The resonance frequency is 29.5 kHz, and the switching frequency is varied from 13 kHz to 66 kHz to control the output power. The output voltage and the current observed at 15 kHz switching frequency is shown in Figure 8a. The output voltage and current waveforms for 30 kHz and 66 kHz are shown in Figure 8b,c, respectively. For a 66 kHz switching frequency, according to Equation (2), $P_0 = 0.82 \times 500 = 410$ W. The rms value of the current at 66 kHz switching frequency is 6.4 A and P_0 is $6.4^2 \times 10 = 410$ W. The input power (P_{in}) is $80 \times 6.4 = 512$ W. The efficiency for 66 kHz is 80.1%, and it is presented for other switching frequencies in Table 3. It is observed that efficiency is maximum for the resonance condition, and the efficiency is reduced both below and above the resonance frequency due to switching losses. The switch (S_1) voltage and current waveforms for 15 kHz are shown in Figure 8d. It is observed that hard switching is realized as switching frequency is far from the resonance frequency. As the switch is operated in hard switching mode, there exists a ringing effect on the output voltage waveform, which results in more switching loss and stress. Hence, it advisable to operate the inverter switch in resonant operating mode to reduce switching loss and stress. The temperature variation with respect to time is shown in Figure 8e. The heat rate for various switching frequencies are shown in Figure 8f and FLIR thermal image for various switching frequencies for a fixed time period (5 min) is illustrated in Figure 8g. The temperature varies with respect to the switching frequencies.As the switching frequency is varied, the inverter is operated in hard switching mode, which results in more switching loss. In addition, the frequency ringing effect, heat conversion losses, experimental coil design constraints, and capacitor selection restrict the efficiency to 86.9% in real-time implementation.

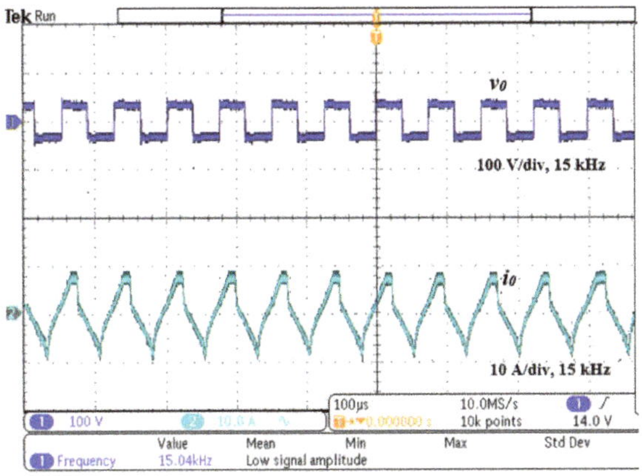

(**a**)

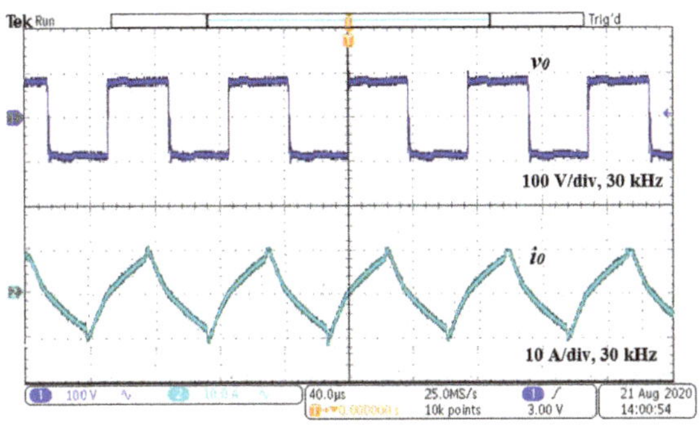

(**b**)

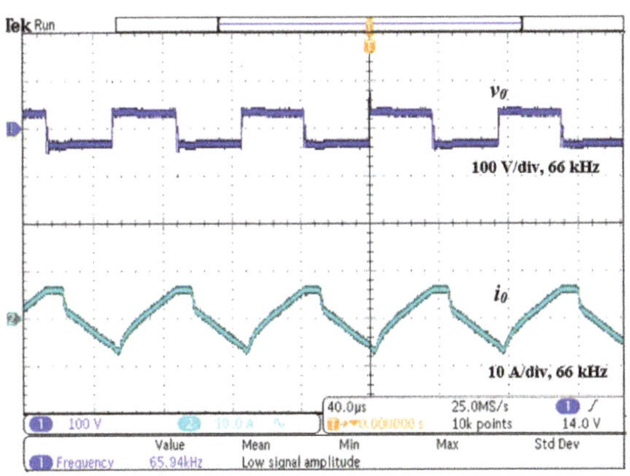

(**c**)

Figure 8. *Cont.*

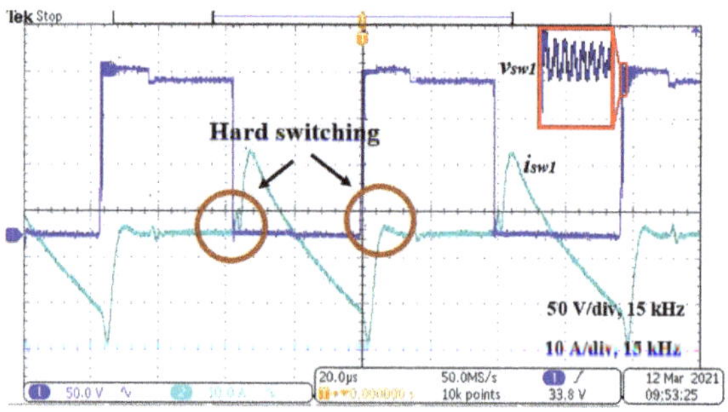

(d)

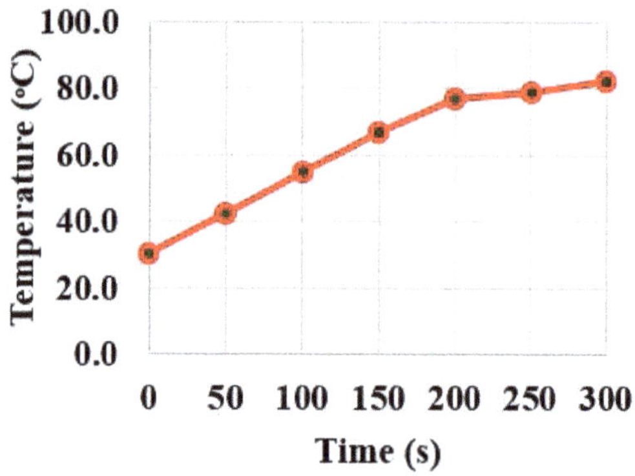

(e)

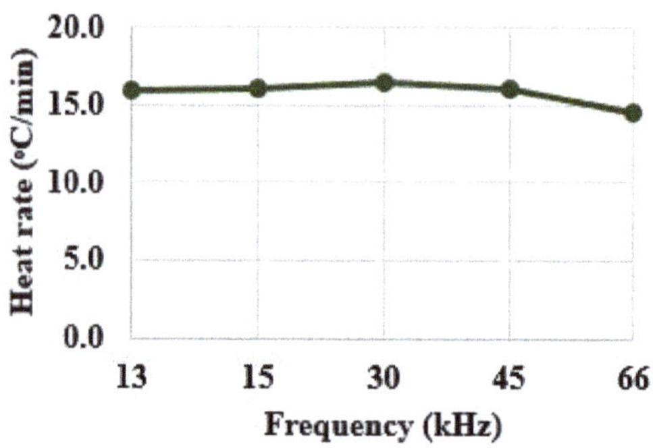

(f)

Figure 8. *Cont.*

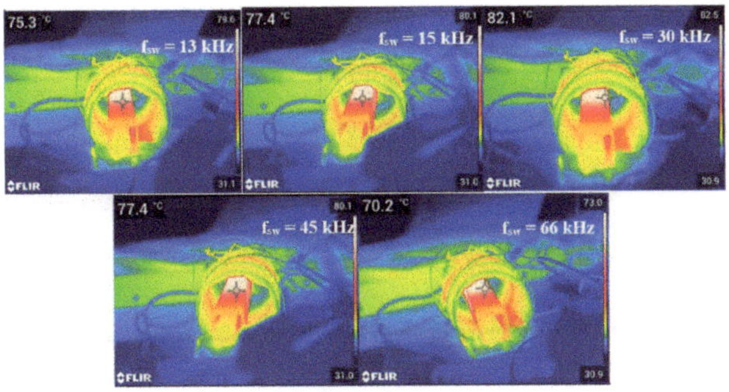

(g)

Figure 8. Variable frequency control: output voltage and current waveforms for (**a**) 15 kHz, (**b**) 30 kHz, and (**c**) 66 kHz switching frequency. (**d**) Switch voltage and current waveforms for 15 kHz switching frequency. (**e**) Temperature variation. (**f**) Heat rate for various switching frequencies. (**g**) FLIR thermal image for various switching frequencies at t = 5 min.

Table 3. Efficiency of variable frequency control.

S. No	Switching Frequency (kHz)	Input Voltage (V)	Input Current (A)	Input Power (W)	RMS Value of Output Voltage (V)	RMS Value of Output Current (A)	Output Power (W)	Efficiency (%)
1	13	80	7.29	583	69.21	6.92	479	82.2
2	15	80	7.27	581.6	69.57	6.96	484	83.2
3	30	80	6.97	558	70.71	7.07	500	89.6
4	45	80	7.27	581.6	69.57	6.96	484	83.2
5	66	80	6.4	512	64.03	6.4	410	80.1

4.3. Phase Shift Control

In phase shift control, the switching pulses are shifted with respect to the adjacent leg of the inverter. The variation in rms values of the output voltage and current changes the output power. The experimental results of PSC are shown in Figure 9. The switching pulses, and output voltage and current waveforms for ϕ = 20° are shown in Figure 9a,b, respectively. For ϕ = 20°, according to Equation (5), P_0 is $500 * cos^2(20/2)$ = 485 W. The rms value of the output current is 6.96 A and $P_0 = 6.96^2 \times 10$ = 484 W. The input power is 80×6.96 = 557 W. The efficiency for various phase angle is given in Table 4. The output voltage and current waveforms for ϕ = 60° is represented in Figure 9c. The soft-switching range for phase shift control is shown in Figure 9d. It is inferred that inverter switches operate in soft switching for ϕ ranging from 0° to 168°. The switch (S_1) voltage and current waveforms for ϕ = 170° are shown in Figure 9e, where the switch is operated in hard switching mode. The temperature variation with respect to time is shown in Figure 9f. The heat rate for various phase angles are shown in Figure 9g and FLIR thermal image for various phase angles for a fixed time period (5 min) is illustrated in Figure 9h. As the rms value of output voltage varies, the temperature decreases for an increase in phase angle.

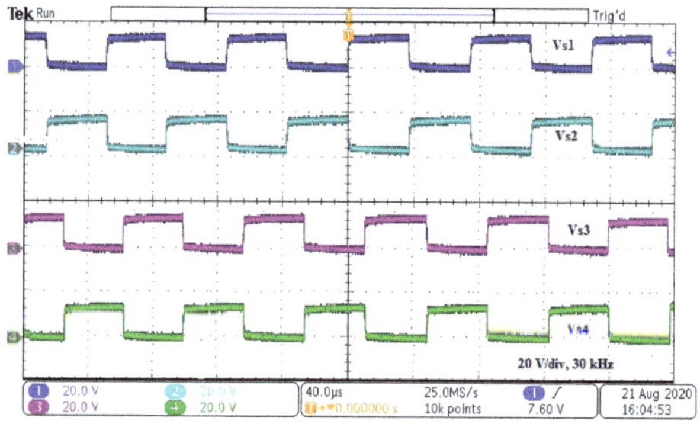

(a)

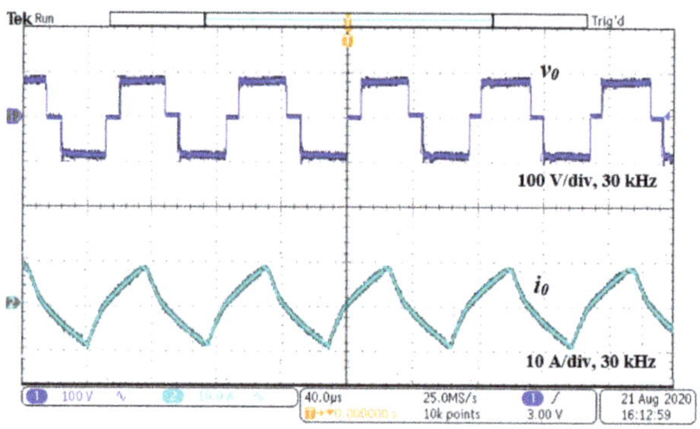

(b)

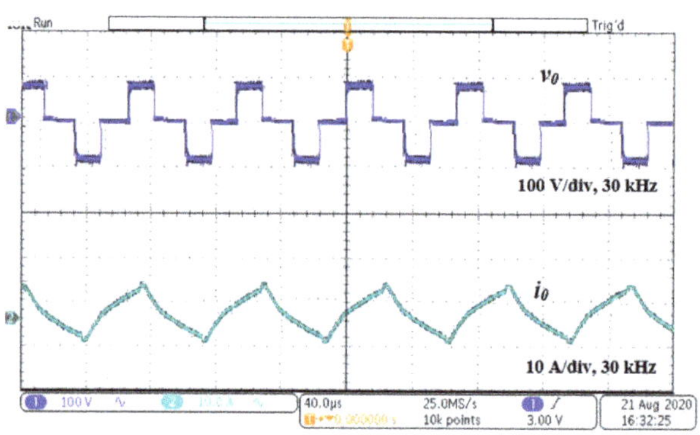

(c)

Figure 9. *Cont.*

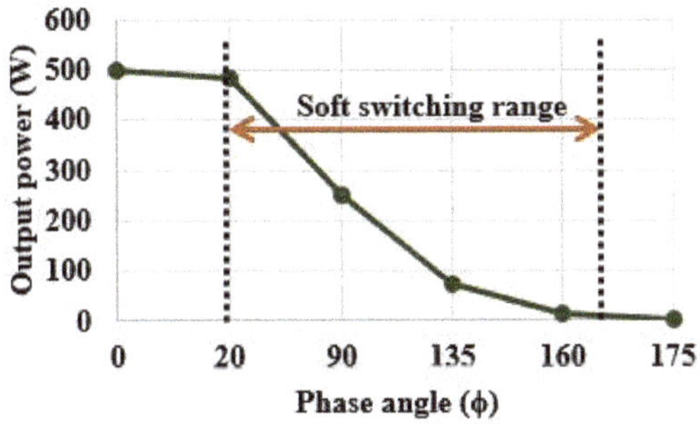

(d)

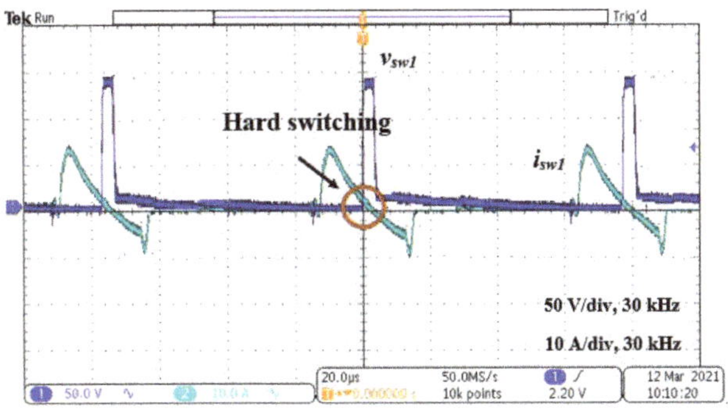

(e)

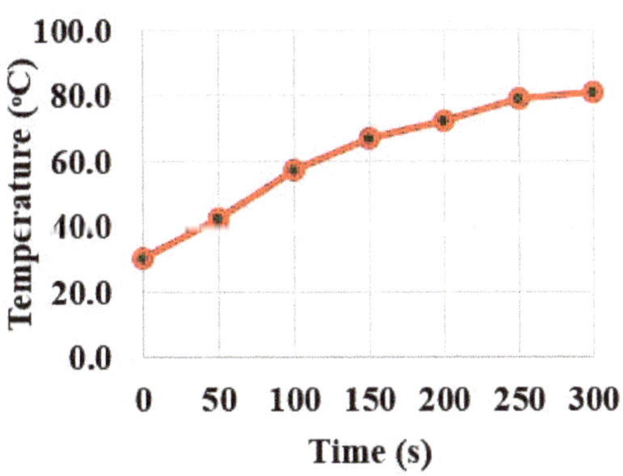

(f)

Figure 9. *Cont.*

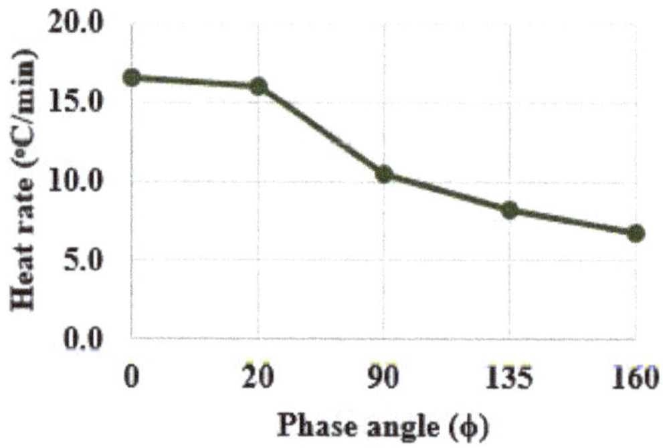

(g)

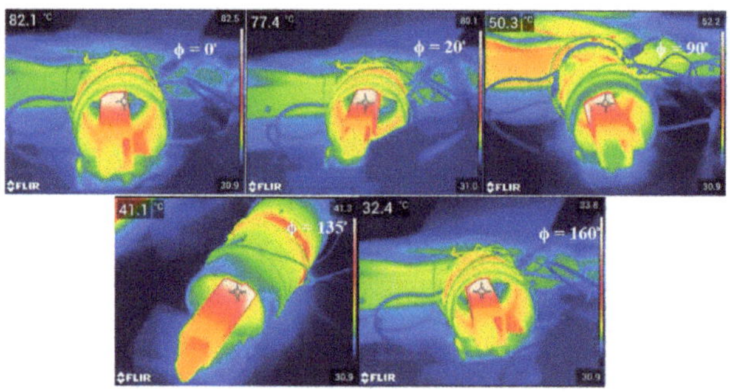

(h)

Figure 9. Phase shift control. (**a**) Switching pulses for ϕ = 20°. (**b**) Output voltage and current waveforms for ϕ = 20°. (**c**) Output voltage and current waveforms for ϕ = 60°. (**d**) Soft switching range. (**e**) Switch voltage and current waveforms for ϕ = 170°. (**f**) Temperature variation. (**g**) Heat rate for various phases angles. (**h**) FLIR thermal image for various phase angles at t = 5 min.

Table 4. Efficiency of phase shift control.

S. No	Phase Angle (°)	Input Voltage (V)	Input Current (A)	Input Power (W)	RMS Value of Output Voltage (V)	RMS Value of Output Current (A)	Output Power (W)	Efficiency (%)
1	0	80	6.97	557.6	70.71	7.07	500	89.6
2	20	80	6.96	557	69.57	6.96	484	87
3	90	80	3.77	301.6	50	5	250	83
4	135	80	1.12	89.6	27.09	2.71	73.4	82
5	160	80	0.23	18.4	12.25	1.22	15	81.5

4.4. Asymmetric Duty Cycle Control

The experimental waveforms corresponding to ADC for 30 kHz switching frequency is shown in Figure 10. The switching pulses and output voltage and current waveforms for α = 30° are shown in Figure 10a,b. respectively. For α = 30°, according to Equation (7), $P_0 = 500 * cos^2(30/2) = 466.5$ W. The rms value of the output current is 6.8 A and P_0 is $6.8^2 \times 10 = 462.4$ W. The input power is $80 \times 6.7 = 536$ W. The efficiency for various control angles is given in Table 5. The output voltage and current waveforms for α = 140° are shown in Figure 10c. As there is a dead band for zero crossing of current, the switching losses are reduced by making the inverter operate in soft switching mode. The switch (S_1) voltage and current waveforms for α = 30° are shown in Figure 10d, where the switch is operated in soft switching mode. The output voltage is controlled for α ranging from 0° to 165°. The harmonic spectrum for α = 140° is shown in Figure 10e. It is inferred that there exist even harmonics due to the asymmetric voltage waveform. The variation in temperature with respect to time is shown in Figure 10f. The heat rate for various control angles is shown in Figure 10g, and an FLIR thermal image for various phase angles for a fixed time period (5 min) is illustrated in Figure 10h. As the rms value of the output voltage varies, the temperature decreases with an increase in control angle. The ADC technique holds for a rated power operation and variation in the α results in even harmonics in the output voltage due to asymmetric voltage on either half cycle. This results in a reduction in the efficiency.

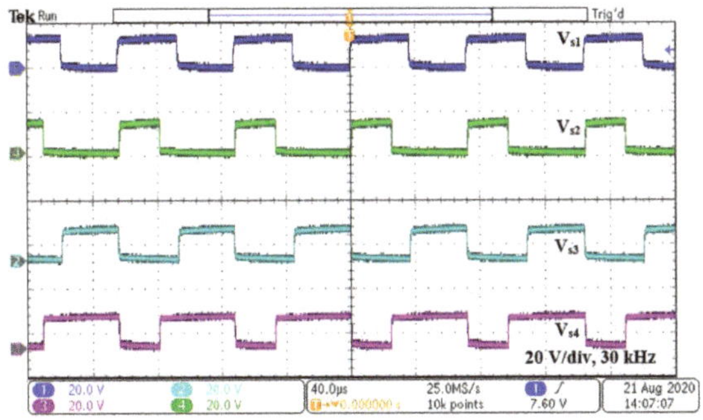

(**a**)

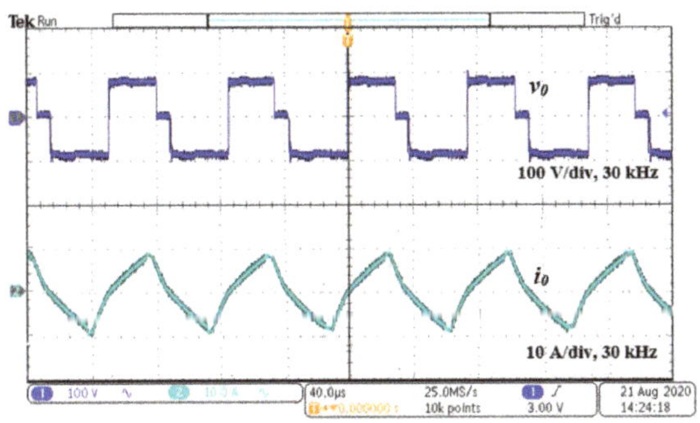

(**b**)

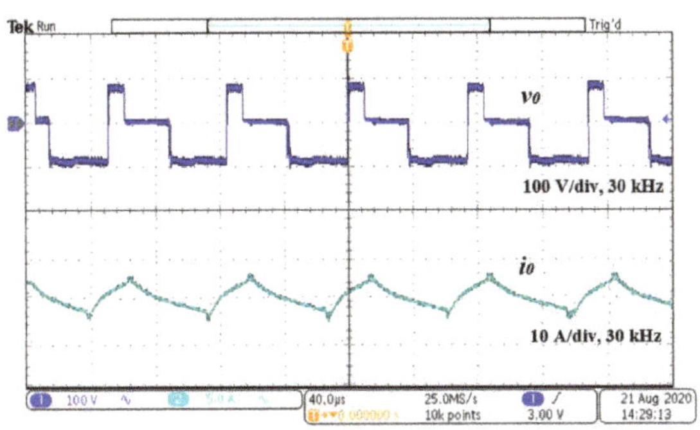

(**c**)

Figure 10. *Cont.*

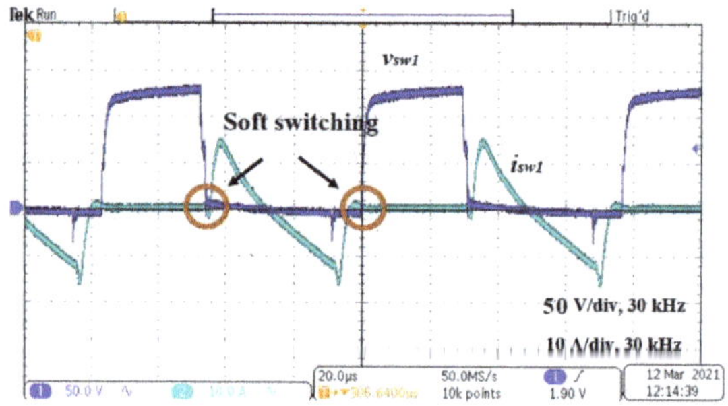

(d)

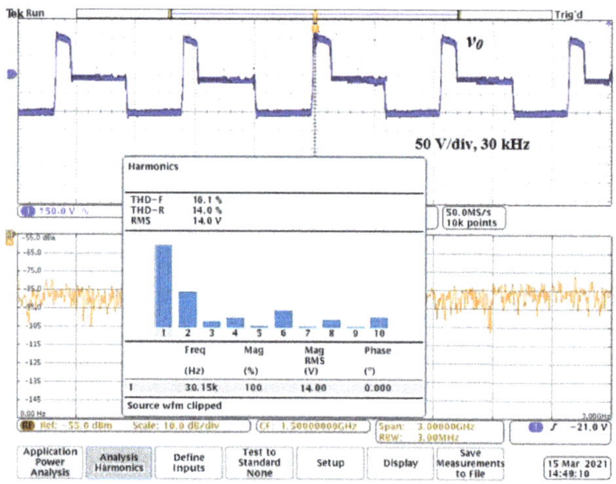

(e)

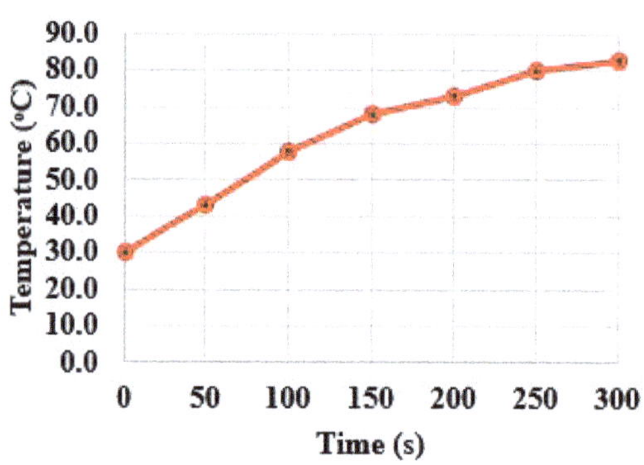

(f)

Figure 10. *Cont.*

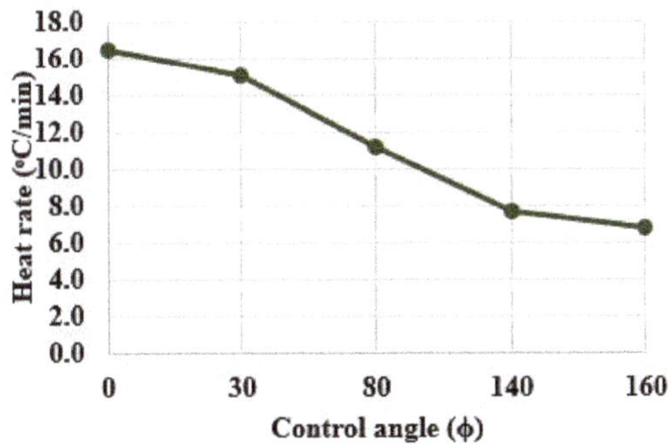

(g)

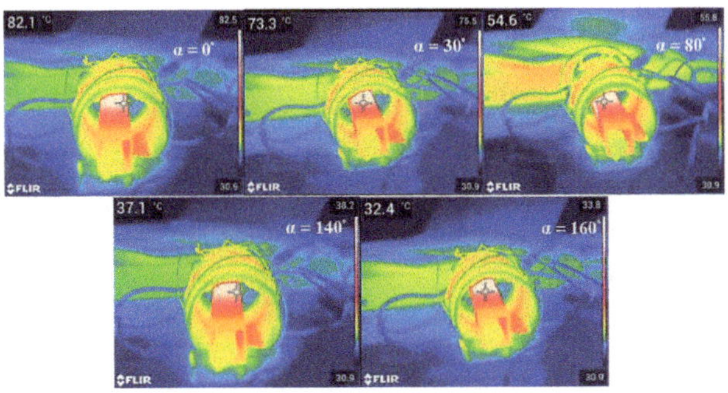

(h)

Figure 10. Asymmetrical duty control. (**a**) Switching pulses for $\alpha = 30°$. (**b**) Output voltage and current waveforms for $\alpha = 30°$. (**c**) Output voltage and current waveforms for $\alpha = 140°$. (**d**) Switch voltage and current waveforms for $\alpha = 30°$. (**e**) Harmonics spectrum for $\alpha = 140°$. (**f**) Temperature variation. (**g**) Heat rate for various control angles. (**h**) FLIR thermal image for various control angles at $t = 5$ min.

Table 5. Efficiency of asymmetrical duty control.

S. No	Control Angle (°)	Input Voltage (V)	Input Current (A)	Input Power (W)	RMS Value of Output Voltage (V)	RMS Value of Output Current (A)	Output Power (W)	Efficiency (%)
1	0	80	6.97	557.6	70.71	7.07	500	89.6
2	30	80	6.7	536	68	6.8	462.4	86.2
3	80	80	4.26	341	54.1	5.41	292.6	86
4	140	80	0.87	70	24.2	2.42	58.5	84
5	160	80	0.23	18.4	12.25	1.22	15	81.5

4.5. Pulse Density Modulation Control

One of the most efficient control techniques is the pulse density modulation. In this technique, as the switching frequency and its duty cycle are not varied, the soft-switching operation is retained for the entire power variation. In this technique, a low-frequency signal (20 Hz) is compared with high-frequency switching pulses (30 kHz) to produce the pulse pattern in a manner such that, during the on period of a low-frequency signal, high-frequency pulses are applied to the inverter for powering the load. By controlling the density of high-frequency pulses, the output power is controlled.

The experimental waveforms for PDM-based control is shown in Figure 11. The switching pulses and output voltage and the current for D_{PDM} = 60% are shown in Figure 11a,b, respectively. For 60% D_{PDM}, according to Equation (9), the calculated $P_0 = 500 \times 0.6 = 300$ W. The rms value of the output current is 5.48 A, and P_0 is $5.48^2 \times 10 = 300$ W. The input power is $80 \times 4.36 = 348.8$ W. The efficiency for various D_{PDM} is given in Table 6. The output voltage and current waveforms for D_{PDM} = 90% are shown in Figure 11c. As switching frequency is not varied, resonance is maintained for the entire range of operation. The output power is controlled for the D_{PDM} ranging from 0% to 100%. The variation of temperature is shown in Figure 11d. The heat rate for various D_{PDM} is shown in Figure 11e and FLIR thermal image for various D_{PDM} for a fixed time period (5 min) is shown in Figure 11f. The temperature varies with respect to D_{PDM}.

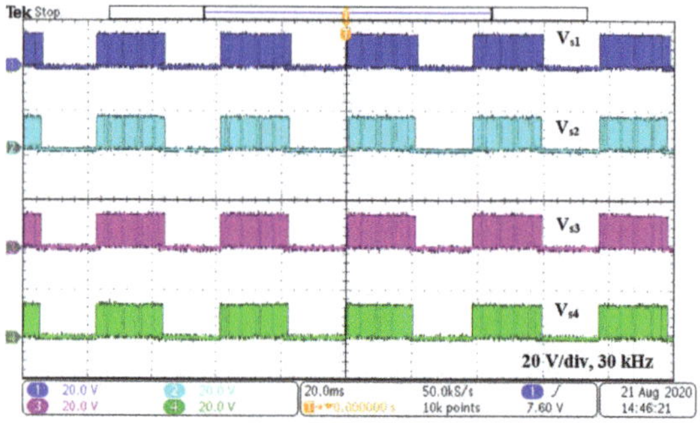

(**a**)

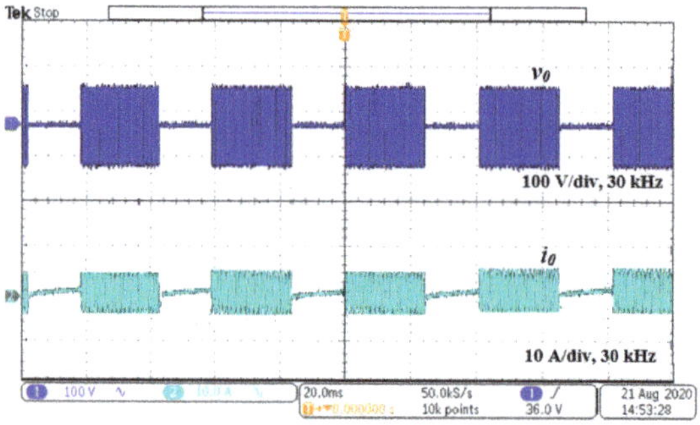

(**b**)

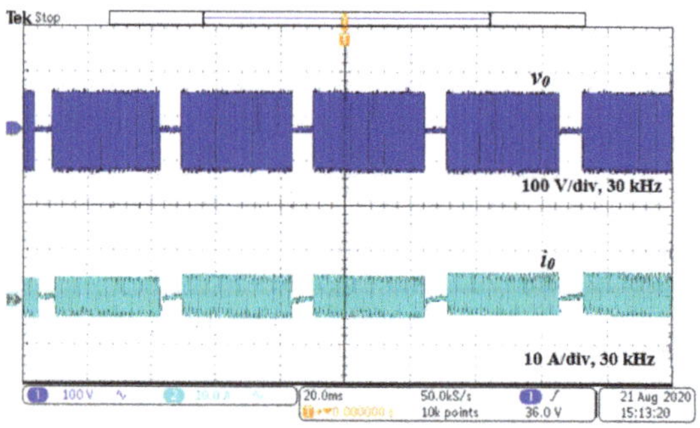

(**c**)

Figure 11. *Cont.*

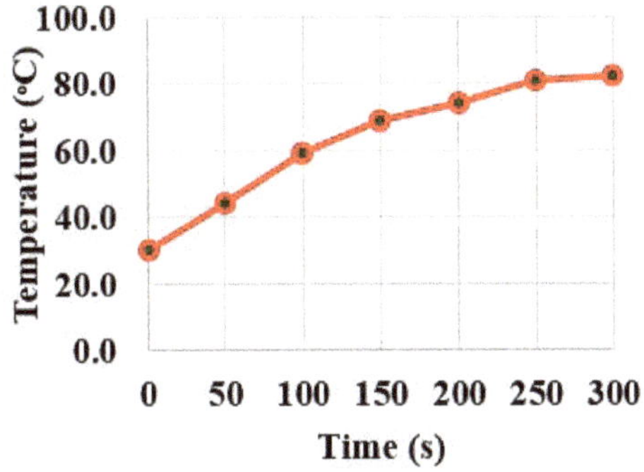

(d)

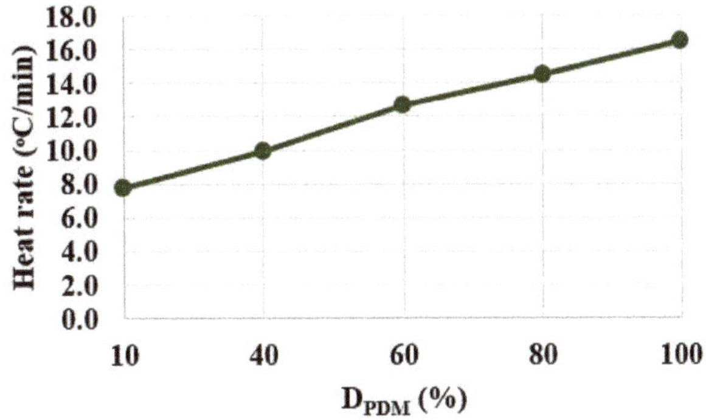

(e)

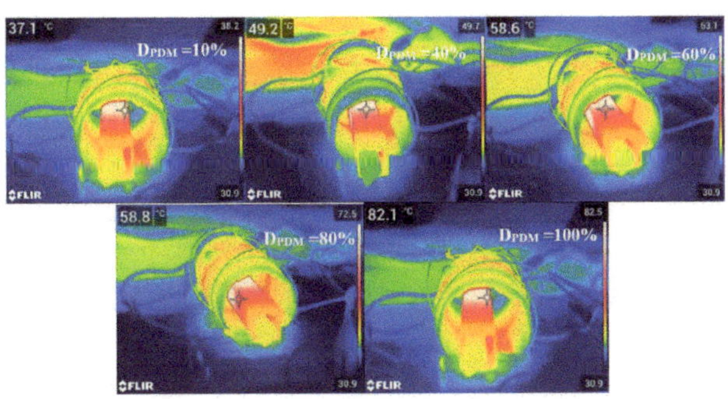

(f)

Figure 11. Pulse density modulation control. (**a**) Switching pulses for D_{PDM} = 60%. (**b**) Output voltage and current waveforms for D_{PDM} = 60%. (**c**) output voltage and current waveforms for D_{PDM} = 90%. (**d**) Temperature variation. (**e**) Heat rate for various D_{PDM}. (**f**) FLIR thermal image for various D_{PDM} at t = 5 min.

Table 6. Efficiency of pulse density modulation control.

S. No	D_{PDM} (%)	Input Voltage (V)	Input Current (A)	Input Power (W)	RMS Value of Output Voltage (V)	RMS Value of Output Current (A)	Output Power (W)	Efficiency (%)
1	10	80	0.75	60	22.36	2.24	50	83
2	40	80	2.94	235	44.72	4.47	200	85
3	60	80	4.36	339	54.77	5.48	300	86
4	80	80	5.71	457	63.25	6.32	400	87.5
5	100	80	6.97	557.6	70.71	7.07	500	89.6

5. Summary of Discussions on the Experimental Results

In this paper, the various power control techniques used for bifilar coil-based IH systems are discussed. The modular losses are calculated using the formula given in [19]. The modular losses (conduction and switching losses) are calculated to evaluate the best control technique. Figure 12 shows the comparison of total modular power loss with respect to output power. It is inferred that the losses are the same at the rated operating power (500 W) for various control methods. In the PDM control technique, the amount of input power drawn from the supply is at a minimum, which results in a minimum total modular loss as compared with other techniques for the same output power. Similarly, the variation in efficiency with respect to the output power is illustrated in Figure 13. The efficiency of the bifilar coil-based IH system varies between 60% to 89.6% for output power variation. Among the tested control techniques, the PDM control scheme yields higher efficiency (84% to 89.6%) with linear variations in the output power (0% to 100% of rated power).

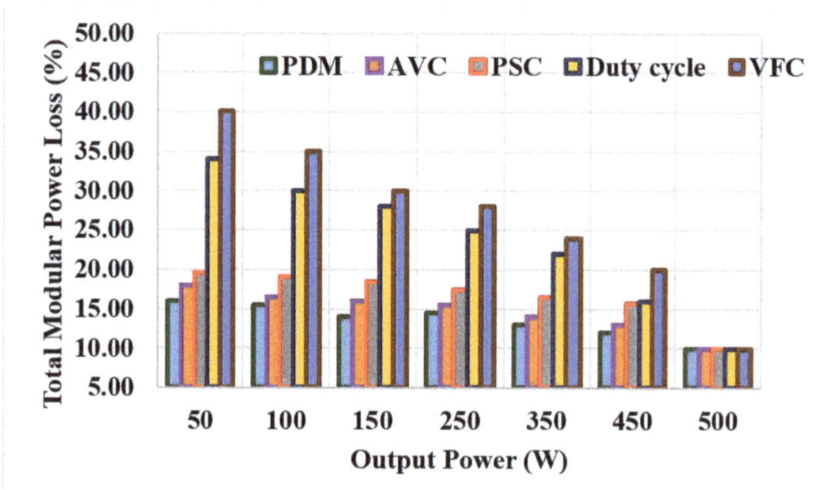

Figure 12. Total modular power loss.

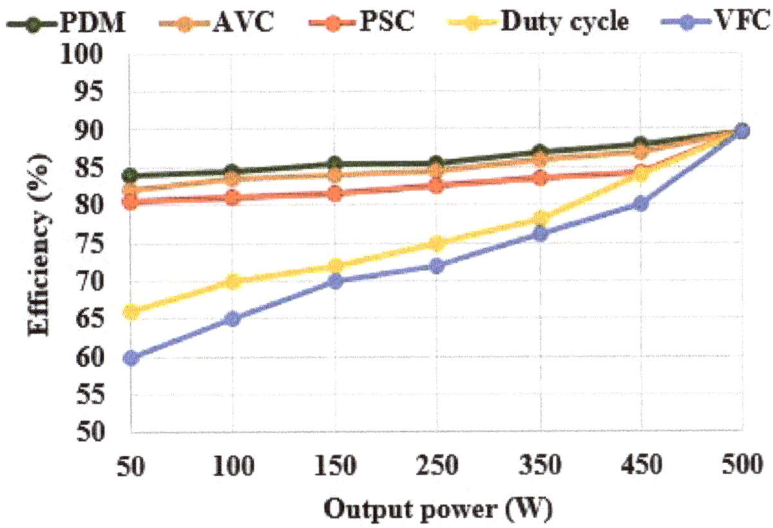

Figure 13. Efficiency of various control techniques.

6. Conclusions

In this work, various control techniques used for IH systems are investigated for bifilar coil-based IH systems used for melting applications. The performance of the system is examined using a full-bridge series-resonant inverter at 500 W output power. The contributions are summarized as follows.

- The pulse density modulation-based control scheme shows optimal performance with respect to heat rate, output variation, and efficiency (maximum of 89.6%). This trend is similar to that of the work published for cooking applications [17,19,31–33].
- Asymmetric duty cycle control can be applied to a system where a wide range of ZVS operations are required. The system possesses even harmonics for various values of α.
- Phase shift control can be deployed for an IH system where power control needs to be varied from 0% to 100% of the rated power. The resonance operation is maintained for ϕ, ranging from 0° to 165°.
- Frequency control is suitable for induction melting applications, since it requires differential heating over the surface and depth of the material.
- A pulse duty cycle-based control scheme is preferred when delivering power for a rated load.

All modulation techniques were implemented using a single PIC 16F877A micro controller, providing a cost effective solution for the prototype developed.

Author Contributions: Conceptualization, A.S., P.V. and R.G.; methodology, P.V.; software, A.S.; validation, P.V., R.G. and S.R.; formal analysis, P.V. and R.G.; investigation, P.V.; resources, B.N.; data curation, S.R.; writing—original draft preparation, P.V.; writing—review and editing, A.S., R.G., S.R. and B.N.; visualization, B.N.; supervision, R.G. and S.R.; project administration, B.N.; funding acquisition, B.N. All authors have read and agreed to the published version of the manuscript.

Funding: This research received no external funding.

Institutional Review Board Statement: Not applicable.

Informed Consent Statement: Not applicable.

Data Availability Statement: Not applicable.

Conflicts of Interest: The authors declare no conflict of interest.

References

1. Vishnuram, P.; Ramachandran, G.; Ramasamy, S.; Dayalan, S. A comprehensive overview of power converter topologies for induction heating applications. *Int. Trans. Electr. Energy Syst.* **2020**, *30*, e12554. [CrossRef]
2. Lozinskii, M.G. *Industrial Applications of Induction Heating*; Pergamon: Oxford, UK, 1969.
3. Moreland, W.C. The Induction Range: Its Performance and Its Development Problems. *IEEE Trans. Ind. Appl.* **1973**, *IA-9*, 81–85. [CrossRef]
4. Acero, J.; Burdio, J.M.; Barragan, L.A.; Navarro, D.; Alonso, R.; Garcia, J.R.; Monterde, F.; Hernandez, P.; Llorente, S.; Garde, I. The domestic induction heating appliance: An overview of recent research. In Proceedings of the 2008 Twenty-Third Annual IEEE Applied Power Electronics Conference and Exposition, Austin, TX, USA, 24–28 February 2008; pp. 651–657. [CrossRef]
5. Lucía, O.; Maussion, P.; Dede, E.J.; Burdío, J.M. Induction Heating Technology and Its Applications: Past Developments, Current Technology, and Future Challenges. *IEEE Trans. Ind. Electron.* **2014**, *61*, 2509–2520. [CrossRef]
6. Fujita, H.; Uchida, N.; Ozaki, K. A New Zone-Control Induction Heating System Using Multiple Inverter Units Applicable Under Mutual Magnetic Coupling Conditions. *IEEE Trans. Power Electron.* **2011**, *26*, 2009–2017. [CrossRef]
7. Fernández, O.; Delgado, J.; Martínez, F.; Correa, J.; Heras, M. Design and implementation of a 120A resonant inverter for induction furnace. In Proceedings of the 2013 IEEE International Autumn Meeting on Power Electronics and Computing (ROPEC), Morelia, Mexico, 13–15 November 2013; pp. 1–6. [CrossRef]
8. Yongyuth, N.; Viriya, P.; Matsuse, K. Analysis of a Full-Bridge Inverter for Induction Heating Using Asymmetrical Phase-Shift Control under ZVS and NON-ZVS Operation. In Proceedings of the 7th International Conference on Power Electronics and Drive Systems, Bangkok, Thailand, 27–30 November 2007; pp. 476–482. [CrossRef]
9. Villa, J.; Navarro, D.; Dominguez, A.; Artigas, J.I.; Barragan, L.A. Vessel Recognition in Induction Heating Appliances—A Deep-Learning Approach. *IEEE Access* **2021**, *9*, 16053–16061. [CrossRef]
10. Shenkman, A.; Axelrod, B.; Berkovich, Y. Single-switch AC-AC converter with high power factor and soft commutation for induction heating applications. *IEEE Proc. Electr. Power Appl.* **2001**, *148*, 469–474. [CrossRef]
11. Sarnago, H.; Lucía, O.; Mediano, A.; Burdío, J.M. Class-D/DE Dual-Mode-Operation Resonant Converter for Improved-Efficiency Domestic Induction Heating System. *IEEE Trans. Power Electron.* **2013**, *28*, 1274–1285. [CrossRef]
12. Ahmed, N.; Nakaoka, M. Boost-half-bridge edge resonant soft switching PWM high-frequency inverter for consumer induction heating appliances. *IEEE Proc. Electr. Power Appl.* **2006**, *153*, 932–938. [CrossRef]
13. Forest, F.; Faucher, S.; Gaspard, J.; Montloup, D.; Huselstein, J.; Joubert, C. Frequency-Synchronized Resonant Converters for the Supply of Multiwinding Coils in Induction Cooking Appliances. *IEEE Trans. Ind. Electron.* **2007**, *54*, 441–452. [CrossRef]
14. Lucia, O.; Burdio, J.M.; Millan, I.; Acero, J.; Puyal, D. Load-Adaptive Control Algorithm of Half-Bridge Series Resonant Inverter for Domestic Induction Heating. *IEEE Trans. Ind. Electron.* **2009**, *56*, 3106–3116. [CrossRef]
15. Lucía, O.; Burdío, J.M.; Barragán, L.A.; Acero, J.; Millán, I. Series-Resonant Multiinverter for Multiple Induction Heaters. *IEEE Trans. Power Electron.* **2010**, *25*, 2860–2868. [CrossRef]
16. Sarnago, H.; Lucía, O.; Burdio, J.M. A Versatile Resonant Tank Identification Methodology for Induction Heating Systems. *IEEE Trans. Power Electron.* **2018**, *33*, 1897–1901. [CrossRef]
17. Meziane, B.; Zeroug, H. Comprehensive Power Control Performance Investigations of Resonant Inverter for Induction Metal Surface Hardening. *IEEE Trans. Ind. Electron.* **2016**, *63*, 6086–6096. [CrossRef]
18. Bi, C.; Lu, H.; Jia, K.; Hu, J.; Li, H. A Novel Multiple-Frequency Resonant Inverter for Induction Heating Applications. *IEEE Trans. Power Electron.* **2016**, *31*, 8162–8171. [CrossRef]
19. Vishnuram, P.; Ramachandran, G.; Ramasamy, S. A Novel Power Control Technique for Series Resonant Inverter-Fed Induction Heating System with Fuzzy-Aided Digital Pulse Density Modulation Scheme. *Int. J. Fuzzy Syst.* **2017**, *20*. [CrossRef]
20. Acero, J.; Carretero, C.; Alonso, R.; Burdio, J.M. Quantitative Evaluation of Induction Efficiency in Domestic Induction Heating Applications. *IEEE Trans. Magn.* **2013**, *49*, 1382–1389. [CrossRef]
21. Wang, C.S.; Stielau, O.; Covic, G. Design considerations for a contactless electric vehicle battery charger. *IEEE Trans. Ind. Electron.* **2005**, *52*, 1308–1314. [CrossRef]
22. Chudnovsky, V.; Axelrod, B.; Shenkman, A.L. An approximate analysis of a starting process of a current source parallel inverter with a high-Q induction heating load. *IEEE Trans. Power Electron.* **1997**, *12*, 294–301. [CrossRef]
23. de Miranda, C.M.; Pichorim, S.F. A Self-Resonant Two-Coil Wireless Power Transfer System Using Open Bifilar Coils. *IEEE Trans. Circuits Syst. II Express Briefs* **2017**, *64*, 615–619. [CrossRef]
24. Miranda, C.; Pichorim, S. On the self-resonant frequency reduction of closed- and open-bifilar coils. *Int. J. Circuit Theory Appl.* **2019**, *47*, 641–653. [CrossRef]
25. Vishnuram, P.; Ramachandran, G. Capacitor-less induction heating system with self-resonant bifilar coil. *Int. J. Circuit Theory Appl.* **2020**, *48*, 1411–1425. [CrossRef]
26. Sarnago, H.; Lucia, O.; Mediano, A.; Burdio, J. High-efficiency parallel quasi-resonant current source inverter featuring SiC metal-oxide semiconductor field-effect transistors for induction heating systems with coupled inductors. *IET Power Electron.* **2013**, *6*, 183–191. [CrossRef]
27. Bojoi, R.; Griva, G.; Profumo, F.; Cesano, M.; Natale, L. Shunt active power filter implementation for induction heating applications. In Proceedings of the Twentieth Annual IEEE Applied Power Electronics Conference and Exposition, Austin, TX, USA, 6–10 March 2005; Volume 3, pp. 1674–1679. [CrossRef]

28. Barragan, L.; Burdio, J.; Artigas, J.; Navarro, D.; Acero, J.; Puyal, D. Efficiency optimization in ZVS series resonant inverters with asymmetrical voltage-cancellation control. *IEEE Trans. Power Electron.* **2005**, *20*, 1036–1044. [CrossRef]
29. Burdio, J.; Barragan, L.; Monterde, F.; Navarro, D.; Acero, J. Asymmetrical voltage-cancellation control for full-bridge series resonant inverters. *IEEE Trans. Power Electron.* **2004**, *19*, 461–469. [CrossRef]
30. Ahmed, N.A. High-Frequency Soft-Switching AC Conversion Circuit With Dual-Mode PWM/PDM Control Strategy for High-Power IH Applications. *IEEE Trans. Ind. Electron.* **2011**, *58*, 1440–1448. [CrossRef]
31. Vishnuram, P.; Ramachandran, G. A simple multi-frequency multiload independent power control using pulse density modulation scheme for cooking applications. *Int. Trans. Electr. Energy Syst.* **2021**, *31*, e12771. [CrossRef]
32. Vishnuram, P.; Ramachandran, G.; Thanikanti, S.B.; Nastasi, B. Induction Heating in Domestic Cooking and Industrial Melting Applications: A Systematic Review on Modelling, Converter Topologies and Control Schemes. *Energies* **2021**, *14*, 6634. [CrossRef]
33. Vishnuram, P.; Dayalan, S.; Thanikanti, S.B.; Balasubramanian, K.; Nastasi, B. Single Source Multi-Frequency AC-AC Converter for Induction Cooking Applications. *Energies* **2021**, *14*, 4799. [CrossRef]

energies

Review

Extended Reality for Smart Building Operation and Maintenance: A Review

Marco Casini

Department of Planning, Design, and Technology of Architecture (PDTA), Sapienza University of Rome, 00185 Roma, Italy; marco.casini@uniroma1.it

Abstract: The operation and maintenance (O&M) of buildings and infrastructure represent a strategic activity to ensure they perform as expected over time and to reduce energy consumption and maintenance costs at the urban and building scale. With the increasing diffusion of BIM, IoT devices, and AI, the future of O&M is represented by digital twin technology. To effectively take advantage of this digital revolution, thus enabling data-driven energy control, proactive maintenance, and predictive daily operations, it is vital that smart building management exploits the opportunities offered by the extended reality (XR) technologies. Nevertheless, in consideration of the novelty of XR in the AECO sector and its rapid and ongoing evolution, knowledge of the specific possibilities and the methods of integration into the building process workflow is still piecemeal and sparse. With the goal to bridge this gap, the article presents a thorough review of virtual reality (VR), augmented reality (AR), and mixed reality (MR) technologies and applications for smart building operation and maintenance. After defining VR, AR, and MR, the article provides a detailed review that analyzes, categorizes, and summarizes state-of-the-art XR technologies and their possible applications for building O&M along with their relative advantages and disadvantages. The article concludes that the application of XR in building and city management is showing promising results in enhancing human performance in technical O&M tasks, in understanding and controlling the energy efficiency, comfort, and safety of building and infrastructures, and in supporting strategic decision making for the future smart city.

Keywords: building operation and maintenance; extended reality; virtual reality; augmented reality; mixed reality; immersive technologies; digital twins; metaverse

Citation: Casini, M. Extended Reality for Smart Building Operation and Maintenance: A Review. *Energies* 2022, 15, 3785. https://doi.org/10.3390/en15103785

Academic Editor: Fernando Morgado-Dias

Received: 29 March 2022
Accepted: 18 May 2022
Published: 20 May 2022

1. Introduction

In the last five years, a more widespread adoption of digitalization and innovative technologies and systems has started making the digital and physical worlds become more deeply interconnected and interrelated, thus transforming the way that buildings and infrastructure are managed with the goal to provide a more energy efficient, comfortable, sustainable, and profitable built environment [1–3].

This transformation is enabling buildings to exchange, process, and exploit data and information, communicate with users, and share their assets with those of cities. Unprecedented possibilities are coming from the spread of key technologies such as building information modeling (BIM), artificial intelligence (AI), big data, and the Internet of Things (IoT), which are making the built environment become smarter, allowing building performance and user experiences to be greatly improved [4–10]. At the urban level, in particular, the integration of intelligent building systems with those of the city can allow for smart energy management capable of taking into account power demand and availability in real time [11–13].

At the center of this revolution is the possibility of overlapping (combining, mirroring) the life cycle of the building with a "digital twin building life cycle" able to interact bidirectionally with the physical world [1]. The integration of sensing technologies into the real world can generate a detailed and continuous flux of data that can power realistic and accurate digital replicas of the physical objects (digital twins) that can be freely analyzed and configured, knowing that every informed decision made in the digital world will be valid, and automatically implemented, in the physical world in turn [14,15]. In this way, the huge amount of data collected in both the construction and management phase of the building can be easily analyzed by artificial intelligence and machine learning algorithms to identify patterns and to create a data model to make predictions and support decision making along the entire value chain (resource optimization, cost prediction, risk analyses, predictive maintenance, etc.) [16].

The transition towards this digital ecosystem has accentuated the need for a more effective visualization of such a wealth of information, highlighting the limitations of traditional visual techniques throughout the life cycle of the building and requiring more dynamic tools capable of creating immersive virtual experiences of the real world, and thus allowing a greater understanding of the built environment. As real spaces are becoming more and more connected and sensor enabled, the demand is to combine real-time 3D digital representations with information from heterogeneous data sources, including BIM models and IoT sensors, and to improve the user experience with human interfaces with clear information dashboards and intuitive controls such as vocal and gesture commands.

In this picture, the extended reality (XR) technology, encompassing virtual, augmented, and mixed reality, is proving revolutionary in supporting building operation and maintenance, providing users with enhanced visualization thanks to the ability to show superimposed instructions, technical schemes, or sensor data right in their field of view, as well as allowing remote control and hands-free communication [17].

XR is the tool that sits between the digital twin and real-world information, essentially helping people to complete the marriage of the virtual and physical worlds and improve both. The application of XR technology shows promising results in enhancing human performance in carrying out operational tasks, as well as improving maintenance activities and supporting strategic decision making [18]. Furthermore, in a world situation where remote work has become the standard for companies in every industry (with over 1 in 4 Americans continuing to work remotely in 2022 [19]), integrated cooperation platforms enable remote connected colleagues and experts to see the live, first-person view captured by the XR viewer and collaborate proactively with the person on field.

Currently, the literature about XR technologies in the AECO sector and their applications in O&M is growing but is still limited in scope and lacking in more operative aspects. In particular, Alizadehsalehi et al. [20] presented a review of the most recent VR, AR, and MR technologies in the design and construction industry and an introduction to the most commonly used wearable XRs on the market. Khan et al. [21] presented a study of 64 papers on the integration level of XR immersive technologies with BIM in the AECO industry, divided into eight domains: client/stakeholder, design exploration, design analysis, construction planning, construction monitoring, construction health/safety, facility/management, and education/training. Sidani et al. [22] presented a review of 24 selected papers on the tools and techniques of BIM-based augmented reality in the AECO sector exploring 6 main application fields: collaboration, construction design, construction management, construction safety, facility management, and worker performance. Delgado et al. [17] presented a study on the usage landscape of AR and VR in the AEC sectors and proposed a research agenda to address the existing gaps in the required capabilities. Cheng et al. [23] presented a state-of-the-art review on MR applications in the AECO industry where 87 journal papers on MR applications were identified and classified into four categories: architecture and engineering, construction, operation, and applications in multiple stages. Delgado et al. [24] presented a systematic study of the factors that limit and drive the adoption of AR and VR in a construction sector-specific context. Albahbah, et al. [25] reviewed seven applications of AR that may benefit the construction industry, namely, safety management, communication and data acquisition, visualization, construction management education, progress tracking, quality management, and facility management. Zhu et al. [26] presented a state-of-the-art review of the application of virtual and augmented reality technologies for emergency management in built environments, while Li et al. [27] reviewed VR and AR applications in construction safety. Coupry et al. [28] presented a study on how XR technologies combined with digital twin technology can improve the maintenance operations in smart buildings. Noghabaei et al. [29] investigated the trends in AR/VR technology adoption in the AEC industry by conducting two user surveys in 2017 and 2018 involving 158 industrial experts. The results showed the potential for a solid growth of AR/VR technologies in the AEC industry in the following 5 to 10 years. The survey also highlighted some limitations of adopting AR/VR in the AEC industry such as "lack of budget," "upper management's lack of understanding of these technologies," and "design teams' lack of knowledge". Prabhakaran et al. [30] carried out a systematic review of scientific publications between the years 2010 and 2019 to understand the state-of-the-art immersive technology applications in AEC, revealing the following nine critical challenges: infrastructure, algorithm development, interoperability, general health and safety, virtual content modelling, cost, skills availability, multi-sensory limitations, and ethical issues.

In consideration of the speed with which this technology has appeared in the AECO sector and the equally rapid technological and application developments, in turn linked to the diffusion of other digital technologies (BIM, IoT, 3D mapping), knowledge of the specific possibilities offered by XR technology and on the methods of integration into the building process workflow is still piecemeal and sparse. Currently, there is no granular study that analyzes how and for what purposes O&M companies are using XR technologies. In particular, the differences between VR, AR, and MR are not yet clear to industry insiders, and there is still insufficient information on which devices are most appropriate to employ and which skills are necessary for their effective use. The study presented in this paper seeks to fill these gaps.

Overall, this article presents a thorough review of virtual reality (VR), augmented reality (AR), and mixed reality (MR) technologies and applications for smart building operation and maintenance. After defining VR, AR, and MR, the article provides a detailed review that analyzes, categorizes, and summarizes the state-of-the-art XR technologies and their possible applications for building O&M along with their relative advantages and disadvantages.

The paper is structured in six sections. After the introduction, Section 2 describes the research method used. Section 3 is dedicated to the definition of extended reality (XR) technology, illustrating the main differences between virtual (VR), augmented (AR), and mixed reality (MR), and presenting an overview of the most recent applications of XR in the AEC industry. Section 4 describes XR technologies and the most commonly used wearable XR on the market in terms of features, ease of use, and specifications. Section 5 highlights and illustrates the main possible applications of XR in buildings' O&M showing the potential benefits in terms of time and cost reductions and performance improvement. Finally, the conclusions are expounded in Section 6.

2. Research Method

This review is based on the following main sources of knowledge:

(i) bibliographic research of articles retrieved from renowned academic journals within the domain of the AECO sector;
(ii) market research of XR hardware and software products available and in development.

2.1. Academic Articles Review

Literature review was conducted on articles retrieved from well-acknowledged academic journals within the domain of the AECO industry to reflect the recent development trend and current situation of XR applications in the building sector.

Scopus, Web of Science, and Google Scholar were used as database search engines. The selected search period ranged from January 2018 up to the end of April 2022. This interval was chosen because of the breakthrough innovations that took place in the last five years on immersive technologies, along with the widespread adoption of BIM and digital Twin in the AECO sector and in O&M in particular. Document and source type was restricted to conference papers, research, and review articles in English language. Subject area was limited to engineering, computer science, environmental science, and energy.

Research method followed two steps. As a first step, articles were selected from the aforementioned databases using in the title, keywords, and in the abstract sections, different combinations of the following keywords: extended reality, virtual reality, augmented virtuality, augmented reality, mixed reality, immersive technologies, digital twin, building, architecture, engineering, design, construction, operation, maintenance, building information modeling (BIM), civil engineering, and facility management (FM).

Based on the findings of the first search round, a second search round was conducted by manually filtering the papers related to XR applications in the AECO industry to remove irrelevant papers. The abstract of each paper was read by the author to ensure that the application of the paper was within the AECO industry. For example, several papers obtained from the first round of searching were about design in manufacturing or engineering in mechanics. Additionally, other papers were removed as they were repeats or duplicates. After two rounds of filtering, 72 articles were selected from 43 journals and 4 proceedings and classified into three application categories, namely: (i) architecture and engineering design, (ii) building construction, (iii) building operation and management.

In order to track down the evolution of XR throughout past years, additional papers before 2018 were considered and examined, and 14 articles were added to the list for a total of 86 (see Tables 1 and 2).

Table 1. Selected literature for review.

XR Application Field	Selected Articles
Architecture and engineering design	[17,21,23,29–46]
Building construction	[17,20–25,27,29–31,47–56]
Building operation and management	[17,18,21,23,26,28,54,57–104]

Table 2. Number of selected articles by journal and year of publication.

Academic Journals	B. 2018	2018	2019	2020	2021	2022	Total
Advanced Engineering Informatics	1	1		1		2	**5**
Applied Ergonomics			1				**1**
Applied Sciences			2	3	6		**11**
Architectural Science Review		1					**1**
Automation in construction	3	7	4	6	3	1	**24**
BIO Web of Conferences	1						**1**
Building and Environment					1		**1**
Buildings				1	1	2	**4**
Business Horizons				1			**1**
Computers in Industry		1		1			**2**
Data				1			**1**
Digital Applications in Archaeology and Cultural Heritage			1				**1**
Energies				2			**2**
IEEE Transactions of Haptics			1				**1**
Information and Software Technology			1				**1**
International Journal of Human-Computer Interaction			1				**1**
Journal of Computing in Civil Engineering	1		1				**2**
Journal of Construction Engineering					1		**1**
Journal of Construction Engineering and Management				2			**2**
Journal of Building Engineering					1		**1**
Journal of Computational Design and Engineering			1				**1**

Table 2. *Cont.*

Academic Journals	B. 2018	2018	2019	2020	2021	2022	Total
Journal of Destination Marketing & Management	1						**1**
Journal of Industrial Information Integration				1			**1**
Journal of Information Technology in Construction	1						**1**
Journal of Manufacturing Systems				1			**1**
Journal of safety science and resilience					1		**1**
Multimodal Technol. Interact.	1						**1**
Procedia CIRP	1			2			**3**
Procedia Engineering	1						**1**
Proceedings	1	2	1				**4**
Robotics and Autonomous Systems						1	**1**
Robotics and Computer Integrated Manufacturing		1					**1**
Safety Science					1		**1**
Sustainable Cities and Society	1						**1**
Teknik Dergi			1				**1**
Virtual Reality & Intelligent Hardware						1	**1**
Visualization in Engineering	1						**1**
Total	**14**	**13**	**15**	**22**	**15**	**7**	**86**

Selected articles concerning XR application in building operation and management were further divided into VR and AR/MR categories, each articulated in different domains according to their possible use in building facility management as shown in Table 3.

Table 3. Domains of O&M application of selected papers.

XR Technology	Domains of O&M applications	Selected Papers
VR	• Maintainability design	[61–64]
	• Immersive building visualization and monitoring	[65–68]
	• Human-centered building management	[69–78]
	• Personnel training	[26,63,79–81]
	• Vehicles and robot's teleoperation	[82]
AR and MR	• Building monitoring and management	[83–90,96]
	• Maintenance and repair operations	[18,23,28,54,57–60,84,91,92,97–101]
	• Renovations and retrofit works	[17,102]
	• Personnel training and workplace safety	[26,93–95,103,104]

2.2. XR Hardware and Software Research

Along with academic articles review, a thorough analysis of XR hardware and software technology on the market was carried out in order to understand the state-of-the-art technology of the industry in terms of performance, availability, and accessibility of products as well as the trends shown by in-development prototypes and proofs of concept.

VR and AR devices database such as VRcompare [105] and infinite.cz [106] were used to conduct a wide view research of the market offering, which identified 70 different VR headsets and 34 AR/MR headsets released in the period 2018–2022 (see Table 4).

Table 4. XR headsets released on the market in the period (2018–2022).

XR Headsets	VR	AR/MR
Number of brands	32	24
Number of models	70	34

All XR devices were then investigated by examining the technical sheets and documentation available from the manufacturers themselves, with particular focus on display characteristics and virtual space fruition. In particular, relevant characteristics considered included typology of the device (head-mounted, smartphone, or tablet based), type of 3D tracking, degrees of freedom (DoF), screen resolution, field of view (FoV), and, in the case of VR, whether the device was tethered to a PC or self-contained. Such information allowed the most relevant products to be identified to compare and present to the reader in the VR, AR, and MR chapters of the article.

Likewise, starting from online databases such as Sourceforge [107] and Capterra [108], software on the market to develop and use XR contents were investigated, studying the different possible applications in the building process. In addition to commercial ready-to-use XR software investigated in the following sections of this article, most used development toolkits for AR, VR, and MR were examined to better contextualize limits, requirements, and opportunities of current and future applications in the building process.

In particular, resources from main AR software development kits (SDK) for smartphones application, such as AR Core by Google (Mountain View, CA, USA), ARKit by Apple (Cupertino, CA, USA), Vuforia (PTC, Boston, MA, USA), and Wikitude (Salzburg, Austria) were consulted to understand the means of AR visualization and interaction (environment understanding and motion tracking). Web-based AR tools such as AR.js and A-Frame were also considered.

Concerning VR, research covered both 3D and BIM model creation tools (Autodesk Revit and 3ds Max, Graphisoft Archicad, SketchUp Pro, Cinema 4D, Rhinoceros) as well as the main graphic engines supporting VR visualization and interaction (Unity Reflect, Unreal Engine 4, Enscape, CryEngine). These engines are compatible with the majority of VR devices on the market: Meta Quest (Meta, Cambridge, MA, USA), Valve Index (Valve Corporation, Bellevue, WA, USA), HTC Vive (HTC, Taoyuan, Taiwan), PlayStation VR (Sony Interactive Entertainment Inc., San Mateo, CA, USA), etc.) out of the box, i.e., they can read input from sensors and controllers, process this data, and produce the correct instructions for visualization. Additional functions allowed by SDKs provided by the manufacturers of VR headsets were considered.

Regarding MR, the main resource tool for importing 3D models and integrating MR functionalities investigated was Microsoft Mixed Reality Toolkit (MRTK, Redmond, WA, USA) available both for Unity and for Unreal Engine 4 graphical engines, which can be further customized using the OpenXR standard.

3. Extended Reality

The terms extended reality, "X-reality", or XR, refer to the use of different technologies to create immersive digital experiences, and include various combinations of computer-generated content and reality, including virtual reality (VR), augmented reality (AR), and mixed reality (MR) [109]. XR is used as an umbrella category that encompasses all real–virtual combined environments as well as man–machine interactions through computer technology and wearables [20,21].

These different technologies within the XR domain are set apart according to their immersivity in the virtual environment, to the level of interaction between the real and virtual worlds, and to which hardware is required for fruition (see Figure 1).

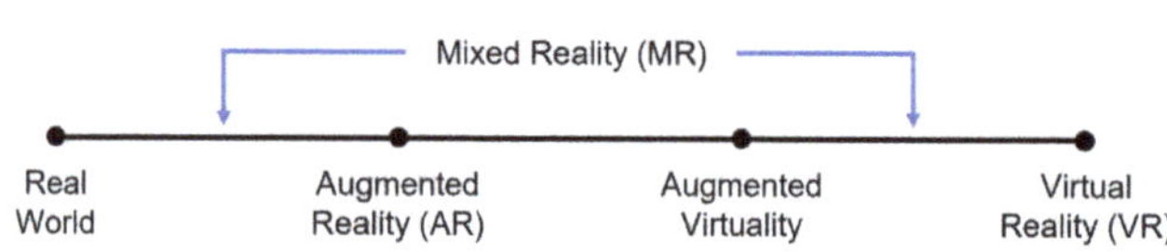

Figure 1. Extended reality and virtual continuum Redrawn from [110].

Virtual reality (VR) includes all immersive content that is completely digital and computer-generated (CG) and experienced through a VR headset or head-mounted display (HMD). In VR, the user is isolated from the real world and its surroundings as the current reality is replaced aurally and visually with a new 3D digital environment.

Augmented reality (AR) instead superimposes CG content over the real world, which remains tangible. This CG overlay can also superficially interact with the environment in real time. AR is primarily experienced through wearable goggles or HMDs or via smartphone and tablet screens.

Finally, mixed reality (MR), also known as immersive media, hybrid reality, or spatial computing, produces a CG overlay in which virtual 3D elements can integrate, enrich, and interact with the real-world environment beneath. MR wearable devices combine several technologies and commonly feature transparent lenses to allow CG overlay. The definition of MR often includes the concept of augmented virtuality, another take of XR in which the scenario is a virtual world to which real contents are added, as in the case of a virtual classroom where the lesson from a teacher in the flesh can be attended, compared to conventional AR in which the context is the physical world to which digital contents are overlaid.

The diffusion of all XR technologies is on the rise with an expected market of USD 60.55 billion by 2023 [22] and an expected CAGR of 57.91% over the forecast period 2022–2027 [111]. At the same time, the worldwide market for AR and VR headsets grew 92.1% year over year in 2021, with global shipments reaching 11.2 million units [112].

After being considered in the top 10 Gartner strategic technology trends for 2019 [113], since 2020, XR has been rapidly approaching a much more mature state and becoming more integral to business and IT. Several companies are currently developing supporting technologies that may help commercialize XR technologies globally, and MR-based applications were one of the top 10 ranked ICT technologies in 2020 [114]. The increasingly pervasive adoption of geo-localized apps, supported by an extended network and massive digitization that involves both the built environment and processes, along with the intersection with data mining, deep learning, and machine learning, is creating an ever-growing fertile ground for XR development and experimentation. Finally, the increasing adoption of gesture-based computing is also aiding the market growth by opening up several applications and opportunities.

Currently, XR technology represents one of the most promising innovations within the digitalization process of the construction industry, showing remarkable potential in increasing quality, efficiency, productivity, and profitability in all phases of the AECO process—design, construction, operation, maintenance, and real estate management.

XR can support designers, facility managers, and technicians by enhancing their physical realities with digital data and making the combined content readily shareable, allowing them to improve the quality and efficiency of work, to take better decisions faster, as well as to collaborate and communicate in a much more effective way. Using XR, they can visualize, explore, and comprehend blueprints, models, and site conditions more conveniently, greatly improving workflows throughout the lifecycle of buildings and infrastructure (see Table 5). Overall, the principal uses of XR in AECO activities include stakeholder engagement, design support and review, construction planning, progress monitoring, construction safety, operative support, operation and management, as well as workmen training [1].

3.1. Virtual Reality

Virtual reality (VR) refers to the use of computer hardware and software to generate realistic aural, visual, and other sensations, creating an immersive environment and simulating the user's physical presence in it by enabling real-time interactions via sensorimotor channels.

As a concept, VR first became mainstream in the 1990s as several industries were influenced by videogames and by the first immersive human–computer interaction prototype, the "Man-Machine Graphical Communication System". A second wave of VR emerged after 2005 and saw more effective applications in different fields, including engineering, design, architecture, construction, medicine, mental health, military, education and training, fine arts, entertainment, business, communication and marketing, and travel. Today, research and development on VR devices, hardware and software products, and user interfaces is progressing rapidly, and technology adoption is quickly reaching many leading players in manufacturing, including several in the AECO industry.

Table 5. Main XR applications in AECO sector.

XR	Design	Construction	Operation
VR	• Design interpretation and visualization • Design collaboration • Design maquettes development • User-centered design • Human behavior investigation • Education • Communication and client engagement	• Site planning • Virtual construction interpretation and visualization • Safety planning • Personnel training • Vehicles and robot's teleoperation	• Maintainability design • Immersive building visualization and monitoring • Human-centered building management • Personnel training • Vehicles and robot's teleoperation
AR	• On-site superimposed 3D models for building renovation interventions • 1:1 preview of furniture and interior decorations • Virtual tours and in situ walkthroughs • On-site measurement and floor plan creation	• Superimposed data visualization • Remote collaboration and assistance • Contextualization of 3D models at actual scale • On-site measurement and floor plan creation • Safety on the worksite	• Augmented building monitoring and management • Augmented maintenance and repair operations • Augmented renovations and retrofit works • Personnel training and workplace safety
MR	• Design collaboration • Tabletop interactive virtual 3D models • Interactive virtual tours and in situ walkthroughs • On-site measurement and 3D model creation	• Superimposed data visualization • Remote collaboration and assistance • Contextualization of 3D models at actual scale • On-site measurement and 3D model update • Safety on the worksite	• Collaborative augmented building monitoring and management • Collaborative augmented maintenance and repair operations • Collaborative augmented renovations and retrofit works • Personnel training and workplace safety

To provide users with effective and productive VR experiences, VR content creators carefully design and translate 3D models, 2D images, and spatial sounds into a machine-operable format that considers the intended final use, the VR device employed, and the degree of interaction expected. Specifically, VR experiences can be either static (i.e., limited to the immersive visualization of a 360° spherical image or video from a predetermined point of view, with no possibility of free movement) or dynamic (allowing the user freedom of movement and interaction within the VR environment).

Dynamic VR grants much more freedom to the viewer compared to static VR, allowing them to wander unrestricted inside the CG environments, interact with reproduced objects, and even alter the scene directly from within the virtual dimension. Static images are not a viable solution for dynamic VR, in fact all content must be generated continuously according to the user's input and movements using Real Time 3D (RT3D) rendering tools. However, rendering time must be fast enough to be imperceptible by the viewer, to ensure that the virtual reproduction of reality is as authentic as their analog experience and to provide them with precise control and feedback over their movement and interaction. Such level of performance cannot be achieved by conventional 3D and BIM rendering software, hence VR usually employs specific optimized graphic engines that often borrow their technology from the videogame industry. These can be integrated with user–environment interactions using their own libraries and VR platform software development kits (SDKs).

With its advanced capabilities of immersive and interactive visualization, VR has been advocated to facilitate design, engineering, construction, and management for the built environment [17,31,115].

As VR provides a fully immersive experience, it can prove a particularly effective tool for designers, contractors, and owners to review plans and fully comprehend operation sites before projects are put into motion. The VR application controls the entire environment that the user sees, creating a uniquely detailed and dynamic individual experience. VR can also improve collaboration among stakeholders [32], enable a better understanding of complex designs [33], help identify design issues [34], support outdoor [35] and indoor [36] lighting design, realistically recreate building geometry so that users can fully comprehend the project and reach a better design decision [37], and aid collaborative decision making [38]. Regarding the construction phase, evidence suggests that VR technologies can be effectively deployed to support construction safety training [47], project schedule control [48], and optimization of the construction site layout [49].

3.2. Augmented Reality

Augmented reality (AR) can be defined as the technology that combines the physical and digital worlds by superimposing CG content—text, images, and interactive graphics—onto real-world objects. It is experienced through HMDs, smartphones, or tablets able to provide users with both the CG video feed and the direct view of the real world. Superimposed information can be either constructive (i.e., additive to the natural environment) or destructive (i.e., masking the natural environment). Differently from VR, therefore, AR is neither immersive, as the viewer continues to see the real world around them without any feeling of being elsewhere, nor exclusive, as the user maintains their capacity of interaction with objects and people around them. As such, the main benefit of AR consists in the possibility of providing content or information associated with the context in which users actually are, for instance, integrating in the real environment objects that are not present in reality, greatly facilitating the transfer of information from abstract representations to the physical world. AR can provide relevant information right in the context where it is needed, proving enormous potential in facilitating manual tasks, enabling time saving, and improving quality and efficiency, as already shown by evaluations in numerous projects.

AR systems commonly determine the positioning of the CG content in space using geolocation (GNSS, Wi-Fi triangulation, beacons, etc.) in situations where the approximation can be large, or, when higher precision is required, via using tags, markers, or anchors in the real world, usually consisting of 2D prints similar to QR codes, which can be recognized by the camera acting as a trigger for displaying the CG content as well as a reference point to define its position, size, and perspective.

Nonetheless, in all the aforementioned cases, the CG content cannot recognize any physical objects within the real-world environment, meaning that CG and real-world content are not able to respond to one another. Even when the 3D virtual object is correctly placed, scaled, and rotated in the space environment, it is solely anchored to the camera view and is always displayed in the foreground with respect to the real objects in the environment. For instance, when the user sees a CG image of a chair on the floor in front of them, the latter appears in the foreground even if a real object is introduced in between, disrupting the illusion. Indeed, AR does not allow digital objects to be occluded behind real objects, thus preventing true interaction between the two worlds: the information of objects overlaid onto the real environment cannot be interacted with, and that is where mixed reality comes into play.

Researchers have proposed several ways to use AR for architecture, engineering and construction, and facility management (AEC/FM) projects that can yield many advantages for enhancing and improving representation techniques on a job site [50,51].

Thanks to their ability to precisely overlay CG objects and information on the real world, in the AECO industry, AR and MR certainly show their most effective deployment in the phases of construction and maintenance of buildings, where they act as true human augmentation tools [52–54].

Nevertheless, AR and MR, the latter in particular, are finding numerous and promising applications in the design phase as well [17,39,40], concerning, in particular, building renovations, infrastructure design, urban planning [17], digital fabrication in architecture [41], virtual tours, and in situ walkthroughs (for instance overlaying virtual reconstruction on heritage sites to enrich cultural tours) [42–44], or for an augmented user experience in art and architecture exhibitions (such as for example the Exploring SongEun Art Space exhibition in Seoul by Herzog and de Meuron of November 2021), engaging attendees in a combination of physical and virtual worlds in which virtual agents can be interacted with as one would with real world counterparts [45].

3.3. Mixed Reality

Mixed reality can be described as an upgrade of AR that further bridges the virtual and real world together, providing a more connected experience where both virtual and real elements can fully interact with each other. As a term, "mixed reality" was first defined in 1994 by Paul Milgram and Fumio Kishino as "anywhere between the extrema of the virtuality continuum", with the virtuality continuum extending from the completely virtual throughout to the completely real environments [110].

Mixed reality can effectively remove the boundaries between real world and virtual content by properly implementing occlusion: CG objects can be correctly obscured from the user's point of view by objects in the physical environment, effectively becoming part of the real world. A CG ball will bounce off real tables and walls, or disappear under a real couch. MR allows users to manipulate virtual elements as they would in the real world, with the digital content behaving and reacting accordingly. For instance, one could turn a virtual object around using physical gestures to inspect it from all angles.

Mixed reality can be achieved either by integrating CG content into the real world or by adding elements from the real world into the virtual environment (such as streaming video from physical spaces, as through a webcam). Therefore, MR must be experienced through semi-transparent lenses or MR headsets equipped with a camera in order to film and display the user's environment. During their operation, MR devices continuously scan the surrounding environment to update the associated 3D model: this allows digital content to be placed over the real world and enables users to interact with it seamlessly, thus enabling the mixed reality experience [114].

In the design phase, in addition to the possibilities offered by AR, MR can provide virtual collaborative environments in which several users from different places in the world can meet to inspect and interact with a CG maquette as they would with a physical one actually present in the room, communicating the design intent to stakeholders without needing expensive real models. For instance, SketchUp Viewer is an MR application designed for multiple platforms, the main features of which are a tabletop display of a 3D model and of a 1:1 scaled version of the model itself, as well as a user interface (UI) to edit or view information on the tabletop model and a navigation interface for the 1:1 model.

Carrasco et al. [46] assessed the effectiveness of design review using MR compared to traditional 2D methods. The results showed that MR-based design review can effectively communicate 85% of the information to the client instead of 70% provided by 2D media. At the same time, it showed the potential to enhance the client's comprehension of the aesthetic characteristics of materials, giving the possibility to replace physical samples or mockups during the finishing stage of construction.

Through MR, clients can see what is being constructed, where before the physical elements would have been uncertain, acquiring a level of confidence in the end product that they would not traditionally have. Using programs such as Holoview MR, stakeholders can view a building as if completed in its actual location, accurately interpreting plans and their spatial relationship to the surrounding physical environment. BIM/CAD models overlay real world construction environments for when works have already started. Contractors can easily see how their upcoming activities are impacted by existing works. As the model overlay is extremely accurate (to 1 cm), engineers and contractors can confidently identify design mistakes and resolve clashes and constructability issues early. The new design scope can be projected into the construction environment so that all stakeholders can assess the impact that changes would have on current and future works. Furthermore, MR allows plans to be reviewed and issues to be corrected, to remotely undertake progress walkthroughs, and to undertake code compliance inspections for all elements included in the design model.

4. XR Technologies

XR can be visualized through portable devices such as smartphones and tablets, or through HMDs that can be worn on the head or integrated into helmets (akin to helmet-mounted displays for aviation pilots). HMDs contain a display and lens assembly in front of either one (monocular HMD) or both eyes (binocular HMD). The employed display technologies include liquid-crystal displays (LCDs), organic light-emitting diodes (OLED), liquid crystal on silicon (LCos), or multiple micro-displays to increase total resolution and field of view.

Virtual reality HMDs can only display computer-generated imagery (CGI) and feature an electronic inertial measurement unit (IMU) that uses a combination of accelerometers, gyroscopes, and sometimes magnetometers to keep track of their specific acceleration, angular rate, and orientation.

AR and MR headsets, instead, must overlay CGI onto the view of the real world; therefore, they feature optical head-mounted displays (OHMDs). OHMDs employ optical mixers, which consist of partly silvered mirrors that let the user look through the lens while reflecting artificial imagery produced by the device.

Currently, portable XR devices still have limited capacity in terms of the size and complexity of displayed files because the performance required for continuous 3D rendering in terms of CPU, RAM, and storage is ill-suited for miniaturization and low battery consumption without resorting to an external computing unit such as a high-end PC. The current limitations of XR technology, namely the compromise between desktop or portable solutions in terms of performance, fidelity, and mobility, can be effectively addressed by resorting to cloud computing, meaning that a key role will be played by the progress in wireless communication, in both local (802.11ax or Wi-Fi 6) and mobile networks (5G).

4.1. VR Technologies

VR systems normally include more than one hardware device to allow full operation. Generally speaking, the hardware components of any VR system can be classified into three categories according to their function, namely, display, controllers, and motion capture (mocap) devices. The display outputs stereoscopic images to users and is an essential element of the VR system. Commonly used display types include HMDs, mobile devices, and display walls. A typical VR platform can generally be classified into three categories, namely head-based, stationary, and hand-based operation.

Head-based VR devices consist of helmets or HMDs in which CGI is displayed on the internal screen or pair of screens, one for each eye, with an embedded position-tracking sensor that keeps track of where the user is looking (see Figure 2). Conversely, stationary VR platforms are usually fixed in place and employ projectors and/or large screens to display CGI to viewers. Lastly, hand-based VR devices are held by the viewers up to their eyes with their own hands and include smartphones or tablets.

Concerning head-mounted systems, three types of VR headsets on the market are suitable for the AECO sector (see Table 6): tethered VR (also known as desktop VR or PC VR), standalone VR (also known as wireless VR or all-in-one VR), and smartphone VR headsets (also known as mobile VR or VR viewers).

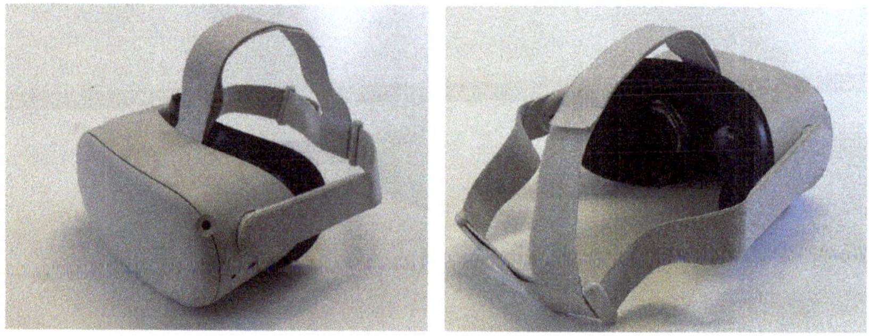

(a) (b)

Figure 2. Meta Quest 2 VR headset. (**a**) Front view; (**b**) Rear view.

Table 6. Main VR tethered and standalone headsets on the market.

Brand	Type	Display	FOV (Horizontal)	Resolution per Eye	Refresh Rate	Release Date
HTC Vive Focus 3	standalone	2 × LCD binocular	116°	2448 × 2448	90 Hz	2021
Meta Quest 2	standalone	Single Fast switch LCD binocular	97°	1832 × 1920	120 Hz	2020
Pico Neo 3 Pro	standalone	Single LCD binocular	98°	1832 × 1920	90 Hz	2021
Lenovo Mirage VR S3	standalone	Single LCD binocular	101°	1920 × 2160	75 Hz	2020
HP Reverb G2	tethered	2 × LCD binocular	98°	2160 × 2160	90 Hz	2020
HTC VIVE Pro 2	tethered	2 × LCD binocular	116°	2448 × 2448	120 Hz	2021
Valve Index	tethered	2 × LCD binocular	108°	1440 × 1600	144 Hz	2019
Varjo Aero	tethered	2 × Mini LED binocular	102°	2880 × 2720	90 Hz	2022

Tethered VR headsets (HTC VIVE Pro 2, HP Reverb G2 (HP, Palo Alto, CA, USA) and Varjo AERO (Varjo Technologies Oy, Helsinki, Finland)) are designed to be connected to the PC, with or without wires, to exploit the resources inside them. They need compatibility with operating systems and specific hardware requirements. Standalone VR headsets (Meta Quest 2 (Meta, Cambridge, MA, USA), HTC VIVE Focus 3/Plus (HTC, Taoyuan, Taiwan), and Pico Neo 3 Pro (VR Expert, Utrecht, The Netherlands)) are completely independent viewers designed to operate without the need for other peripherals and without wires. They are equipped with internal memory and integrate all functions. Bluetooth or a smartphone connection may be required for configuration. Smartphone VR headsets are designed to be connected to the smartphone, which can be physically inserted into the viewer or connected via Bluetooth. In this regard, we have viewers designed exclusively for a smartphone brand or universal VR viewers that adapt to multiple models and operating systems (BNEXT VR PRO, https://www.aniwaa.com/product/vr-ar/bnext-vr-pro/ (accessed on 15 March 2022)).

Notwithstanding the type of headset, the immersivity level that can be achieved depends on different characteristics of the device, which include the field of view (FoV) and the quality of the display (pixel density, color accuracy, dynamic range, and brightness), the refresh rate (or frame rate) of the CGI, the number of movements allowed to the user (degree of freedom, DoF), the accuracy tracking system, the presence of controllers, and the audio system. Generally speaking, a satisfactory immersive experience requires a FoV of a minimum of 100 degrees (for reference, the human eye has about 220 degrees FoV), a refresh rate between 90 Hz and 120 Hz, six degrees of freedom (3DoF rotational freedom to

360 degrees head rotation and 3DoF positional freedom to allow up/down, left/right, and forward/backward movement), as well as an accurate motion tracking system.

Wide FoV displays enable viewers to experience the virtual environment in a more lifelike way, i.e., focusing on what is in front of them while also perceiving peripheric objects. Higher refresh rates and low latency are instead recommended to avoid motion sickness symptoms (so-called cybersickness). 6DoF controllers allow for more advanced interactions compared to less sophisticated point-and-click controllers (limited to 3DoF).

Motion tracking is a crucial function in the VR system because it ensures user movements and orientation are effectively replicated on screen, in turn enabling a satisfactory interaction with the immersive VR environment [116]. VR Mocap systems continuously capture and process real-world motions of the user in order to track their current view and provide positioning for interaction with the virtual environment. A more precise motion tracking translates to a more seamless and lifelike immersion in VR. Likewise, any perceived gap or lag between the user's actions in real life and their reproduction in VR may greatly disrupt the immersive experience. Positional tracking can be either external or internal.

External tracking (also known as outside-in tracking) employs external sensors and/or cameras to keep track of the VR headset's position and orientation within a user's defined space (room-scaling). For full room scale, more than two sensors are installed to avoid any occlusion. Internal tracking (also known as inside-out tracking) uses one or more front-facing cameras embedded into the VR headset to detect its position and may function with the support of external markers. Internal tracking generally has poorer performance and is less accurate than external tracking; however, it is much more convenient to set up.

Mocap technology commonly employs optical, inertial, mechanical, and magnetic sensors. In portable VR systems, sensors are usually embedded in the headset itself and their data feed is processed by algorithms to provide motion tracking. Other solutions may rely on more complex systems including volumetric capture (MS Azure Kinect, Microsoft, Redmond, WA, USA), hand tracking and haptics (Ultraleap, Leap Motion, San Francisco, CA, USA), eye tracking, or even full body tracking thanks to special suits.

Finally, interactive controllers (joysticks or wands, data gloves, and haptic devices) are an equally important media for enhancing the reality of VR environments as they are responsible for how users interact with objects in the virtual environment and for which sensory feedback they receive, including haptic and auditory feelings.

4.2. AR and MR Technologies

Similarly to VR, hardware components required for augmented and mixed reality fruition include processing units, displays, several sensors, and dedicated input devices [54].

On the other hand, both AR and MR technologies are context-aware instead of immersive: their key characteristic is the capacity to combine the reality that users see with their own eyes with CG objects that are seamlessly overlaid in their specific position, differently from VR in which users are completely isolated from the real world upon entering the virtual environment. Therefore, a crucial function of context-aware technologies is geolocation, that is, ensuring the correct alignment of the virtual environment onto the real world in order for the device to properly display the virtual content in its expected position.

In particular, AR and MR devices can calculate the coordinates of their actual 3D position in the real world by processing the spatial relationship between themselves, external markers, and key points through the method of Simultaneous Localization and Mapping (SLAM). As soon as the AR/MR device turns on, its sensor equipment (cameras, gyroscope, and accelerometer) scan the surroundings and feed their data to an algorithm able to reconstruct a 3D model of the real-world environment and then position itself within it. Following this process, the system understands its environment well enough to be able to display CG objects that are realistically placed, oriented, and illuminated to feel part of the real world, with the viewer able to move close and inspect them from multiple directions. Differently from AR, after the SLAM process is complete and the CG content is properly positioned in the real space, MR additionally allows the virtual objects to be occluded from the view when they would be obscured by real ones (such as walls, floors, and columns that stand between the viewer and their expected position). MR also allows this occlusion to be controlled to display, for example, the pipes beneath a floor or wall surface as an X-ray view by regulating the transparency of the related shaders in the graphical engine.

In particular, there are several methods to display concealed utilities in AR/MR, which mainly differ in how they provide the perception of depth of the virtual object in relation to the viewer, and therefore in how accurate and intuitive the AR/MR scene is to the user. Muthalif et al. [57] focused on the AR/MR visualization of underground pipe utilities and investigated six main display techniques, namely X-ray view, topographic view, shadow view, image rendering, transparent view, and cross-sectional view.

Among these, the X-ray view, which superimposes a cutout of the ground—a virtual "excavation box"—in which utilities are shown in their correct depth and position, was found to be the most accurate and intuitive for the viewer, allowing them to distinguish even multiple utilities in the same view. However, this method requires an accurate and detailed 3D model (a BIM file or even a 3D point cloud of a previous excavation), as well as pre-captured data to operate at best, and the virtual excavation box may end up covering a large part of the real-world view, which may pose hazards in working conditions.

The topographic view instead superimposes a 2D map of underground utilities directly on the ground akin to traditional paint markings on street and paved surfaces. Such a technique is very intuitive, leaves most of the real-world view non-occluded, and does not require any 3D model; however, it lacks any information on depth and may prove confusing if more utilities are shown at once. The shadow view integrates depth information into the topographic view by means of shadow lines and projections on the surface, but this adds complexity to the scene.

Image rendering improves the 3D visualization of objects in space by integrating additional reference points in context, e.g., adding virtual edges to the real world and masking virtual objects beneath real ones (occlusion). This improves the understanding of the scene at the expense of additional computing power and 3D model accuracy needed for localization. Lastly, the transparency view and cross-sectional view see little application in the field.

Table 7 shows the different static and dynamic methods with which virtual content can be superimposed to the real-world view.

Table 7. Static and dynamic methods to overlay CGI to the physical world.

CG Content Overlay Mode	Description
Screen-fixed, static CG overlay	CG content is always fixed to the same position on the screen: recommended for elements that should always be visible, e.g., menus or return prompts, or the cockpit of a vehicle.
World-fixed, static CG overlay	CG content is locked in its expected position in the real world: recommended for labels or graphics that complement a specific object or marker in space.
Screen-flexible, dynamic CG overlay	CG content can be moved on the screen, allowing users to place elements in specific areas: recommended for target-based or drag and drop assets.
World flexible, dynamic 3D CG content	CG content can be freely moved in the real world: recommended for educational purposes or to fully understand 3D models and break down their components.
World proportionate, dynamic 3D CG content	CG content is realistically placed in its real world position with actual lighting and measurements taken into account.

In order to support human activities effectively, AR and MR devices should be, preferably, head mounted, as this allows to hands-on interaction to be eliminated with the device, should be equipped with semi-transparent lenses or optical displays, to allow CGI to be superimposed on the real-world view, and should feature cameras and sensors to scan the real environment continuously and allow mixed reality experiences. AR MR headsets can also be equipped with directional speakers and active noise reduction microphones, with more advanced products allowing vocal control by virtual assistants.

Multiple models of AR and MR devices are currently on the market [55], designed to meet different needs (see Table 8).

Table 8. Main AR and MR devices on the market.

Brand	Category	Type	FOV	Resolution per Eye	Release Date
Epson Moverio BT-40	AR	Tethered	34° diagonal	1920 × 1080	2020
Google Glass Enterprise Edition 2	AR	Standalone	80° horizontal	640 × 360	2019
Magic Leap 1	MR	Standalone	40° horizontal	1280 × 960	2018
Magic Leap 2	MR	Tethered	44° horizontal	1536 × 1856	Announced 2021
Microsoft HoloLens 2	MR	Standalone	43° horizontal	1440 × 936	2019
Varjo XR-3	VR/MR	Tethered	115° horizontal	2880 × 2720	2021
Vuzix M4000	AR	Standalone	28° diagonal	854 × 480	2020

Overall, AR can be implemented according to four methods: optical see-through, video see-through, eye multiplexed, and projection based. The former two are the most widely adopted in AR headsets available on the market [56]. In optical see-through systems, AR is achieved by superimposing virtual images over the direct view of the real world, commonly by projecting CG content through half mirrors or prisms. With this method, the real-time view of the world is maintained while seeing AR content. In video see-through systems, on the other hand, the camera of the AR device continuously captures the real world in front of it, processes each frame by adding CG content, and finally displays the AR image to the viewer on the device's screen. Concerning the types of display featured in AR devices, there are three main solutions: monocular (a screen in front of one eye), binocular (the same screen in front of both eyes), or dichoptic (a different screen in front of each eye, to enable depth perception).

Regarding AR HMDs, Google Glass Enterprise Edition 2 (see Figure 3) is among the most diffused optical see-through devices in the industry. It is used to access documents on the go, which may include texts, images annotated with detailed instructions, training videos, or quality assurance checklists. It can also connect with other AR devices to communicate, for instance livestreaming one's view to enable real-time collaboration and remote assistance. Control methods include voice commands to launch applications allowing full hands-free operation.

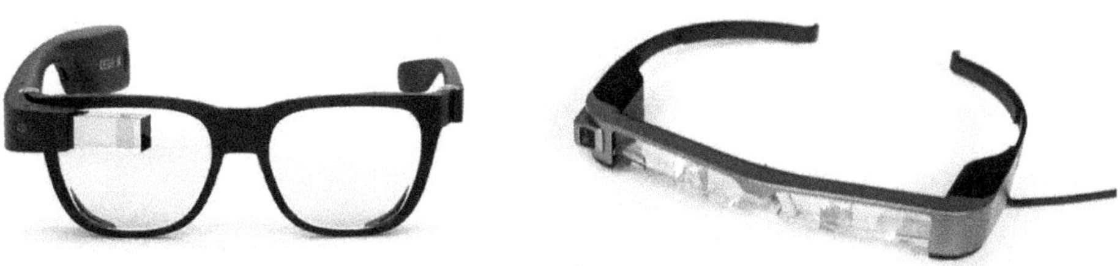

Figure 3. Google Glass Enterprise Edition 2 (**left**) and Epson Moverio AR glasses (**right**).

Moverio BT-300FPV AR glasses by Epson (Suwa City, Japan) are optimized for drone flying: the video feed from the aircraft is displayed in first-person view (FPV) on a transparent screen, which allows the viewer to pilot the UAV, monitor its flight statistics, and keep the UAV always in sight (see Figure 3). In 2022, XYZ (London, UK) released The Atom, a next-generation engineering-grade AR headset for construction. Combining a safety-certified hard hat, augmented reality displays, and the in-built computing power of the platform HoloSite, the device can position holograms of BIM models on-site to millimeter accuracy.

In addition to headsets, smartphones or tablets can provide handheld AR experiences: the user directs the device's camera at the real-world environment and the AR app installed superimposes content such as an image, animation, or data on the main screen. Indeed, most mobile devices on the market feature high resolution cameras as well as accelerometers, a GNSS receiver, solid state compass, and even LiDAR (Measure Australia, Surry Hills, Australia), making them fully equipped for AR operation. AR applications and software development kits (SDKs) are already available from the largest consumer tech companies such as Apple, Facebook, Google, and others, and some AR headsets are explicitly designed to mount smartphones (Mira Prism, Vuzix M300, https://www.aniwaa.com/product/vr-ar/vuzix-m300/, accessed on 15 March 2022).

On the other hand, the market for MR devices is still smaller than AR, and currently consists of Microsoft HoloLens and Magic Leap (West Sunrise Boulevard Plantation, FL, USA) with the recent introduction of Varjo XR-3 (see Figure 4). Indeed, MR goggles are high-performing HMDs, which must include sensors such as gyroscopes, accelerometers, Wi-Fi antennas, digital compasses, GNSS, and conventional and depth sensing cameras [1,2] in order to scan and capture the surroundings and integrate them with fully interactive CG content.

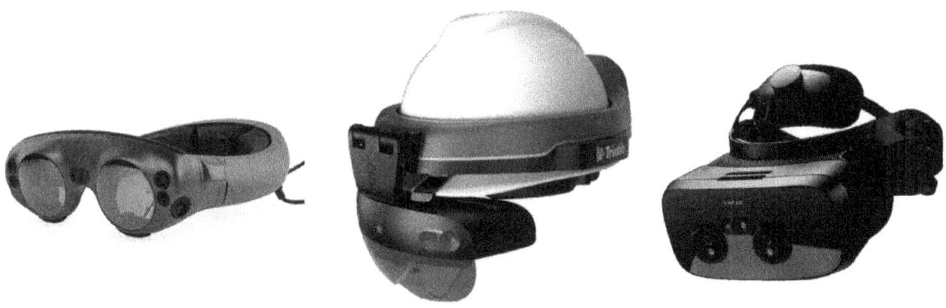

Figure 4. Magic Leap 1 (**left**), MS HoloLens 2 on Trimble (Sunnyvale, CA, USA) XR10 (**center**), and Varjo XR-3 (**right**).

The flagship MR device on the market is arguably MS HoloLens, currently in its second iteration, which takes advantage of eye and hand movement tracking to support the positioning of virtual content and a user interface that does not rely on external controllers. Users can also log into HoloLens seamlessly using eye recognition. Moreover, HoloLens features smart microphones and natural language speech processing algorithms to ensure vocal controls work properly even in noisy industrial environments. Park et al. [58] conducted a literature review to investigate the current status and trends in HoloLens studies published over the past five years (2016–2020), showing a growing use of MS HoloLens multiple fields, from medical and surgical aids and systems, and medical education and simulation, to industrial engineering, architecture, and civil engineering, and other engineering fields.

A recent example of a video pass-through MR headset is represented by Varjo XR3, a tethered HMD that claims the industry's highest resolution (2880 $\times$ 2720) and the widest field of view (115°), and provides depth awareness powered by LiDAR for pixel-perfect real-time occlusion and 3D world reconstruction. The device allows, in addition, a full VR experience.

Apart from standalone MR headsets, several smartphone-based MR headsets are already available or about to enter the market, including Tesseract Holoboard Enterprise Edition, Occipital Bridge (https://www.aniwaa.com/product/vr-ar/occipital-bridge/, accessed on 15 March 2022), and Zappar ZapBox (https://www.aniwaa.com/product/vr-ar/zappar-zapbox/, accessed on 15 March 2022). These can provide MR experiences using the camera and display of the smartphone or semi-transparent mirrors to overlay the CG content to the real world.

5. XR Applications in Building O&M

Operation and Maintenance (O&M) represents a strategic activity in building and infrastructure management and includes a diverse range of services, skills, methods, and tools to ensure that they maintain the expected performance over their lifetime and to reduce energy consumption and upkeep costs at the urban and building scale [117]. Common O&M activities include the maintenance, repair, and replacement of components, energy management of the built asset, management of emergencies, management of changes or relocations of services and uses, security, management of hazardous and non-hazardous waste, ICT, and others [118,119]. Operational costs, energy consumption, indoor comfort, and emissions are some of the aspects that can be significantly affected by suboptimal efficiency and effectiveness of O&M. In particular, optimal planning of maintenance work is paramount to improve the efficiency of building management and minimize O&M costs, thus requiring the extensive adoption of advanced management tools, including building automation systems (BAS) and BIM-integrated facilities management software [59]. BIM has, in fact, numerous potential applications in O&M, both as a central part of an integrated facility management system or as a relevant data repository for the latter [120]. BIM allows facility managers to store, maintain, and access building information, including spatial, technical, warranty, maintenance, and spare parts data, and thereby to conduct their commissioning and O&M activities more efficiently [121,122]. The planning of major repairs, retrofits, and expansions greatly benefits from an up-to-date BIM model. BIM integration can be even more effective when paired with real-time data provided by building sensors, systems, and service robots that can offer detailed information on the asset's condition. In fact, BIM is the foundation for the digital building twin, that is, the digital replica of the built asset that is continuously updated by live data provided by sensors and systems in the real structure replicate, a breakthrough instrument with unprecedented potential in O&M improvement and optimization [14,123,124]. AI- and ML-powered analytics hosted in the cloud can extract meaningful knowledge from data streams provided by the digital twin, and results can be fed back into the system to allow self-learning and optimization [125]. The spread of BIM and digital twins has also paved the way for other technologies such as reality capture (3D laser scanning, point clouds, etc.) and new XR tools [60,126]

In particular, the application of VR, AR, and MR technology in building management is showing promising results in enhancing human performance in building O&M tasks, in improving maintenance operations, and in supporting strategic decision making [18]. Due to the possibility of benefiting from an immersive and collaborative environment and of superimposing a digital model, real environment, data, and information in an integrated way, XR allows, at best, the functioning of the digital twins of buildings and infrastructures, to be exploited.

5.1. VR Applications

Thanks to its advanced capabilities of immersive and interactive visualization, VR can facilitate several building O&M activities. VR applications concern the following main areas:

- design of the building's maintainability, improving its effectiveness by allowing multidisciplinary experts better understanding of the final product and thus reducing reworks down the line;
- immersive building visualization and monitoring, enabling facility managers to interact with any asset at a more immersive and intuitive level than with RT3D visualization;
- human-centered building management, providing a realistic, safe, and fully customizable virtual environment to observe human–building interaction in different situations;
- training for maintenance technicians and managers, supporting or replacing altogether traditional image- and text-based learning methods;

- teleoperated maintenance, allowing effective interventions with robots even in dangerous or difficult to reach locations.

5.1.1. Maintainability Design

Regarding the design of maintainability, VR allows maintenance operations of any equipment or product to be realistically simulated in a fully replicated environment, enabling effective design reviews and identifying potential maintenance-related design flaws preemptively. Design teams can involve facility managers, technicians, and maintenance engineers in a VR simulation, allowing them to enter the project, replicate conventional operations, then return valuable feedback in the early stages of design. Such interaction with the virtual prototypes in the immersive environment enables a more complete and tactile understanding of the practical maintenance operations of any product well before it is commissioned [61].

Data observed during VR interactive simulations can in turn be collected and analyzed by embedded algorithms to allow qualitative and quantitative evaluation of maintainability. VR simulations have already proven effective in maintainability design, supporting the evaluation of ergonomics, accessibility of components, environment factors such as heat or radiation, workers' fatigue, and other human factors [62,63]. For instance, Akanmu et al. [64] presented an automated system that integrated BIM models, Microsoft Azure, and VR to engage facility managers in the design phase regarding the accessibility of building components for maintenance. The system provided a platform for mining and extracting knowledge from feedback provided by facility managers to improve building design tools. The functionality and usability of the system were presented with an example of the lighting and air-conditioning systems.

5.1.2. Immersive Building Visualization and Monitoring

VR technology transforms 2D or 3D models into immersive, fully manipulable virtual environments, enabling facility managers to interact with their assets to a much more tactile level than what is allowed by RT3D visualization. Several plug-ins for BIM authoring software (Revit, BIM 360, Archicad, etc.) such as Arkio, The Wild, Prospect, Sentio VR, Unity Reflect, or dedicated software (Twinmotion, Enscape, Fuzor, BIMx, VRcollab, Lumion, VRex, Resolve, Techviz) are today available on the market allowing 3D models to be brought into VR with a single click, and also importing data (so users can easily access BIM metadata and layer visibility) and enabling remote collaboration.

Differently from RT3D, in fact, VR visualization provides a realistic perception of spatial depth, enabling any project to be explored on an actual 1:1 scale. Although conventional 2D renderings and 3D digital models can allow an abstract understanding of the environment, the full experience of interaction with a physical space cannot be replicated with only a computer screen or sheet of paper. In urban planning, for instance, VR visualization allows the user to realistically experience how a new structure will fit in and relate with the surrounding buildings and environment.

Exploiting digital building twin (DBT) technology, VR can augment both the visualization of the digital model and the operation, management, and maintenance of the actual building [127]. In fact, a digital twin consists of an accurate and continuously updated model a given physical entity, which may be a single building component, e.g., an HVAC element, or an entire building, or even whole cities or districts. It allows users to keep track of the current status of its related physical twin, to simulate alternative scenarios, identify internal and external complexities, discover unusual patterns, monitor performance, and accurately predict ongoing trends to optimize operation [1,65]. The digital twin enables predictions and simulations of what-if scenarios, allowing to test adjustments and variations that would be inconvenient to try out in the real world, providing a larger space for experimentation and trial and error. Finally, AI and machine learning analytics can be integrated into digital twins to make them more intelligent and improve autonomous decision making.

Shahinmoghadam et al. [66] presented a method to monitor in real time the thermal comfort conditions of the digital twins of a building's closures, based on the integration of BIM and live IoT data into a game engine environment. The proposed system would allow facility operators and experts to intuitively monitor the complex and dynamic data associated with actual thermal comfort conditions. Users could navigate the virtual environment remotely and observe actual thermal sensations in real time in every place. Parameters could be changed via the user interface to implement different what-if scenarios and observe the outcomes based on the live monitoring data streaming from the IoT nodes installed within the building. A proposed application is in interior layout planning, in which designers can keep track of any spot that is flagged with discomfort signs.

At the city and district level, smart city digital twins can collect information from different sources on the territory in real time and transfer it to administrators, policymakers, and citizens, putting actual data at the base of more informed and participated-in decisions and enabling a truly knowledge-driven city management [67,128]. Acoustic engineering can take advantage of VR as well to realistically simulate sound sensory input, allowing the creation of urban and building spaces with a lower noise exposure and better acoustic comfort. Sound designers can import the BIM model into a VR engine to simulate the physics of acoustic waves passing through a structure and reflecting on different materials. With VR, users may distinguish how the sound bounces off a tree or a stone surface, or hear the difference between an open window and a closed one.

Regarding augmented visualization, VR enables exploration of the DBT even remotely, including first-person movement, the measurement of distances and angles, the interrogation of single components, querying any element by category, the clustering of objects, and the retrieval of documents (e.g., drawings, data sheets), as well as the direct modification of assets that are synced with the building BIM model. From inside the VR environment, the viewer may replace objects, track steps, take pictures, or sketch three-dimensional notes. Augmented visualization with VR can also be used for computational fluid dynamics (CFD) simulation, providing a more intuitive representation method to comprehend the airflow in a given space [68].

Concerning building management and maintenance, VR allows free navigation inside any environment and full interaction with its components. VR can therefore be used to interrogate single components of the DBT by exploring the 3D scene, facilitating monitoring of various internal and external parameters of the building (e.g., temperature, illumination, air quality), and tracking the condition of systems (e.g., on/off, consumption/production, or open/closed status, power levels, operating time). Additional uses include the remote control of systems, devices, and actuated components (e.g., windows, blinds), and the simulation of energy consumption and internal comfort in different scenarios to apply to the real building according to climate conditions, thermostat settings, operating times of the equipment, etc.

With VR, virtual meeting rooms can be set up to allow different members of the building management team to gather in the same place and time as 3D anthropomorphic avatars, even if they are physically remote. In the virtual space, issues can be discussed and solutions proposed as in any typical problem-solving meeting, with the added possibility for participants immersed in the synchronized environment to jointly manipulate and evaluate the emergence of creative solutions in real time. Compared to other forms of remote collaboration, VR co-presence was found to strengthen the sense of collaboration among team members, especially in terms of social aspects such as dependencies, encouragement, and mutual learning.

5.1.3. Human-Centered Building Management

A VR space can be used as a realistic, safe, and fully manipulable testing environment to observe how users would interact with the building in different scenarios and how any variation may affect the human state of mind and corresponding reactions, promoting a more user-centered building management [69]. Interactions with the virtual environment provide valuable insights to inform decision making [70]. In particular, occupant behavior was found to have a significant impact on the building's performance and energy consumption, and has therefore become a major research area in the human–building interaction field. Human–building interactions that have energy efficiency consequences, i.e., indoor lighting preferences, space layout, daylighting (operational blinds), or the use of appliances, shall be fully comprehended. In this picture, researchers can take advantage of the VR environment to collect behavioral data to analyze the human–building interaction and to assess how human experience and behaviors affect the building's energy performance [70].

Furthermore, VR is an effective tool for behavior investigation and training in emergency situations, especially those concerning fire or earthquake occurrences [71,72]. In particular, VR allows users to construct a stressful environment in emergency evacuation simulations without any short-term or long-term harm to the participants, providing the virtual stimuli to evoke mental and behavioral responses similar to those experienced in real-life situations [73]. Lorusso et al. [74] proposed a VR platform integrated with numerical simulation tools to reproduce an evolutionary fire emergency scenario for an existing school building, including real-time simulation of the crowd dynamic during the evacuation process. The results showed that the proposed VR-based system can be used to help decision makers determine emergency plans and to help firefighters as a training tool to simulate emergency evacuation actions.

Quantitative measure metrics, i.e., movement speed and directions, can also be incorporated in VR-based human–building interaction studies, as well as biometric sensors such as EEG (electroencephalogram), GSR (the galvanic skin response), and PPG (facial- or vision-based electromyography), to assist researchers in gaining insights into how human body functions work and react in the case of different built environment settings, such as spatial cognition, with a more conclusive and complementary measurement result [75,76].

Risk situations and the safety of workers can be effectively assessed with VR, allowing the feasibility and convenience of different design alternatives to be gauged. Specific aspects that imperil the virtual labor can be easily identified during the VR experience and adjusted accordingly, avoiding unnecessary reworks down the line.

Finally, recent global sanitary emergencies have clearly shown that not only the quality and characteristics of spaces, but also the way in which people experience these spaces and interact with each other, can have a direct influence on people's health. The problem of movement and the interaction of people within both closed and open spaces has now become a critical issue for safety linked to the transmission of diseases, equal in importance to established concepts such as systems safety and fire protection and evacuation. Therefore, understanding how to arrange spaces and how to organize the movement of people has now become a central theme of planning at the building and district scale, aimed at minimizing health risks by maximizing well-being and productivity. Spaces where people congregate—offices, construction sites, schools, retail spaces, warehouses—must be designed to reduce the risk of disease transmission. Such planning requires an understanding of mathematics, biomechanics, data science, design, demography, psychology, local regulations, sociology, and geography, among other disciplines, which today can find support in VR technology as well as simulation software and artificial intelligence. In this sense, a study by Pavon et al. [77] utilized BIM possibilities to reduce crowding and facilitate social distancing as a COVID-19 measure in a public building. More recently Mukhopadhyay et al. [78] presented a VR-based DT implementation of a physical office space with the goal of using it as an automatic social distancing measurement system. The VR environment was enhanced with an interactive dashboard showing information collected from physical sensors and the latest statistics on COVID-19.

5.1.4. Personnel Training

VR tools are finding widespread application in personnel training, particularly for larger contractors, as these show numerous benefits compared to conventional training methods concerning the speed and effectiveness of learning, schedule flexibility and better compatibility with work activities, lower travel expenses, lower overall costs, and greater safety for both learners and equipment.

Such advantages are especially meaningful in the O&M sector, in which continuous training and skills certification is required for labor, and several environments and situations may pose a high risk for workers' safety.

In VR, multiple scenarios can be simulated, including extremely dangerous events, allowing learners to "learn by doing"—with the optional support of text or audio information—without any risk for their safety or damage to the equipment. In addition to the end results and decision-making skills, VR-assisted learning allows other aspects to be evaluated such as response times, the precision of movements, and other specific variables, providing a more comprehensive insight on the level of training and competence acquired. Furthermore, VR can considerably reduce training costs for companies and learners, as it does not require real equipment to be rented—especially of large dimensions (cranes, etc.)—nor to replicate dangerous scenarios, while allowing training activities to be carried out on an individual level without affecting working hours or interrupting company activities (as it happens instead of group courses) [79]. The VR immersive environment is demonstrated to allow procedures and operations in a more involved and distraction-free way, with cognitive benefits including both working memory development (information encoding) and subsequent retrieval (information recalling). VR-trained subjects were observed to outperform those trained with conventional 2D and text instructions in both the rapidity and accuracy of maintenance operations [80]. In particular, haptic feedback from VR systems has proven to increase users' situational awareness and overall performance for maintenance applications. Indeed, although VR studies mostly focus on the pursuit of realistic visual effects, as vision accounts for 80% of external information processed by the brain, the possibility to accurately simulate the haptic feedback of the real maintenance process in VR is also very important, as the human–machine interaction in maintenance is mostly carried out by hand [63]. Compared to visual feedback, haptic feedback was found to be more effective in improving the performance of maintenance workers and particularly helpful in increasing situational awareness during telemanipulation tasks, in which the completion time was reduced and the quality of operation improved [81].

5.1.5. Vehicles and Robot's Teleoperation

VR can be successfully employed to enhance teleoperated maintenance operations, including those in hazardous or hard to reach environments such as nuclear plants, underwater facilities, offshore platforms, or emergency situations. Teleoperated vehicles include UAVs and robots equipped with articulated arms and manipulators characterized by impressive dexterity and precision of movement [82].

In VR-assisted teleoperation, drivers and operators wear headsets that display a full view of the working area and reproduce any sounds on-site as well as audible warnings. VR controllers perfectly mimic those on the real machine, providing an experience equal to sitting in the machine's cab, even if it is kilometers away. Verbal-, thought-, and eye-based control inputs are also employed. VR teleoperation also allows users to rotate operators in working shifts (even in different time zones), enabling 24/7 uptime. VR can completely immerse the robot operator in the task at hand, displaying stereoscopic video feed from the machine's cameras and faithfully reproducing commands imparted by VR controllers.

In its most advanced application, VR teleoperation involves biomorphic robots controlled using full-body suits: called VR exoskeletons, these robotic systems allow a machine placed in a separate location to perfectly synch with the movements of an operator as detected by a sensors-equipped exoskeleton [1]. Hand controllers, gloves, and similar allow objects of any size and weight to be manipulated with remarkable dexterity and accuracy. An example of a virtual exoskeleton is the Guardian GT system by Sarcos, consisting of a robotic system with one or two highly mobile arms mounted on a track or wheeled base, controlled by a body suit that allows various tasks (heavy lifting, intricate processes such as welding and joining, or even operating switches and levers) to be performed remotely.

5.2. AR Applications

Augmented reality (AR) is the application of XR technology in real-world scenarios, and has found considerable utilization in building management activities, allowing technicians to superimpose data to the real-world view while on site by using wearable HMDs or a camera-equipped mobile device [52,53].

With the ability to provide real-time information, AR is being used within the industry to increase efficiency, improve safety, streamline collaboration, manage costs, and boost overall project confidence, proving effective in the following building O&M applications: building monitoring and management, maintenance and repair operations, renovations and retrofit works, personnel training, and workplace safety.

AR combines both physical and digital content into one coherent environment by overlaying CGI over the user's view through a mobile device or HMD. AR systems can determine their position using GNSS positioning and cameras, then present the user with real-time context-related data, continuously updating and displaying the necessary information as they move throughout space. Information and documents such as operational status, plans and drawings, and schedules are easily accessible, allowing fully informed decision making in the field. AR solutions can be implemented with either head-mounted displays (HMDs) such as smart goggles or handheld displays (HHS) such as tablets, smartphones, or dedicated devices. AR apps for smartphones or tablets, such as XMReality, Reflekt Remote, or XOi, can be easy installed, do not require any dedicated hardware, and directly overlay graphics and text on the video feed of the devices' cameras. However, these handheld devices are not particularly suitable for supporting manual work as the operator has to interrupt the activities to consult the device, and remote support is equally suboptimal.

Conversely, AR systems that employ HMDs such as Epson Moverio, Google Glass, or Vuzix M-Series, which can also be integrated in PPEs, e.g., helmets (XYZ Atom), allow the operator to retrieve information and to collaborate remotely while keeping their hands on the tools and their attention on the task. Both HHS and HMD AR devices are autonomous and self-packaged, granting sufficient mobility for training, service, and maintenance operations on-site.

Several programs that are available allow AR functions to be created and provided (UpSkill Skylight, ScopeAR Worklink), with visualization methods ranging from displaying text bullet points to interactive 3D animations. The simplest AR interfaces superimpose virtual arrows and text onto the real-world view. Plain text contents are quick to create or update and their overlay does not occlude the view of the user; however, such solutions are less intuitive and thus indicated more for supporting already skilled maintenance workers. Support data can be presented as static 2D images and 3D models as well, overlaying relevant information on the screen to enhance inspections or other interventions. CG content can be animated to display specific processes at each stage more intuitively. Depending on the worker's skills, AR support can guide their actions step by step or just present top-level information. Another function of AR headsets is remote conferencing with coworkers and technicians, allowing colleagues off-site a precise glance on the field; much more effective than traditional vocal communication and picture sharing.

Additionally, AR allows users to erect 3D models over 2D floor or site plans, enabling more intuitive and accurate previews of the work on the field. Mobile AR programs such as GAMMA AR and Trimble's SiteVision (see Figure 5) capture the scene ahead on a smartphone or tablet and overlay it with 3D models, allowing on-site preview of utilities, buildings, and infrastructures presented with cm accuracy.

In particular, Trimble SiteVision is an X-ray view AR tool consisting of a lightweight, handheld, field controller equipped with an integrated GNSS positioning system. The system combines satellite data, to correctly position itself, and measurements from the on-board rangefinder, to determine the position of target points relative to the 3D model. The user is presented with a CG model accurately superimposed on the real-world context scene captured by the camera, and can comprehend intuitively how the 3D model interacts with the environment. As the viewer moves and shifts their point of view, the device continuously adjusts the 3D model on the display to match what the camera is seeing. By previewing the 3D model on the field in its expected position, potential issues can be detected beforehand to enact corrections timely and avoid clashes and unnecessary reworks down the line. In addition, the device allows users to capture photos, take measurements, and attach notes, then directly create and assign tasks to coworkers to ensure productive follow-up. Updates from the job site are transmitted from and archived in the office, allowing to better sync workers, activities, and information.

Figure 5. Trimble SiteVision AR system (**left**) and smartphone-based AR (**right**).

5.2.1. Augmented Building Monitoring and Management

AR allows facility managers to enhance information retrieval and visualization by presenting relevant content right in its real-world context, such as, for instance, displaying real-time sensor readings, process telemetry, or aggregated performance, thus improving the monitoring and control of whole building systems. This can be particularly effective for planning maintenance interventions, using virtual performance panels and dashboards to access real-time data for monitoring and control, and analytics purposes.

AR-enhanced visualization of non-visual data can be a useful cognitive aid for identifying the information needed for decision making in FM, and can highly improve the usability and accessibility of BIM data [83]. Chung et al. [84] presented a study in which AR-based smart FM systems demonstrated faster and easier access to information compared with existing 2D blueprint-based FM systems, with information obtained through AR allowing immediate, more visual, and easier means to express the information when integrated with actual objects. A study by Abramovici et al. [85] proposed an AR-based support system for collaboration among working-level stakeholders involved in the FM process, whereby AR was used not only as a tool for the visualization of maintenance data, but also for communication and alerts, or for displaying other coordination-related aspects for the whole team. Another study by Chu et al. [86] developed, tested, and evaluated mobile systems (Artifact) with cloud-based storage capabilities aimed at integrating BIM and AR to improve information-retrieving processes as well as operational efficiency during construction.

Alonso-Rosa et al. [87] presented a monitoring energy system based on mobile augmented reality (MAR) to visualize in real time the power quality parameters and the energy consumption of home appliances. Tests showed that by simply focusing the smartphone on the home appliance, the system could detect the image targets with a negligible response time to overlap the energy information on the captured frame.

AR systems can also overlay the results of building thermal or fluid dynamics simulations on the virtual model, or project those on the real environment [112,113]. Indeed, the integration of AR with numerical simulations can improve the solution of practical problems, reducing the misinterpretation in spatial and logical aspects by superimposing engineering analysis and simulation results directly on real-world objects [90].

5.2.2. Augmented Maintenance and Repair Operations

During actual interventions, technicians can use AR to augment their view of the physical world with overlaid digital content, which may consist of real-time data related to the task or asset of interest or even guidance on each step to replace a component. Differently from traditional instructions based on static text and images, AR presents information in a context-aware and more comprehensible manner, allowing more effective operations and greater flexibility in workers' deployment. Therefore, AR is especially indicated to support maintenance operations and training in buildings and infrastructure since it enables quick access to contextual information and seamless integration of data right into the operation, in addition to allowing remote support and assistance by colleagues and experts to improve problem solving and decision making in the field [84]. Moreover, machine vision and object recognition algorithms can be integrated in the AR experience to provide real-time feedback on the task: cameras can determine the actual position and orientation of the components, check those against the targets and show alerts to correct the procedure on the go, and prevent mistakes or reworks down the line. Traditional text and picture manuals can be transformed into digital multimedia content and seamlessly integrated into the workflow using AR headsets, allowing operators to completely focus on the task without diverting their attention. As such, AR helps to reduce unnecessary eye and head movements, improve spatial perception, and increase overall productivity [18]. Moreover, today, systems and equipment are growing in complexity and commonly embed several sensors to monitor operations and perform initial diagnosis. AR can retrieve and display valuable data and diagnostic results right beside the object to be maintained, instead of accessing this information with a separate computer.

Manuri et al. [91] proposed an AR-based system to help the user in detecting and avoiding errors during the maintenance process, which consisted of a computer vision algorithm able to evaluate, at each step of a maintenance procedure, if the user had correctly completed the assigned task or not. Validation occurred by comparing the image of the final status of the machinery, after the user had performed the task, and a virtual 3D representation of the expected result. In order to avoid false positives, the system could also identify both motions in the scene and changes in the camera's zoom and/or position, thus enhancing the robustness of the validation phase.

One of the most common applications of AR is certainly that of remote maintenance, also known as "collaborative maintenance" or "remote assistance", which consists of the cooperation of an expert and a technician who are actually in physically different locations. Especially in consideration of the growing technological complexity of building systems, the conventional remote assistance carried out over the telephone is no longer an effective solution. Programs for AR-based collaboration such as VSight or Virtualist, on the other hand, allow the transfer of information and knowledge between the expert and maintainer in real time in a remote "see-what-I-see" collaboration. Colleagues and experts can see the direct view captured by the operator's AR device on-site and send back voice instructions, documents, or superimpose their annotations directly on the technician's display. This enables even unskilled operators to carry out complex tasks on their own, allowing waiting times to be reduced for more experienced teams and overall system downtime [32,92].

Finally, AR solutions can support quality assurance by allowing users to compare the resulting configuration directly with the design model, which is overlaid in the correct location, highlighting inspection-relevant features and deviations right on the single element. Using AR programs such as GAMMA AR, available for Android or iOS, contractors can validate what has been built correctly with the BIM model and stick photos or add virtual notes to specific built or designed elements. Components of the model can be hidden, isolated, and annotated with text, pictures, or voice notes. With AR tools such as VisualLive, it is possible to vary the transparency of 3D models to allow X-ray views. Users must place QR codes on flat or vertical surfaces, then scan them to import and properly align the virtual models.

5.2.3. Augmented Renovations and Retrofit Works

The ability to merge virtual objects with the physical environment makes augmented reality particularly suited for previewing renovations and retrofit interventions. With AR, the user can use the screen of a smartphone or tablet to project a "digital window" that overlays the BIM model of an object, a building, or a road onto the real-world scene, helping the visualization of the positioning of entire structures, as well as specific design features, elements, or equipment. Using AR, a client can enter the construction site before the renovation works begin, walk around, and view the end result in a simulated built view, as the final 3D model is superimposed on the real images.

In building renovations, mobile AR apps such as AirMeasure and MeasureKit enable direct measurement of the height, length, and dimensions of objects directly on the screen of the device. Some programs (Magic Plan, Room Scan and Floor Plan Creator) allow the creation of 2D and 3D floor plans by pointing the device's cameras at the corners of the room and snapping pictures while turning around. AR solutions can also support the scheduling and planning of building renovations, allowing users to preview and virtually test alternative process sequences, such as crane trajectories or vehicle movements, in a spatial context, or evaluate manual activities such as assembly/disassembly in a realistic manner.

5.2.4. Workplace Safety

AR solutions can be effectively integrated in workplace safety procedures by using HMDs or mobile devices to scan tags or labels placed in specific areas or objects. These labels can then bring up text or 3D models to communicate specific safety or hazard information.

Hurtado et al. [93] proposed a method to support the collective protective equipment inspection process by equipping the safety advisor with an AR-based 3D viewer with an intuitive interface. Authors also provided the background for the use of AR in safety inspection processes on construction sites and in offering methodological recommendations for the development and evaluation of these applications.

Chen et al. [94] explored the use of AR technology to facilitate fire safety equipment inspection and maintenance using mobile devices, overcoming the constraints imposed by paper files on these tasks. The demonstration and validation results showed that the proposed AR-based system provided highly comprehensive, mobile, and effective access to fire safety equipment information, facilitating the presentation of information in an immediate, visual, and convenient manner.

Codina et al. [95] presented an implementation of AR for emergency situations in smart buildings by means of indoor localization using sub-GHz beacons. The system should help emergency services that need to move quickly to rescue trapped people in situations where there is no visibility, or in large buildings such as hospitals, by generating a three-dimensional model of the building and facilitating navigation through it.

5.3. MR Applications

Mixed reality (MR) extends AR technology by enabling more direct interaction between reality and virtuality, allowing the user to manipulate virtual elements as they would in the real world with 3D digital content responding and reacting accordingly [129].

Microsoft HoloLens headsets are the most used MR devices in the construction sector as they are already certified as basic protection goggles (see Figure 6) and are available integrated into hard hats such as Trimble XR10 for personnel working in dirty, noisy, and safety-controlled environments. Trimble XR10 are also available with the Trimble HoloTint accessory visor, which is equipped with photochromic lenses for use outdoors or in brightly lit environments.

MR devices allow workers to visualize building plans, to make measurements on the field, to create 3D models from surrounding site scans, and enable technicians and designers to make modifications while on-site, proving effective in the following building

O&M applications: collaborative building management, maintenance and repair operations, renovations and retrofit works, personnel training, and workplace safety.

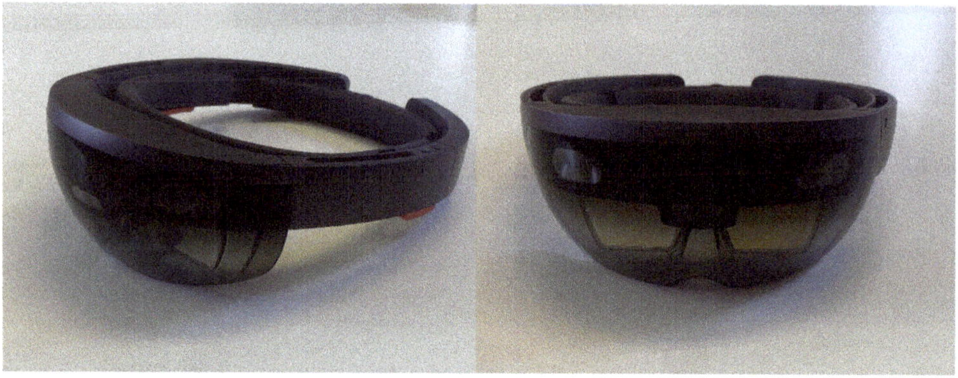

Figure 6. MS HoloLens MR headset.

5.3.1. Collaborative Augmented Building Monitoring and Management

As for AR, MR allows facility managers to enhance data visualization by displaying information right on field, including, for example, the visualization of real-time data from sensors, process parameters, or aggregated performance, therefore improving the monitoring, control, and maintenance of the whole building's components. The improved spatial awareness of MR allows the device to estimate the location and orientation of the user within the 3D BIM model or DT of the building, enabling first-person navigation aids (arrows, directions) inside the real-scale model with the possibility of displaying or interrogating all relevant data, including MEP systems, building components, and data from sensors and IoT devices.

Localization is performed via GNSS positioning and/or by comparing the user's perspective to BIM based on deep learning computation, commonly stream-processed in graphics processing unit (GPU)-enabled servers via transmission control protocol/Internet protocol (TCP/IP). Once the viewer is correctly positioned and oriented in the virtual space, spatial mapping visually fits the object of interest (superimposed text, elements-to-build, concealed pipes) onto the MR image [96]. MR navigation is also particularly useful to locate the target for maintenance and repair operations (electrical cabinets, air handling units, pipes manifolds) and guide the technician to the location even if they are unfamiliar with the premises.

A recent example of the application of MR in residential buildings is that of the Restart4Smart project by Sapienza University of Rome, which participated in the Solar Decathlon Middle East 2018, in which a digital twin of the building was created, and MS HoloLens was used on public tours to show visitors the structure and systems of the house, highlighting specific components such as pipes, tensioning cables, and technical devices [130].

Furthermore, MR wearable devices can enable a collaborative workspace between different professionals: as managing more complex buildings and infrastructure requires a great deal of collaboration between multiple teams, including designers, managers, contractors, users, tenants, supervisors, etc., a successful collaboration is needed at all times even if all the figures involved in a project are not always present at the workplace. MR allows the creation of collaborative environments where several people, who may even be in different places, can literally walk around and at the same time interact with a virtual 3D model as with a physical one actually present in the room, and show the design intent to stakeholders without needing an actual maquette (see Figure 7). Instead of waiting for all parties involved to assemble and review blueprints to make any changes, MR allows users to take notes and share video views of any suspected issues and send the information to remote teams in real time. On-site and remote team members are able to consult with each other and work with the information needed, greatly lowering the time and costs needed to make a decision among teams. MR collaborative environments such as WearCom further increase spatial faithfulness by using spatial cues such as virtual avatars to locate every observer speaking with each other in a conference.

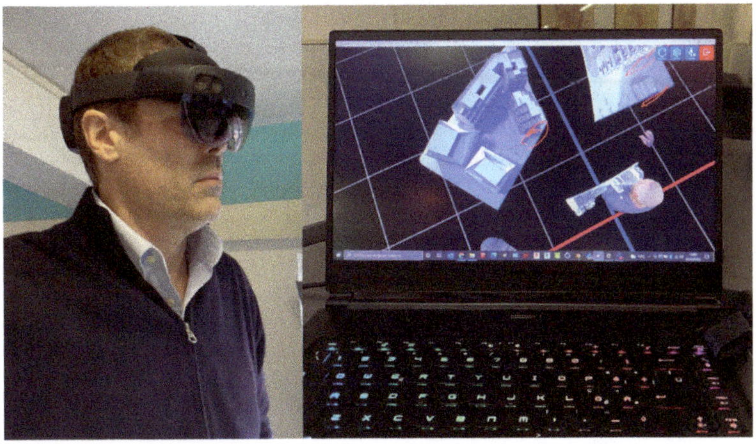

Figure 7. MR collaborative environment with MS HoloLens 2.

5.3.2. Collaborative Augmented Maintenance and Repair Operations

Thanks to the full integration between the virtual and real world, MR represents the best suited technology to support maintenance and repair operations in the building and infrastructure sector. Naticchia et al. [97] merged information from a BIM model with an MR environment to reflect the maintenance workers' benefits. Another MR case study was investigated by Ammari and Hammad [98], who proposed a framework to coordinate BIM and MR for supporting field tasks in the facility management domain.

MR can perform all activities that require AR integration and have been described above, including remote collaboration, with much improved precision and contextual interaction. In addition to visualizing maintenance-related data such as service requests, work order information, real-time IoT readings of the asset being serviced, documentation, or knowledge base items, MR can highlight parts and operations in their correct order (screws, switches, levers, valves, etc.) and virtually expand or explode the 3D image of the machine to visualize how it is built and all its components.

In particular, MR 3D models have been observed to be easier to understand than conventional paper documents, allowing for faster assembly and fewer errors made during construction. Furthermore, MR-assisted construction showed that participants with no previous assembly experience achieved the best times using MR, and they were also faster than the most experienced participants who used traditional paper plans [99], confirming the potential of MR to address the current shortage of experienced or skilled labor in the construction industry.

By using an MR integrated cooperation platform such as Microsoft Dynamics 365 Remote Assist, a remotely connected expert can see the live, first-person view taken by the HMD, freeze any scene of this video feed, and create annotations and 3D holograms, which are then integrated into the real-world view of the person wearing the device, anchored to their position irrespective of head movements [100].

By far, the most promising feature of MR is the ability to combine all digital data and documentation with one's physical view. Information, including actual or expected locations of walls, pipes, outlets, switches, and ventilation, is accessible directly on the site in levels that can be easily turned on and off. Having the precisely placed MEP overlays superimposed on the actual construction site can lead to subsequent work being completed with great precision. For example, contractors have been using robotic total stations and HoloLens to visualize a planned underground piping grid and mark the layout of required excavations accordingly directly on-site, thus minimizing unnecessary earth movements. Software such as BIM Holoview visualize infrastructure components where they will be constructed to evaluate and verify their correct positioning and can also be used as a post-construction tool to view existing infrastructure behind walls. The MR device can also overlay information on the work in progress, highlighting necessary tools and components, or overlaying visual guidelines for masonry or electrical installations [101], up to entire step-by-step assembly instructions.

The ability to accurately overlay BIM model data that would otherwise be hidden, such as built-in wall ducts or overhead ducts, allows spot interventions without wasting time and resources on probing. This is particularly useful for underground infrastructures, where excavation operations often present a high risk of inadvertently damaging the existing subsurface utilities, causing financial loss or even accidental injury [18]. A workflow proposed by Bentley Systems with the ContextCapture tool leverages real-world photographic reconstruction to generate a 3Dmesh in which 2D pipe maps are projected and aligned based on visible surface features such as manholes, valve access covers, and drains. On site, the pipe-augmented 3D mesh is geolocated with the physical world by selecting common control points (doors, lampposts), then turned off to show only the pipe augmentations in the aligned position. This way, it is possible to identify pipes and manholes to service and to trace precise excavation markers minimizing impact on the surrounding environment. At the same time, the user can visually identify discrepancies between the augmentation and the actual location of surface assets and propose changes to the pipe database by adjusting the virtual pipe's position directly on-site.

Regarding building maintenance with MR, ThyssenKrupp developed a HoloLens-based solution for elevator maintenance [131]. The application provides maintenance workers with an interactive UI and 3D holograms of the objects in need of fixing and offers live Skype calls for assistance (see Figure 8): the responder can visualize what the caller is seeing, making it possible to verify or assist the repair work in real time.

Figure 8. ThyssenKrupp MR solution for elevator maintenance.

5.3.3. Collaborative Augmented Renovations and Retrofit Works

In renovation planning, MR can be used to improve information retrieval from BIM models and thus reduce time and errors in related tasks. Indeed, most issues originate normally from poor coordination, incomplete or poor understanding by contractors, inability to detect errors or omissions early, or disconnect between stakeholders and construction teams. MR can mitigate all these problems greatly increasing overall productivity.

Using MR glasses, project team members can walk to the site of the future intervention and digitally mark problems they encounter, such as cracks or misplaced elements. Each "tag" is linked to a 3D model of the space and comes complete with built-in links to requests for information (RFIs), drawing details, additional images, vocal notes, or preset tags, allowing workers to use the same MR headset in a second moment to locate, identify, diagnose, track, and ultimately fix issues.

Furthermore, MR can be used to measure the physical properties of a space, i.e., width, length, and height. Contractors can integrate this information into 3D models of the building, allowing them to generate even more accurate structures and have a more comprehensive view of how the project is being built. Using MR headsets, workers will be able to automatically take measurements of the built components and compare them against the dimensions specified in the as-designed models to identify any inconsistencies in the structures and quickly resolve them to prevent delays or higher costs (see Figure 9).

Figure 9. Example of Trimble XR10 MR application on construction site.

Thanks to the interoperability between different MR and BIM systems, it is possible to automatically update BIM models and construction schedules from the MR device itself [17]. Workers can easily display interior and exterior views of a facility and make changes to virtual floor plans while keeping one view intact, such as removing or repositioning walls or components, or modifying the layout on their headset or mobile MR devices. This allows experts to resolve any errors in a virtual view before applying changes to the physical structure. As this digital data is continuously updated, it takes the guesswork out of any design changes while improving the workflow and preventing material waste. Retrieving BIM information also allows field workers to effectively monitor a project against its construction plan and ensure its successful completion. In addition to this useful information database, MR can allow users to virtually see the building's progress against its schedule, providing an additional level of project management. A Progress monitoring interface can provide color-coded overlays to easily identify sections of the construction site that are ahead or behind schedule [102]. Platforms such Fologram allow users to create interactive models that guide step-by-step construction in real-world environments. Fologram can also use QR codes or ArUco markers to manipulate models in space. For example, with wireframe models, these guides specify how to trace out the dimensions for molds to create building elements and where to place each brick in a wall with millimeter accuracy.

MR can also be used to build mockups that combine digital and physical elements by inserting the digital model over the real one. This is particularly useful in the case of interventions on an urban scale. For example, the City Planning application by HoloPundits can help to graphically visualize a blueprint as a completed project, enabling 3D walkthroughs, to adjust building spaces, to drag and drop props, and to replace the virtual buildings. Taqtile's HoloMaps is a VR/AR app used to navigate through a geographical holographic model by superimposing real-time data, integrating Microsoft Bing 3D with over 250 cities and landmarks as a 3D model of any size. Data source integration provides contextually relevant information (weather, real-time traffic, and Twitter feeds) that can be shared with team members or stakeholders locally or remotely, synchronously or asynchronously, alongside voice and text annotations, and MR photo and video screenshots.

5.3.4. Personnel Training

In O&M training, MR presents the great advantage of allowing the trainee to interact with real-world objects (also having tactile feedback) and to access virtual information at the same time, easily making the mapping between the real task and the instructions without the need to use separate external training materials such as user guides or manuals. According to Fitts' skill acquisition model, MR enables the trainee to learn the basics of the task by observing augmented instructions and to develop behavioral and movement patterns during repeated execution of the instructed tasks, increasing their skills right from initial rehearsals [103].

5.3.5. Workplace Safety

In addition to supporting education and training, MR device applications are also about their ability to boost hazard identification for field safety, improve risk recognition, and enhance real-time communication between managers and workers. The communication of safety information with MR can be enhanced by sharing video feeds between the on-site MR device and off-site PCs or tablets, enabling bi-directional communication between workers and safety coordinators regarding potential hazards, violations, and tips on site, and verbalized annotations and comments that can be tagged and georeferenced on their actual digital location [104]. As MR glasses such as HoloLens 2 continually create and update the 3D map of the environment, it is also possible to provide orientation and navigation of premises in a building or construction site with overlapping arrows, indicators, and information panels that can accurately align to rooms, doors, and equipment.

6. Conclusions

Numerous studies and fieldworks are demonstrating how XR technologies represent a very important tool to best exploit the possibilities offered by the digital transition of the AECO sector. Indeed, the AECO industry is faced with a confluence of trends and technologies that brings the virtual and physical worlds together to create a truly networked world in which intelligent objects communicate and interact with each other.

XR technologies, including VR, AR, and MR, allow immersive digital experiences to be created that permit users to easily visualize, explore, and understand designs, models, and site conditions with many benefits in terms of stakeholder engagement, design support and review, construction planning, progress monitoring, construction safety, and support to operations and management, as well as personnel training.

After providing a detailed review that analyzed, categorized, and summarized state-of-the-art XR technologies and their possible applications for building O&M along with their relative advantages and disadvantages, the article shows that XR can greatly improve building management and maintenance, allowing the optimization of building performance, cost-effectiveness, employee satisfaction, and productivity. VR, AR, and MR can be revolutionary in supporting designers, technicians, and facility managers in their activities thanks to the ability to provide immersive visualization of the asset and to superimpose instructions, sensor data, or technical schemes right in their field of view as well as allowing hands-free communication. Furthermore, integrated cooperation platforms can allow remote connected colleagues and experts to see the live, first-person view taken by the XR headset and collaborate actively with the users.

VR can provide a way to operate the facility remotely in an immersive environment and can be used to simulate alternative scenarios. AR has great potential to support building operation and management because it can provide useful information in context to site workers that operate and maintain the facilities. A combination of both technologies can support field and remote office workers at the same time and improve collaboration.

In particular, VR can support O&M activities through maintainability design, immersive building visualization and monitoring, human-centered building management, personnel training, and vehicles or robots' teleoperation. AR and MR can improve building performance and maintenance activities by enabling collaborative augmented building monitoring and management, collaborative augmented maintenance and repair operations, collaborative augmented renovations and retrofit works, personnel training, and workplace safety.

The main challenges facing the use of both AR and VR for facility management are the lack of integration with other facility management systems, the low accuracy and speed for updating information across several systems, and the difficulty to archive and revisit AR and VR experiences. Furthermore, XR headsets need to become less constrained by battery life, image rendering speeds need to improve, and siloed software platforms need the privacy controls and infrastructure support to integrate more seamlessly in the cloud. Interactions with the virtual space will become more intuitive, supporting gestures and eye-gaze to aid in sketching, prototyping, and animation. Geospatial tags that sync real-world location data with building models will be automated and achieve greater precision.

Today's market trends relating to the diffusion of XR, and the convergence with other sectors such as gaming and video entertainment, suggest a rapid resolution of the present barriers to the application of these technologies to O&M, currently represented by the still high costs of the devices (MR in particular), the level of digital alphabetization required, and the sparse adoption of BIM 7D systems in facility management. On the other hand, the type of technology and the required digital literacy could attract and develop new and more skilled and qualified talents, addressing the labor shortage problems typical of the sector.

Finally, given the novelty of the subject, this study allowed the identification of a lack of scientific literature on the measured costs and benefits deriving from the application of XR technologies in the field of O&M, thus identifying this limitation as a need for further development in this area of research and for a larger diffusion of immersive technologies in the AECO sector.

Funding: This research received no external funding.

Institutional Review Board Statement: Not applicable.

Informed Consent Statement: Not applicable.

Data Availability Statement: Not applicable.

Conflicts of Interest: The author declares no conflict of interest.

References

1. Casini, M. *Construction 4.0. Advanced Technology, Tools and Materials for the Digital Transformation of the Construction Industry*, 1st ed.; Woodhead Publishing Series in Civil and Structural Engineering; Elsevier: Cambridge, UK, 2022; p. 677.
2. Sawhney, A.; Riley, M.; Irizarry, J. (Eds.) *Construction 4.0 an Innovation Platform for the Built Environment*, 1st ed.; Routledge: New York, NY, USA, 2020.
3. Bolpagni, M.; Gavina, R.; Ribeiro, D. (Eds.) *Industry 4.0 for the Built Environment. Methodologies, Technologies and Skills*; Springer: Boston, MA, USA, 2022.
4. Autodesk. *Constructing with the Power of Digital*; Autodesk: San Rafael, CA, USA, 2016.
5. Autodesk. *Connected Construction: A better Way to Build, Together*; Autodesk: San Rafael, CA, USA, 2021.
6. Roland Berger. *Hot Trends in Construction. A New Era of Opportunities*; Roland Berger: Paris, France, 2018.

7. McKinsey. *Construction and Building Technology—Poised for a Breakthrough?* McKinsey & Company: New York, NY, USA, 2020.
8. WEF (World Economic Forum). *Shaping the Future of Construction. A Breakthrough in Mindset and Technology*; WEF: Cologny, Switzerland, 2016.
9. ECSO (European Construction Sector Observatory). *Digitalisation in the Construction Sector*; Analytical Report; ESCO: Brussels, Belgium, April 2021.
10. PEEB (Programme for Energy Efficiency in Buildings). *Smart and Efficient—Digital Solutions to Save Energy in Buildings*; PEEB: Paris, France, 2019.
11. JRC (Joint Research Centre Institute for Energy and Transport). *Digital Transformation in Transport, Construction, Energy, Government and Public Administration*; JRC: Brussels, Belgium, 2019.
12. JRC. *The Future of Cities*; JRC: Brussels, Belgium, 2019.
13. McKinsey. *Smart Cities: Digital Solutions for a More Livable Future*; McKinsey & Company: New York, NY, USA, 2018.
14. Opoku, D.-G.J.; Perera, S.; Osei-Kyei, R.; Rashidi, M. Digital twin application in the construction industry: A literature review. *J. Build. Eng.* **2021**, *40*, 102726. [CrossRef]
15. Jones, D.; Snider, C.; Nassehi, A.; Yon, J.; Hicks, B. Characterizing the digital twin: A systematic literature review. *CIRP J. Manuf. Sci. Technol.* **2020**, *29*, 36–52. [CrossRef]
16. Hong, T.; Wang, Z.; Luo, X.; Zhang, W. *State-of-the-Art on Research and Applications of Machine Learning in the Building Life Cycle*; Lawrence Berkeley National Laboratory: Berkeley, CA, USA, 2020.
17. Delgado, J.M.D.; Oyedele, L.; Demian, P.; Beach, T. A research agenda for augmented and virtual reality in architecture, engineering and construction. *Adv. Eng. Inform.* **2020**, *45*, 101122. [CrossRef]
18. Palmarini, R.; Erkoyuncu, J.A.; Roy, R.; Torabmostaedi, H. A systematic review of augmented reality applications in maintenance. *Robot. Comput. Integr. Manuf.* **2018**, *49*, 215–228. [CrossRef]
19. Statistics on Remote Workers That Will Surprise You. 2022. Available online: https://www.apollotechnical.com/statistics-on-remote-workers/ (accessed on 15 March 2022).
20. Alizadehsalehi, S.; Hadavi, A.; Huang, J.C. From BIM to extended reality in AEC industry. *Autom. Constr.* **2020**, *116*, 103254. [CrossRef]
21. Khan, A.; Sepasgozar, S.; Liu, T.; Yu, R. Integration of BIM and immersive technologies for AEC: A scientometric-SWOT analysis and critical content review. *Buildings* **2021**, *11*, 126. [CrossRef]
22. Sidani, A.; Matoseiro Dinis, F.; Duarte, J.; Sanhudo, L.; Calvetti, D.; Baptista, J.S.; Poças Martins, J.; Soeiro, A. Recent tools and techniques of BIM-Based Augmented Reality: A systematic review. *J. Build. Eng.* **2021**, *42*, 102500. [CrossRef]
23. Cheng, J.C.; Chen, K.; Chen, W. State-of-the-art review on mixed reality applications in the AECO industry. *J. Constr. Eng. Manag.* **2020**, *146*, 03119009. [CrossRef]
24. Delgado, J.M.D.; Oyedele, L.; Beach, T.; Demian, P. Augmented and Virtual Reality in Construction: Drivers and Limitations for Industry Adoption. *J. Constr. Eng. Manag.* **2020**, *146*, 04020079. [CrossRef]
25. Albahbah, M.; Kıvrak, S.; Arslan, G. Application areas of augmented reality and virtual reality in construction project management: A scoping review. *J. Constr. Eng.* **2021**, *14*, 151–172. [CrossRef]
26. Zhu, Y.; Li, N. Virtual and augmented reality technologies for emergency management in the built environments: A state-of-the-art review. *J. Saf. Sci. Resil.* **2021**, *2*, 1–10. [CrossRef]
27. Li, X.; Yi, W.; Chi, H.-L.; Wang, X.; Chan, A.P. A critical review of virtual and augmented reality (VR/AR) applications in construction safety. *Autom. Constr.* **2018**, *86*, 150–162. [CrossRef]
28. Coupry, C.; Noblecourt, S.; Richard, P.; Baudry, D.; Bigaud, D. BIM-Based Digital Twin and XR Devices to Improve Maintenance Procedures in Smart Buildings: A Literature Review. *Appl. Sci.* **2021**, *11*, 6810. [CrossRef]
29. Noghabaei, M.; Heydarian, A.; Balali, V.; Han, K. Trend Analysis on Adoption of Virtual and Augmented Reality in the Architecture, Engineering, and Construction Industry. *Data* **2020**, *5*, 26. [CrossRef]
30. Prabhakaran, A.; Mahamadu, A.-M.; Mahdjoubi, L. Understanding the challenges of immersive technology use in the architecture and construction industry: A systematic review. *Autom. Constr.* **2022**, *137*, 104228. [CrossRef]
31. Zhang, Y.; Liu, H.; Kanga, S.-C.; Al-Hussein, M. Virtual reality applications for the built environment: Research trends and opportunities. *Autom. Constr.* **2020**, *118*, 103311. [CrossRef]
32. Alizadehsalehi, S.; Hadavi, A.; Huang, J.C. *Virtual Reality for Design and Construction Education Environment, AEI 2019*; American Society of Civil Engineers: Reston, VA, USA, 2019; pp. 193–203.
33. Sutcliffe, A.G.; Poullis, C.; Gregoriades, A.; Katsouri, I.; Tzanavari, A.; Herakleous, K. Reflecting on the design process for virtual reality applications. *Int. J. Hum.-Comput. Interact.* **2019**, *35*, 168–179. [CrossRef]
34. Romano, S.; Capece, N.; Erra, U.; Scanniello, G.; Lanza, M. On the use of virtual reality in software visualization: The case of the city metaphor. *Inf. Softw. Technol.* **2019**, *114*, 92–106. [CrossRef]
35. Scorpio, M.; Laffi, R.; Masullo, M.; Giovanni Ciampi, G.; Rosato, A.; Maffei, L.; Sibilio, S. Virtual Reality for Smart Urban Lighting Design: Review, Applications and Opportunities. *Energies* **2020**, *13*, 3809. [CrossRef]
36. Krupiński, R. Virtual Reality System and Scientific Visualisation for Smart Designing and Evaluating of Lighting. *Energies* **2020**, *13*, 5518. [CrossRef]
37. Bille, R.; Smith, S.P.; Maund, K.; Brewer, G. Extending building information models into game engines. In Proceedings of the 2014 Conference on Interactive Entertainment, Newcastle, Australia, 2–3 December 2014; pp. 1–8.

38. Du, J.; Zou, Z.; Shi, Y.; Zhao, D. Zero latency: Real-time synchronization of BIM data in virtual reality for collaborative decision-making. *Autom. Constr.* **2018**, *85*, 51–64. [CrossRef]
39. Russo, M. AR in the Architecture Domain: State of the Art. *Appl. Sci.* **2021**, *11*, 6800. [CrossRef]
40. Lee, J.G.; Seo, J.O.; Abbas, A.; Choi, M. End-Users' Augmented Reality Utilization for Architectural Design Review. *Appl. Sci.* **2020**, *10*, 5363. [CrossRef]
41. Song, Y.; Koeck, R.; Luo, S. Review and analysis of augmented reality (AR) literature for digital fabrication in architecture. *Autom. Constr.* **2021**, *128*, 103762. [CrossRef]
42. Luna, U.; Rivero, P.; Vicent, N. Augmented Reality in Heritage Apps: Current Trends in Europe. *Appl. Sci.* **2019**, *9*, 2756. [CrossRef]
43. Tom Dieck, M.C.; Hyungsoo Jung, T. Value of augmented reality at cultural heritage sites: A stakeholder approach. *J. Destin. Mark. Manag.* **2017**, *6*, 110–117. [CrossRef]
44. Bozzelli, G.; Raia, A.; Ricciardi, S.; De Nino, M.; Barile, N.; Perrella, M.; Tramontano, M.; Pagano, A.; Palombini, A. An integrated VR/AR framework for user-centric interactive experience of cultural heritage: The ArkaeVision project. *Digit. Appl. Archaeol. Cult. Herit.* **2019**, *15*, e00124. [CrossRef]
45. Margetis, G.; Apostolakis, K.C.; Ntoa, S.; Papagiannakis, G.; Stephanidis, C. X-Reality Museums: Unifying the Virtual and RealWorld Towards Realistic Virtual Museums. *Appl. Sci.* **2021**, *11*, 338. [CrossRef]
46. Carrasco, M.D.O.; Chen, P.-H. Application of mixed reality for improving architectural design comprehension effectiveness. *Autom. Constr.* **2021**, *126*, 103677. [CrossRef]
47. Han, Y.; Yang, J.; Yongsheng, D.; Jin, R.; Guo, B.; Adamu, Z. Process and outcome-based evaluation between virtual really-driven and traditional construction safety training. *Adv. Eng. Inform.* **2022**, *52*, 101634. [CrossRef]
48. Fu, M.; Liu, R. The application of virtual reality and augmented reality in dealing with project schedule risks. In *Construction Research Congress 2018: Construction Information Technology, Proceedings of the Construction Research Congress 2018, New Orleans, LA, USA, 2–4 April 2018*; Wang, C., Harper, C., Lee, Y., Harris, R., Berryman, C., Eds.; American Society of Civil Engineers 1801 Alexander Bell Drive: Reston, VA, USA, 2018; pp. 429–438.
49. Muhammad, A.A.; Yitmen, I.; Alizadehsalehi, S.; Celik, T. Adoption of virtual reality (VR) for site layout optimization of construction projects. *Tek. Dergi* **2019**, *31*, 9833–9850. [CrossRef]
50. Rankohi, S.; Waugh, L. Review and analysis of augmented reality literature for construction industry. *Vis. Eng.* **2013**, *1*, 9. [CrossRef]
51. Shin, D.H.; Dunston, P.S. Identification of application areas for augmented reality in industrial construction based on technology suitability. *Autom. Constr.* **2008**, *17*, 882–894. [CrossRef]
52. Behzadan, A.H.; Dong, S.; Kamat, V.R. Augmented reality visualization: A review of civil infrastructure system applications. *Adv. Eng. Inform.* **2015**, *29*, 252–267. [CrossRef]
53. Kwiatek, C.; Sharif, M.; Li, S.; Haas, C.; Walbridge, S. Impact of augmented reality and spatial cognition on assembly in construction. *Autom. Constr.* **2019**, *108*, 102935. [CrossRef]
54. Röltgen, D.; Dumitrescu, R. Classification of industrial Augmented Reality use cases. *Procedia CIRP* **2020**, *91*, 93–100. [CrossRef]
55. Danielsson, O.; Holm, M.; Syberfeldt, A. Augmented reality smart glasses in industrial assembly: Current status and future challenges. *J. Ind. Inf. Integr.* **2020**, *20*, 100175. [CrossRef]
56. Kim, D.; Choi, Y. Applications of Smart Glasses in Applied Sciences: A Systematic Review. *Appl. Sci.* **2021**, *11*, 4956. [CrossRef]
57. Muthalif, M.Z.A.; Shojaei, D.; Khoshelham, K. A review of augmented reality visualization methods for subsurface utilities. *Adv. Eng. Inform.* **2022**, *51*, 101498. [CrossRef]
58. Park, S.; Bokijonov, S.; Choi, Y. Review of Microsoft HoloLens Applications over the Past Five Years. *Appl. Sci.* **2021**, *11*, 7259. [CrossRef]
59. Krystyna, A. Digital technologies in facility management—The state of practice and research challenges. *Procedia Eng.* **2017**, *196*, 1034–1042.
60. Silvestri, L.; Forcina, A.; Introna, V.; Santolamazza, A.; Cesarotti, V. Maintenance transformation through Industry 4.0 technologies: A systematic literature review. *Comput. Ind.* **2020**, *123*, 103335. [CrossRef]
61. Wong, J.K.W.; Ge, J.; He, S.X. Digitisation in facilities management: A literature review and future research directions. *Autom. Constr.* **2018**, *92*, 312–326. [CrossRef]
62. Guo, Z.; Zhou, D.; Chen, J.; Geng, J.; Lv, C.; Zeng, S. Using virtual reality to support the product's maintainability design: Immersive maintainability verification and evaluation system. *Comput. Ind.* **2018**, *101*, 41–50. [CrossRef]
63. Guo, Z.; Zhou, D.; Zhou, Q.; Zhang, X.; Geng, J.; Zeng, S.; Lv, C.; Hao, A. Applications of virtual reality in maintenance during the industrial product lifecycle: A systematic review. *J. Manuf. Syst.* **2020**, *56*, 525–538. [CrossRef]
64. Akanmu, A.A.; Olayiwola, J.; Olatunji, O.A. Automated checking of building component accessibility for maintenance. *Autom. Constr.* **2020**, *114*, 103196. [CrossRef]
65. Lu, Q.; Xie, X.; Parlikad, A.K.; Schooling, J.M. Digital twin-enabled anomaly detection for built asset monitoring in operation and maintenance. *Autom. Constr.* **2020**, *118*, 103277. [CrossRef]
66. Shahinmoghadam, M.; Natephra, W.; Motamedi, A. BIM- and IoT-based virtual reality tool for real-time thermal comfort assessment in building enclosures. *Build. Environ.* **2021**, *199*, 107905. [CrossRef]

67. Deng, T.; Zhang, K.; Shen, Z.-J.M. A systematic review of a digital twin city: A new pattern of urban governance toward smart cities. *J. Manag. Sci. Eng.* **2021**, *6*, 125–134. [CrossRef]
68. Yan, J.; Kensek, K.; Konis, K.; Noble, D. CFD Visualization in a Virtual Reality Environment Using Building Information Modeling Tools. *Buildings* **2020**, *10*, 229. [CrossRef]
69. Heydarian, A.; Carneiro, J.P.; Gerber, D.; Becerik-Gerber, B.; Hayes, T.; Wood, W. Immersive virtual environments versus physical built environments: A benchmarking study for building design and user-built environment explorations. *Autom. Constr.* **2015**, *54*, 116–126. [CrossRef]
70. Niu, S.; Pan, W.; Zhao, Y. A virtual reality integrated design approach to improving occupancy information integrity for closing the building energy performance gap. *Sustain. Cities Soc.* **2016**, *27*, 275–286. [CrossRef]
71. Lovreglio, R.; Gonzalez, V.; Feng, Z.; Amor, R.; Spearpoint, M.; Thomas, J.; Trotter, M.; Sacks, R. Prototyping virtual reality serious games for building earthquake preparedness: The Auckland City Hospital case study. *Adv. Eng. Inform.* **2018**, *38*, 670–682. [CrossRef]
72. Kinateder, M.; Warren, W.H.; Schloss, K.B. What color are emergency exit signs? Egress behavior differs from verbal report. *Appl. Ergon.* **2019**, *75*, 155–160. [PubMed]
73. Zou, H.; Li, N.; Cao, L. Emotional response-based approach for assessing the sense of presence of subjects in virtual building evacuation studies. *J. Comput. Civ. Eng.* **2017**, *31*, 04017028. [CrossRef]
74. Lorusso, P.; De Iuliis, M.; Marasco, S.; Domaneschi, M.; Cimellaro, G.P.; Villa, V. Fire Emergency Evacuation from a School Building Using an Evolutionary Virtual Reality Platform. *Buildings* **2022**, *12*, 223. [CrossRef]
75. Ergan, S.; Radwan, A.; Zou, Z.; Tseng, H.; Han, X. Quantifying human experience in architectural spaces with integrated virtual reality and body sensor networks. *J. Comput. Civ. Eng.* **2019**, *33*, 04018062. [CrossRef]
76. Erkan, İ. Examining wayfinding behaviours in architectural spaces using brain imaging with electroencephalography (EEG). *Archit. Sci. Rev.* **2018**, *61*, 410–428. [CrossRef]
77. Pavón, R.M.; Arcos Alvarez, A.A.; Alberti, M.G. Possibilities of BIM-FM for the Management of COVID in Public Buildings. *Sustainability* **2020**, *12*, 9974. [CrossRef]
78. Mukhopadhyay, A.; Reddy, G.S.R.; Saluja, K.S.; Ghosh, S.; PEÑA-RIOS, A.; Gopal, G.; Biswas, P. Virtual-reality-based digital twin of office spaces with social distance measurement feature. *Virtual Real. Intell. Hardw.* **2022**, *4*, 55–75. [CrossRef]
79. Song, H.; Kim, T.; Kim, J.; Ahn, D.; Kang, Y. Effectiveness of VR crane training with head-mounted display: Double mediation of presence and perceived usefulness. *Autom. Constr.* **2021**, *122*, 103506. [CrossRef]
80. Shi, Y.; Du, J.; Worthy, D.A. The impact of engineering information formats on learning and execution of construction operations: A virtual reality pipe maintenance experiment. *Autom. Constr.* **2020**, *119*, 103367. [CrossRef]
81. Oosterhout, J.V.; Heemskerk, C.J.M.; Boessenkool, H.; Baar, M.R.d.; Helm, F.C.T.v.d.; Abbink, D.A. Haptic assistance improves tele-manipulation with two asymmetric slaves. *IEEE Trans. Haptics* **2019**, *12*, 141–153. [CrossRef] [PubMed]
82. Moniruzzaman, M.D.; Rassau, A.; Chai, D.; Islam, S.M.S. Teleoperation methods and enhancement techniques for mobile robots: A comprehensive survey. *Robot. Auton. Syst.* **2022**, *150*, 103973. [CrossRef]
83. Meža, S.; Turk, Ž.; Dolenc, M. Component based engineering of a mobile BIM-based augmented reality system. *Autom. Constr.* **2014**, *42*, 1–12. [CrossRef]
84. Chung, S.; Cho, C.-S.; Song, J.; Lee, K.; Lee, S.; Kwon, S. Smart Facility Management System Based on Open BIM and Augmented Reality Technology. *Appl. Sci.* **2021**, *11*, 10283. [CrossRef]
85. Abramovici, M.; Wolf, M.; Adwernat, S.; Neges, M. Context-aware maintenance support for augmented reality assistance and synchronous multi-user collaboration. *Procedia CIRP* **2017**, *59*, 18–22. [CrossRef]
86. Chu, M.; Matthews, J.; Love, P.E. Integrating mobile building information modelling and augmented reality systems: An experimental study. *Autom. Constr.* **2018**, *85*, 305–316. [CrossRef]
87. Alonso-Rosa, M.; Gil-de-Castro, A.; Moreno-Munoz, A.; Garrido-Zafra, J.; Gutierrez-Ballesteros, E.; Cañete-Carmona, E. An IoT Based Mobile Augmented Reality Application for Energy Visualization in Buildings Environments. *Appl. Sci.* **2020**, *10*, 600. [CrossRef]
88. Fukuda, T.; Yokoi, K.; Yabuki, N.; Motamedi, A. An indoor thermal environment design system for renovation using augmented reality. *J. Comput. Des. Eng.* **2019**, *6*, 179–188. [CrossRef]
89. Lin, J.-R.; Cao, J.; Zhang, J.-P.; van Treeck, C.; Frisch, J. Visualization of indoor thermal environment on mobile devices based on augmented reality and computational fluid dynamics. *Autom. Constr.* **2019**, *103*, 26–40. [CrossRef]
90. Li, W.; Nee, A.; Ong, S. A State-of-the-Art Review of Augmented Reality in Engineering Analysis and Simulation. *Multimodal Technol. Interact.* **2017**, *1*, 17. [CrossRef]
91. Manuri, F.; Pizzigalli, A.; Sanna, A. A State Validation System for Augmented Reality Based Maintenance Procedures. *Appl. Sci.* **2019**, *9*, 2115. [CrossRef]
92. Mourtzis, D.; Siatras, V.; Angelopoulos, J. Real-Time Remote Maintenance Support Based on Augmented Reality (AR). *Appl. Sci.* **2020**, *10*, 1855. [CrossRef]
93. Ramos-Hurtado, J.; Muñoz-La Rivera, F.; Mora-Serrano, J.; Deraemaeker, A.; Valero, I. Proposal for the Deployment of an Augmented Reality Tool for Construction Safety Inspection. *Buildings* **2022**, *12*, 500. [CrossRef]
94. Chen, Y.-J.; Lai, Y.-S.; Lin, Y.-H. BIM-based augmented reality inspection and maintenance of fire safety equipment. *Autom. Constr.* **2020**, *110*, 103041. [CrossRef]

95. Codina, M.; Castells-Rufas, D.; Carrabina, J.; Salmon, I.; Ayuso, N.; Alfonso Guerendiain, A.; Alvarez, G. Augmented Reality for Emergency Situations in Buildings with the Support of Indoor Localization. *Proceedings* **2019**, *31*, 76.
96. Baek, F.; Ha, I.; Kim, H. Augmented reality system for facility management using image-based indoor localization. *Autom. Constr.* **2019**, *99*, 18–26. [CrossRef]
97. Naticchia, B.; Corneli, A.; Carbonari, A.; Bonci, A.; Pirani, M. Mixed reality approach for the management of building maintenance and operation. In Proceedings of the ISARC 2018—35th International Symposium on Automation and Robotics in Construction and International AEC/FM Hackathon: The Future of Building Things, Berlin, Germany, 20–25 July 2018.
98. El Ammari, K.; Hammad, A. Remote interactive collaboration in facilities management using BIM-based mixed reality. *Autom. Constr.* **2019**, *107*, 102940. [CrossRef]
99. Chalhoub, J.; Ayer, S.K. Using mixed reality for electrical construction design communication. *Autom. Constr.* **2018**, *86*, 1–10. [CrossRef]
100. Vorraber, W.; Gasser, J.; Webb, H.; Neubacher, D.; Url, P. Assessing augmented reality in production: Remote-assisted maintenance with HoloLens. *Procedia CIRP* **2020**, *88*, 139–144. [CrossRef]
101. Fazel, A.I. An interactive augmented reality tool for constructing free-form modular surfaces. *Autom. Constr.* **2018**, *85*, 135–145. [CrossRef]
102. Golparvar-Fard, M.; Peña-Mora, F.; Savarese, S. D4ar: A 4-dimensional augmented reality model for automating construction progress monitoring data collection, processing and communication. *J. Inf. Technol. Constr.* **2009**, *14*, 129–153.
103. Webel, S.; Bockholt, U.; Engelke, T.; Peveri, M.; Preusche, M.O.C. Augmented reality training for assembly and maintenance skills. *BIO Web Conf.* **2011**, *1*, 00097. [CrossRef]
104. Dai, F.; Olorunfemi, A.; Peng, W.; Cao, D.; Luo, X. Can mixed reality enhance safety communication on construction sites? An industry perspective. *Saf. Sci.* **2021**, *133*, 105009.
105. VRcompare. Available online: https://vr-compare.com/ (accessed on 15 March 2022).
106. Infinite. Available online: https://www.infinite.cz/projects/HMD-tester-virtual-reality-headset-database-utility (accessed on 15 March 2022).
107. Sourceforge. Available online: https://sourceforge.net/software/vr/windows/ (accessed on 15 March 2022).
108. Capaterra. Available online: https://www.capterra.com/augmented-reality-software/s/windows/ (accessed on 15 March 2022).
109. Rauschnabel, P.A.; Felix, R.; Hinsch, C.; Shahab, H.; Alt, F. What is XR? Towards a Framework for Augmented and Virtual Reality. *Comput. Hum. Behav.* **2022**, *133*, 107289. [CrossRef]
110. Milgram, P.; Colquhoun, H. A taxonomy of real and virtual world display integration. *Mix. Reality. Merging Real Virtual Worlds* **1999**, *1*, 1–26.
111. Mordor Intelligence. *Extended Reality (XR) Market-Growth, Trends, COVID-19 Impact, and Forecasts (2022–2027)*; Mordor Intelligence: Telangana, India, 2022.
112. International Data Corporation (IDC). *Worldwide Quarterly Augmented and Virtual Reality Headset Tracker 2021 Q4*; IDC: Needham, MA, USA, 2022.
113. Panetta, K. *Gartner Top 10 Strategic Technology Trends for 2019*; Gartner: Stamford, CT, USA, 2018; Available online: https://www.gartner.com/smarterwithgartner/gartner-top-10-strategic-technology-trends-for-2019 (accessed on 24 June 2021).
114. Rokhsaritalemi, S.; Sadeghi-Niaraki, A.; Choi, S.-M. A Review on Mixed Reality: Current Trends, Challenges and Prospects. *Appl. Sci.* **2020**, *10*, 636. [CrossRef]
115. De Regt, A.; Barnes, S.J.; Plangger, K. The virtual reality value chain. *Bus. Horiz.* **2020**, *63*, 737–748. [CrossRef]
116. Kloiber, S.; Settgast, V.; Schinko, C.; Weinzerl, M.; Fritz, J.; Tobias Schreck, T.; Preiner, R. Immersive analysis of user motion in VR applications. *Vis. Comput.* **2020**, *36*, 1937–1949. [CrossRef]
117. Rondeau, E.P.; Brown, R.K.; Lapides, P.D. *Facility Management*, 2nd ed.; Wiley: New York, NY, USA, 2006.
118. Lewis, B.T. *Facility Manager's Operation and Maintenance Handbook*, 1st ed.; McGraw-Hill: New York, NY, USA, 1999.
119. Roper, K.O.; Payant, R.P. *The Facility Management Handbook*, 4th ed.; AMACOM Division of American Management Association: New York, NY, USA, 2014.
120. Aziz, N.D.; Nawawi, A.H.; Ariff, N.R.M. Building information modelling (BIM) in facilities management: Opportunities to be considered by facility managers. *Procedia—Soc. Behav. Sci.* **2016**, *234*, 353–362. [CrossRef]
121. Gao, X.; Pishdad-Bozorgi, P. BIM-enabled facilities operation and maintenance: A review. *Adv. Eng. Inform.* **2019**, *39*, 227–247. [CrossRef]
122. Matarneh, S.T.; Danso-Amoako, M.; Al-Bizri, S.; Gaterell, M.; Matarneh, R. Building information modeling for facilities management: A literature review and future research directions. *J. Build. Eng.* **2019**, *24*, 100755. [CrossRef]
123. Errandonea, I.; Beltra'n, S.; Arrizabalaga, S. Digital twin for maintenance: A literature review. *Comput. Ind.* **2020**, *123*, 103316. [CrossRef]
124. Ozturk, G.B. Digital twin research in the AECO-FM industry. *J. Build. Eng.* **2021**, *40*, 102730. [CrossRef]
125. Pan, Y.; Zhang, L. Roles of artificial intelligence in construction engineering and management: A critical review and future trends. *Autom. Constr.* **2021**, *122*, 103517. [CrossRef]
126. Love, P.E.D.; Matthews, J. The 'how' of benefits management for digital technology: From engineering to asset management. *Autom. Constr.* **2019**, *107*, 102930. [CrossRef]
127. Arup. *Digital Twin. Towards a Meaningful Framework*; Arup: London, UK, 2019.

128. Russom, M.B. Connecting Urban Environments with IoT and Digital Twins. Microsoft Azure. 2020. Available online: https://azure.microsoft.com/it-it/blog/connecting-urban-environments-with-iot-and-digital-twins/ (accessed on 25 March 2021).
129. Marr, B. *Tech Trends in Practice: The 25 Technologies that are Driving the 4th Industrial Revolution*, 1st ed.; John Wiley & Sons: Hoboken, NJ, USA, 2020.
130. Casini, M. A positive energy building for the middle east climate: ReStart4Smart solar house at solar decathlon middle east 2018. *Renew. Energy* **2020**, *159*, 1269–1296. [CrossRef]
131. Thyssenkrupp. Available online: https://www.thyssenkrupp.com/en/newsroom/press-releases/thyssenkrupp-unveils-latest-technology-to-transform-the-global-elevator-service-industry--microsoft-hololens--for-enhancing-interventions--1567.html (accessed on 15 March 2022).

energies

Article

Creating Comparability among European Neighbourhoods to Enable the Transition of District Energy Infrastructures towards Positive Energy Districts

Axel Bruck [1,2,*,†], Luca Casamassima [2,†], Ardak Akhatova [2,†], Lukas Kranzl [2] and Kostas Galanakis [3]

1 Departamento Energías Renovables, Instituto Tecnológico de Canarias, Playa de Pozo Izquierdo s/n, 35119 Santa Lucía de Tirajana , Spain
2 Energy Economics Group, Technische Universität Wien, Karlsplatz 13, 1040 Vienna, Austria; casamassima@eeg.tuwien.ac.at (L.C.); akhatova@eeg.tuwien.ac.at (A.A.); kranzl@eeg.tuwien.ac.at (L.K.)
3 Nottingham Business School, Nottingham Trent University, 50 Shakespeare Street, Nottingham NG1 4QU, UK; kostas.galanakis@ntu.ac.uk
* Correspondence: abruck@itccanarias.org
† These authors contributed equally to this work.

Abstract: Planning the required energy infrastructure for the energy transition is a crucial task for various neighbourhood concepts, such as positive energy districts. However, energy planning often comes with the challenges of data shortages and a lack of comparability among solutions for different districts. This work aims to enable this comparability by introducing an approach for categorising districts according to parameters that are relevant for the planning of neighbourhood energy infrastructures. Four parameters (climate, floor space index, heating demand and share of residential buildings) and their respective ranges (bands) were derived from the literature. Additionally, this work visualised the combination of all parameter bands across Europe to conveniently showcase districts that are comparable according to the selected parameters. This approach and its visualisation could be used in urban planning to share knowledge from existing energy district projects with those planned in comparable districts.

Keywords: positive energy district; district energy infrastructure; decarbonisation of neighbourhoods; GIS; energy transition

Citation: Bruck, A.; Casamassima, L.; Akhatova, A.; Kranzl, L.; Galanakis, K. Creating Comparability among European Neighbourhoods to Enable the Transition of District Energy Infrastructures towards Positive Energy Districts. *Energies* **2022**, *15*, 4720. https://doi.org/10.3390/en15134720

Academic Editors: Andrea Mauri and Benedetto Nastasi

Received: 17 May 2022
Accepted: 18 June 2022
Published: 28 June 2022

Publisher's Note: MDPI stays neutral with regard to jurisdictional claims in published maps and institutional affiliations.

1. Introduction

Cities are responsible for about 70% of global CO_2 emissions, to which the largest contributor is the use of fossil fuels for buildings and transportation [1]. The concept of positive energy districts (PEDs), which are districts or neighbourhoods that have net-zero carbon emissions and positive annual energy balances, was proposed by the Strategic Energy Technology Plan (SET-Plan) Action 3.2 in 2018 as a cornerstone for the creation of carbon-neutral cities in Europe [2]. This action plan has led to various other initiatives across Europe (e.g., Making City [3], +cityXchange [4], IEA EBC Annex 83 (International Energy Agency's Energy in Buildings and Communities programme) [5] and Pocityf [6]). In academia, PED-based research has also been increasing [7–9]. PEDs have been developed using different approaches and techniques from different actors. According to the Joint Programming Initiative (JPI) Urban Europe [10], there were two operating PEDs at the beginning of 2020, with 19 others in the implementation stage and 8 in the planning stage. Furthermore, 32 other projects have not declared ambitions to become PEDs but present characteristics of interest for PED development [10]. These projects span across Europe and provide cases for a wide variety of climates, political contexts, social contexts and national and local energy production mixes and infrastructure. These cases generate plenty of data that other local planners could potentially use for initial assessments and replications. In this context, comparability and replicability are crucial.

The transition towards positive energy districts needs appropriate infrastructure, as defined in Section 2. However, the kind of infrastructure and its capacity in terms of electricity or heating and cooling depend on characteristics that vary across Europe. For example, climate attributes, such as solar irradiation and temperature, affect the potential for photovoltaic (PV) generation, as well as heating and cooling systems. Many district energy system transformation projects rely on case studies for data specific to their locations [3–6]. However, it is not easy to know whether an analysis made for one area would apply to another area within the EU. Therefore, our work aimed to develop a methodology for classifying European districts based on parameters that are helpful in the initial evaluation and planning phase of energy infrastructures that could lead to PEDs.

The method developed in this study clusters existing neighbourhoods within Europe according to key categorisation parameters. This approach can provide benefits for both practice and theory. For example, categorising districts according to their energy infrastructure requirements is of practical use for regional decision-making in the initial stages of planning district energy infrastructures. It simplifies comparing solutions and approaches from other districts that fall into the same district category. The categorisation also supports learning from challenges and opportunities that arose from existing or studied areas, for example, in the planning stage. Furthermore, this categorisation could be of particular interest to academia within the field of energy modelling. The accessibility of data can be a significant burden for energy modelling-related research. Using the presented categorisation approach could justify using data relevant to the infrastructure requirements of another district with similar characteristics.

The remainder of the article begins by narrowing down the scope of the categorisation and introducing important background information in Section 2. Subsequently, Section 3 introduces the overall methodology of the work. The results are presented in Section 4, which also shows the procedure for identifying the relevant parameters for the district categorisation for infrastructure comparability, as well as their ranges (Section 4.1). This is followed by the application of this method to the creation of comparable districts using visualisation maps of different parameter combinations (Section 5). The work is finalised by an elaboration on the application, limitations and prospects of this method in Section 6.

2. Theoretical Background

The positive energy goal, i.e., positive renewable annual energy balance, is barely achievable without the district energy infrastructures undergoing a major transformation [11]. Building on the works of Fulmer et al. [12] and Brozovsky et al. [13], this study defined a district energy infrastructure as the physical components of building and energy infrastructure systems (i.e., heating networks, electricity grids, generation, etc.) that provide commodities and services (e.g., hot water, electricity, etc.) that are essential for enabling, sustaining and enhancing societal living conditions.

According to JPI Urban Europe [2], each PED is supposed to find its own optimal balance between three main elements: the energy efficiency of the infrastructure, local renewable energy production and energy flexibility within the district (see Figure 1). These three elements are relevant to different parts of a district's energy infrastructure, e.g., energy efficiency primarily relates to building envelopes and heating and cooling systems. Thermal insulation, more efficient decentralised boilers and district heating systems are crucial in colder climates. Buildings in warmer climates require insulation and ventilation to reduce cooling demands during hot summer periods. Once the energy efficiency limit is achieved, local energy supply from renewable sources, such as PV or wind power, is deployed to cover the local energy demands. Finally, the flexibility of an energy system can be provided by storage technologies and emerging services, such as dynamic charging for battery electric vehicles (BEV). Although this example has advanced and grown in importance over recent years, this work did not focus on BEVs and their charging. Smart charging and vehicle-to-grid technology are not yet widely implemented. Furthermore, readers can infer relevant information regarding mobility from some of the parameters selected for the classification efforts (Section 4), i.e., the residential share of the total gross floor area and the floor space index. Hence, necessary transformations are the measures (and their ambitions) that need to be implemented to achieve each "sub-goal" (i.e., element). For example, the "transformation" of a building envelope (e.g., insulating walls) or heating system (e.g., installing heat pumps) is necessary for energy efficiency, as energy efficiency is defined by the conditions or characteristics of those infrastructures.

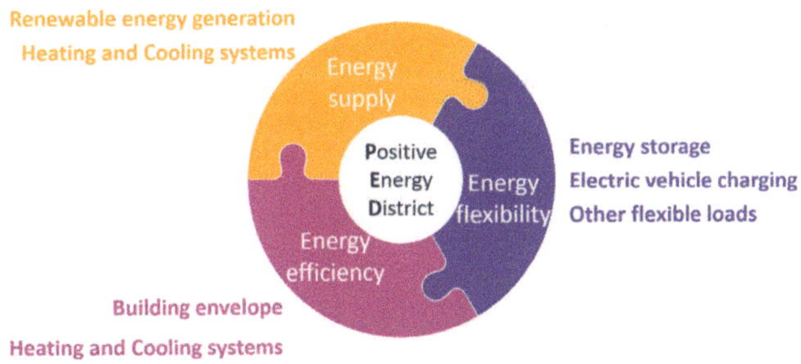

Figure 1. The key concepts/elements of positive energy districts (adapted from [2]).

Hence, for the purpose of this study, a PED infrastructure encompasses the following technology: renewable energy generation, energy storage, charging technology, building envelopes and heating and cooling systems. Each technology also has parametric and installed capacity requirements. Parametric requirements refer to, for example, the supply temperature needed for space heating. Finally, the installed capacity indicates the size of each system, e.g., the installed capacity of renewable generation technologies or the necessary capacity of district heating generation. Such parametric and capacity requirements are usually estimated using energy models [14].

Previously, several studies have attempted to analyse infrastructure requirements by creating district typologies. A German study presented a method for estimating infrastructure costs (for regional spatial planning) based on the structural type of the settlement [15]. Such typology-based approaches are based on the assumption that urban areas with similar building typologies and urban structures are similar in terms of infrastructure configurations and demands. Another approach for creating a district typology was proposed by [16] for estimating heat demands and thermal gains (and losses) within a district. The classification of districts was based on the building typologies and urban forms; thus, parameters such as building shape, district density and building age were used to categorise the districts. Other studies have not created archetypes but have instead analysed the urban characteristics of districts that could help create a sustainable neighbourhood [17,18]. As observed, district categorisation is highly dependent on the location; hence, it is usually not meant to be applied outside its original scope and region.

3. Methodology

This section presents the approach taken in this work to develop the district categorisation method, which was based on infrastructure requirements. This work incorporated three stages, as shown in Figure 2.

First, based on a critical review of the background literature, this work selected district-level energy models and identified the commonly used input parameters from those selected models. Appendix A shows the list of selected models, and information relevant to this study. The models covered the analysis and planning of the energy infrastructures, as discussed in Section 2. The next step was synthesising the findings to define the parameters for district categorisation according to energy infrastructures. Defining the approximate ranges of the values for each parameter was based on the relevant literature. Furthermore, this study showed that this conceptual method could be realised and applied in practice by extracting the raster layers of the defined parameters from Hotmaps, which is a validated open-source application for heating and cooling planning at various spatial levels [19–21]. These raster files of the parameters, which are essential for strategic heat planning, are estimated data. Hence, the limitations of these data should be taken into account (discussed further in Section 5). The Hotmaps data focus on residential buildings; therefore, using these data in non-residential areas must be performed with care. However, since PEDs are a highly residential concept, this limitation was justifiable. The Hotmaps data were further manipulated using the QGIS software. As a result, this study demonstrates similar districts across Europe that are likely comparable regarding infrastructure requirements. Figure 2 illustrates the overall methodology of our work.

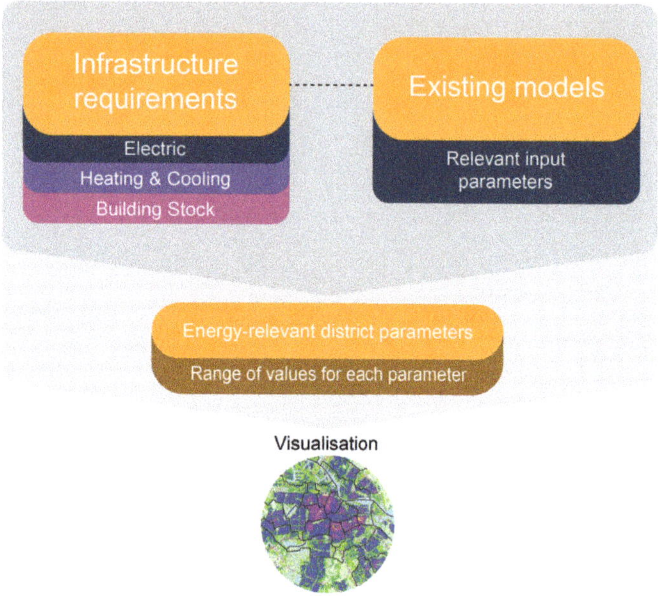

Figure 2. Our methodological approach to classifying districts from an energy modelling point of view.

The first stage of this work (Section 4.1) looked at the existing review papers that deal with models and tools that are used within the energy system modelling domain for various purposes. The article by [14] served as a starting point. The authors listed previous review articles that categorised or analysed energy system models and tools. Many review articles have focused on models that are suitable for analysing local-, community-, district- and neighbourhood-scale energy systems [22–28]. Hence, from the known models, we selected several models that comply with the following criteria, which we extrapolated from our energy infrastructure definition in Section 2:

(a) A district or neighbourhood geographical scale (or any cluster of buildings);
(b) A time resolution from hourly to seasonal;
(c) An infrastructure that was within the scope of this work, i.e., renewable energy generation technology, energy storage and EV charging technology, heating and cooling systems and building envelopes;
(d) Aims that are relevant for energy planning (i.e., not frequency regulation or power sector specifics).

The next step after model selection was to summarise the information that was used as the inputs and outputs of the models, as well as the modelling approach (i.e., method), spatial and temporal scales and the covered infrastructure types. Then, we analysed and generalised the input parameters (along with the other information that was collected) to derive the parameters for our district categorisation (Section 4.1). As specific and granular data are not always available, using generalised district parameters could help with the research for and planning of PEDs (or other related district energy concepts). The extracted district categorisation matrix is illustrated in Section 5, which shows the district typologies that are present in Europe. The cities of Amsterdam, Frankfurt and Torres Vedras, along with selected districts, were used to showcase the use of the final parameter map. These cities were selected in line with the case study cities of the PED-focused "Smart-BEEjS" project. Furthermore, to adequately showcase the comparison map, two cities/districts within the same climate zone were needed (Frankfurt and Amsterdam).

4. Model Details

4.1. Input Parameters for District Categorisation

This study revised several district energy models with the aim of identifying their input requirements. The results of the review are presented in Table A1. These comprehensive and detailed input requirements were grouped into smaller sets of representative and comparable parameters that could be used to classify European neighbourhoods (Table 1). To achieve a feasible categorisation, four parameters that represent the most important input requirements were chosen. The summarised results of the energy modelling review, which are shown in Table 1, were in line with the literature research that was carried out by [29,30].

Table 1. The synthesis of parameters for the district categorisation.

Matching Data Requirements		Representative Parameters
Meteorological data, renewable energy supplies, weather data and climatic characteristics	⟶	Climate Zone
Demand profiles, building envelopes, U-values, insulation and household equipment	⟶	Heating Demand
Available area, building type, building height, building archetype and building geometry	⟶	Floor Space Index
Occupancy behaviour, time of use, net energy demands and PV production	⟶	Share of Residential Buildings

The "climate zone" parameter in Table 1 refers to meteorological and weather- and climate-related data. This input influences heat demands and renewable energy supplies and, therefore, is crucial for the selection of local energy infrastructures. The most widely used climate classification is the Köppen–Geiger (KG) classification [31], which divides the world into five regions and 30 sub-regions according to the threshold values and seasonality of monthly air temperature and precipitation [32]. This classification scheme has been used in the PV community to analyse regions of interest easily [33]. For example, the European project PVSites applied the Köppen–Geiger classification, together with the parameters of the European heat index and European cooling index, to create a zoning map for nearly zero energy buildings (nZEBs) [34].

The "heating demand" parameter indicates the levels of demand for space heating in buildings, which mainly depends on the climate and the energy efficiency state of the buildings. The values and ranges for the annual heat demand were identified from the available data, as explained further in Section 5. This parameter mainly influences the building envelopes and heating and cooling systems.

The "floor space index" parameter (FSI) can also be an indication of the space that is available for renewable energy generation in relation to the number of people living and consuming energy in that specific area, as described by [35]. The FSI (also called the floor area ratio) is defined by Equation (1) [36]:

$$FSI = \frac{gross\,floor\,area}{plot\,area} \quad (1)$$

The FSI ranges indicate the type of the settlements within a neighbourhood: very rural settlements (<0.25), single-family houses (0.25–1), row housing (1–2) and block housing to very dense urban settlements (>2) [35]. A lower FSI could indicate that a larger area (roofs) is available for renewable energy generation, while the overall energy demand density stays low. Thus, a low FSI indicates the increased technical ease of achieving the PED energy balance requirements. However, it does not indicate anything about economic aspects.

The "share of residential buildings" parameter is an indication of the type of energy consumer that is present in a neighbourhood. The type of consumer affects the final net energy consumption of building operations and the load distribution over time [37]. More specifically, the ratio of residential to commercial consumers influences the final net consumption of buildings, as commercial buildings have higher energy requirements (especially supermarkets) [37]. In addition, the electric load distribution of residential buildings is significantly different from that of non-residential buildings, with the residential load being higher during morning and evening hours [38]. Additionally, potential EV charging schedules also depend on the residential share of buildings, as cars are usually available for charging at night in residential districts and during working hours in non-residential districts.

All parameters are shown with their respective ranges in Figure 3. The ranges of the parameters were derived based on the available data, as described in Section 5.

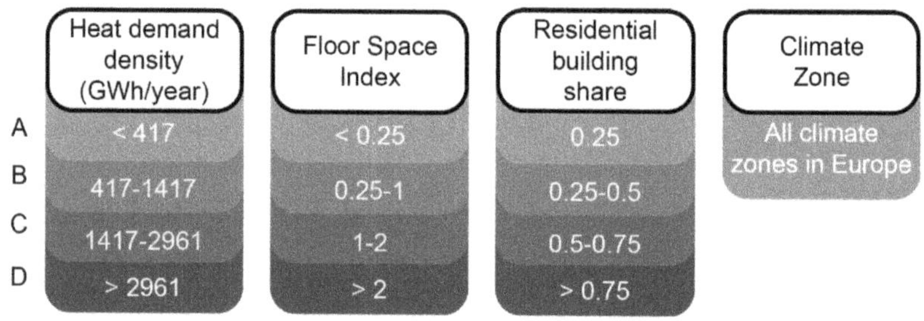

Figure 3. The district categorisation matrix.

The procedure for visualising of district types across Europe is illustrated in Figure 4. First, a raster file of each selected parameter (except for the climate zone parameter) was extracted from the Hotmaps library [20]. The raster layers had a resolution of one hectare; therefore, each hectare equalled one pixel of the raster layer and was associated with one relevant parameter value (e.g., heating demand). To limit the combination of possibilities, this study reduced the available values of the raster layers to our predefined ranges, according to Figure 3. The multiplication of all the Hotmaps-derived raster files resulted in one raster layer with a maximum number of n^x possible district typologies, where n is the number of parameters and x is the number of values that each parameter could obtain. Furthermore, we imported a Köppen–Geiger climate raster file [32] of Europe and overlaid it on the output file (i.e., the district typology map). The climate layer was not multiplied with the remaining parameter layers as it would significantly increase the number of possible combinations that were available. Therefore, the climate layer functioned as an initial filter to find districts within the same climate zone before advancing with the remaining layers. The final two-layer map showed all available combinations of the heat demand, FSI and residential share parameters (as defined in Section 4) on one layer and the climate zone parameter on another layer. Therefore, the map enabled the easy comparison of different districts according to the predefined parameters.

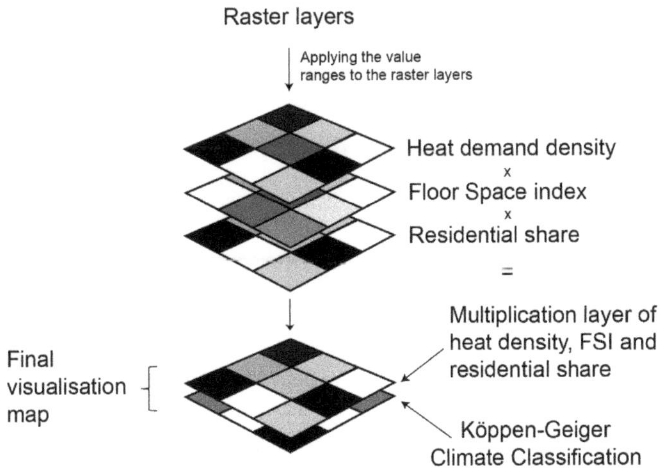

Figure 4. Our methodological approach for the visualisation of the selected parameter combinations on a map.

Defining the parameter value ranges was essential to the categorisation effort. The number of bands had to be kept low to decrease the possible combinations of the values but high enough to guarantee sufficient detail. As the FSI was unavailable as a raster file, this study approximated it using the gross floor area per hectare. The heating demand and the FSI value distributions followed an F-distribution. Furthermore, the heating demand distribution was strictly related to the FSI; thus, the quantiles of the FSI thresholds were transferred to the heat demand values, as defined in Section 4.1, to generate the thresholds that are shown in Figure 3. The residential share distribution showed two peaks on the left- and right-hand sides of the graph, with a valley in the middle. This distribution showed that the most common districts had either a low or high residential share of the gross floor area. Nonetheless, many districts still had a residential share between these two peaks. Applying the exact quantiles used for the FSI and heat demand parameters meant considering a large proportion of the districts to be highly residential, which distorted the categorisation efforts. Consequently, this work divided the residential share values into four equidistant bands. For the first three parameters in Figure 3 (heat demand, FSI and residential share), the letters A to D were used to classify the bands, with A being the lowest and D being the highest value. All of the prevalent climate zones in Europe, according to [32] were used. This work only included heating demand instead of additionally including the cooling demand. Arguably, across one climate zone and one FSI category, the heating demand and cooling demand would be negatively correlated, meaning that heating demand band A would correlate to cooling demand D and vice versa. Thus, adding this additional layer would have increased complexity without adding much additional value. The final district categorisation matrix with the selected parameters and their relevant bands is illustrated in Figure 3.

Figure 5 shows the general application process of the final district typology map (openly accessible). Firstly, two or more districts located in the same KG climate zone need to be chosen for the analysis. After using this climate filter, the climate map can then be disabled to only visualise the districts' typology created by the heat demand density, the FSI and the residential share. The analysis of this can be visual or statistical by extracting the values of the raster layers belonging to each district. If the districts have similar typologies, a high potential for a similar approach towards energy infrastructure requirements can be deducted. Thus, districts that are very similar according to the defined parameters and fall into the same climate zone can initiate knowledge exchange within the planning teams.

On the other hand, if the district typologies vary significantly, the potential for a similar approach is lower.

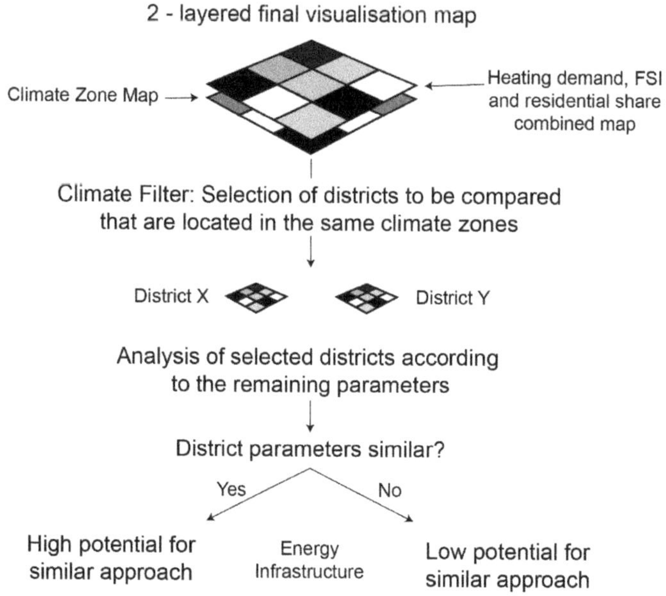

Figure 5. The application process of the final district visualisation map.

5. Visualisation of Results

This section shows the application of the approach presented in Section 3 to three European districts. Figure 6 shows a map of the different climate zones according to the Köppen–Geiger classification across Europe, which was used to filter out districts of interest located in the same climate zone. For further discussion, three partner cities of the "Smart-BEEjS" project for PEDs were selected [39]. Amsterdam and Frankfurt are located in the same climate zone of Cfb (temperate oceanic climate). In contrast, Torres Vedras in the south of Europe is characterised by the Csb climate zone (warm-summer Mediterranean climate). Thus, according to the first layer of diversification, districts in Amsterdam and Frankfurt could be used for comparison. The different climate in Torres Vedras might already have different implications for the energy infrastructures of potential districts; therefore, Torres Vedras could not be compared to the two other cities.

Figure 7 zooms into the three cities of Frankfurt, Amsterdam and Torres Vedras to visualise their district typologies in further detail. Areas that are presented in purple, red or even light orange have increasingly dense heat demands, higher floor space indices and larger shares of residential building usage. Conversely, green areas are less dense and indicate mixed-use or even low residential districts. As this map aimed to compare (but not rank) districts, the order of district combinations in the legend is of little importance. It is simply the result of the prime number approach described in Appendix B. The prime numbers were back-calculated to the respective alphabetical code. The colour code just provides an indication of the density. Additionally, not all of the 64 theoretically possible combinations appear in Europe.

Frankfurt and Amsterdam are similar in size and also show a comparable picture in terms of the district typologies that are present. Both cities show a highly dense centre that is shaded in red/violet, but the central area in Amsterdam is larger and denser. The dense, red-shaded areas are surrounded by blue and then green zones. Frankfurt appears to have multiple smaller settlements around the core city, while Amsterdam seems to be more connected. At the edges of the city and in the canal zones, Amsterdam is mainly shaded in light green. This most likely shows highly industrial areas and ports. Torres Vedras, on the other hand, is predominantly shaded in green and only peaks at light blues in the city centre. This, combined with the warmer climate zone, suggests advantages in terms of the required energy infrastructure for a PED as the heating demand is lower, the potential solar gains are higher, and the available space for local energy generation is larger.

These observations were supported by histograms of the district typologies of the cities. Figure 8 shows the size of each of the ten most common district typologies (in hectares) across the whole cities (Torres Vedras only has seven). The three most common typologies in Frankfurt and Amsterdam are very similar. Both show medium-low heat demands and floor space indices. Amsterdam, however, seems to be slightly less residential and has more highly dense areas within the ten most common categories, such as CCB. Torres Vedras, on the other hand, shows an entirely different picture. The city has a low heat demand thanks to its climate and also a very low floor space index in most areas. The town is very residential but has highly non-residential zones surrounding it, which are most likely industrial or commercial areas.

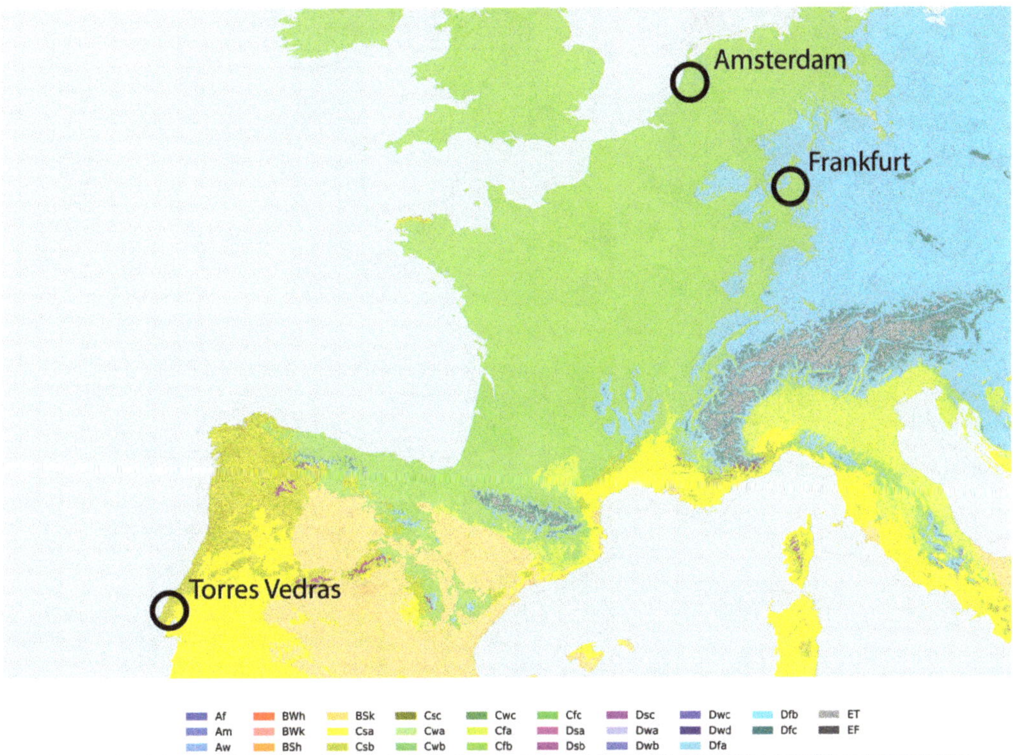

Figure 6. A map of Central Europe, with climate zones according to the Köppen–Geiger classification [32]. Amsterdam, Frankfurt and Torres Vedras are marked on the map.

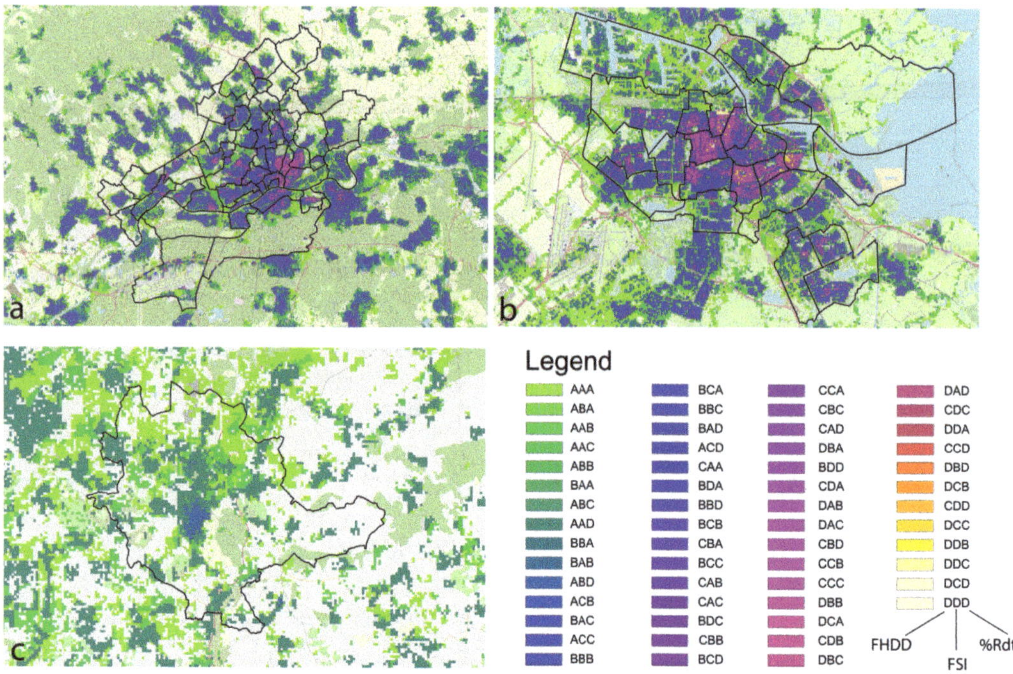

Figure 7. The visualisation of the district typologies in Frankfurt am Main (**a**), Amsterdam (**b**) and Torres Vedras (**c**): FHDD, final heat demand density; *FSI*, floor space index; %Rdt., share of residential buildings.

Torres Vedras is not located within the same climate zone as Frankfurt and Amsterdam (Figure 6) and, therefore, was filtered out by the climate layer as it could not be compared to the other two cities. As Amsterdam and Frankfurt are located within the same climate zone, two example districts were selected to compare according to their district typologies. A PED project called Atelier is located in the south of Buiksloterham in Amsterdam [40]. In Frankfurt, the Griesheim-Mitte district also has major renovation ambitions [41]. Figure 9 shows the distributions of the available district typologies in Frankfurt (Griesheim-Mitte) and Amsterdam (southern Buiksloterham). First of all, Griesheim-Mitte is significantly larger than southern Buiksloterham. Secondly, both districts have predominantly low to medium heating demands and *FSI* values. However, buildings in Griesheim-Mitte have higher levels of residential usage than those in southern Buiksloterham. Furthermore, southern Buiksloterham shows many areas with low heating demands and *FSI* values, with AAB, AAA and AAC being among the five most common areas within the district. On the contrary, Griesheim-Mitte has few areas with higher heating demands and *FSI* values, with CCC and CCB being in the top five categories. As southern Buiksloterham has, on average, lower heating demands and *FSI* values, there is more available space for energy generation, in the form of solar PV or solar thermal panels, in relation to the energy demand. Furthermore, the lower share of residential buildings could be beneficial for the self-supply of the district as commercial energy demands may be more aligned with sun hours and, thus, with renewable energy generation. The comparison indicated a relatively low similarity between the districts and, therefore, limited potential for knowledge transfer among the districts.

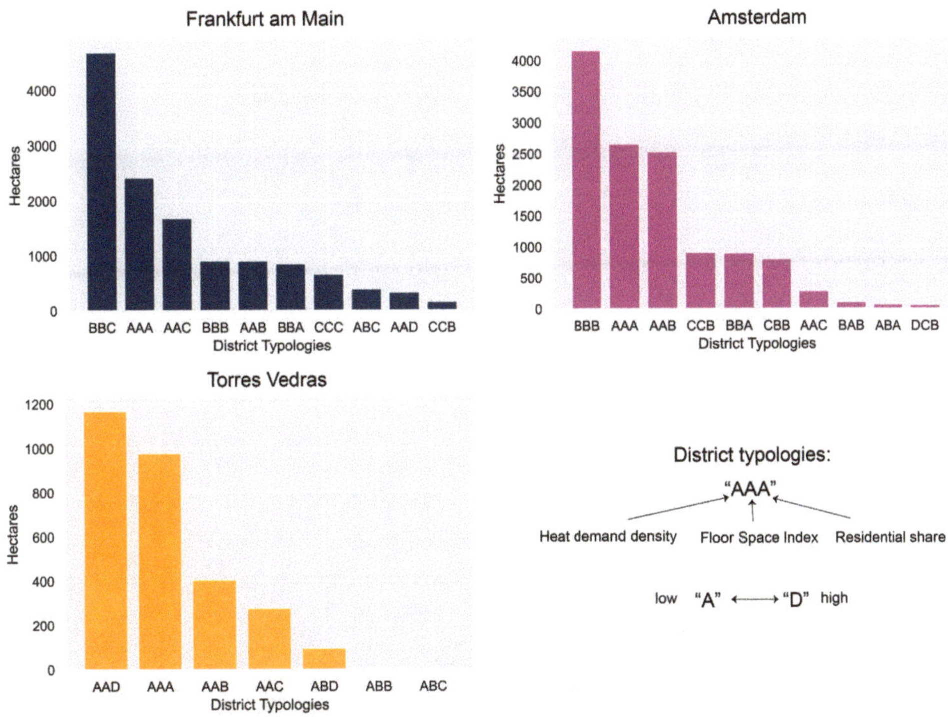

Figure 8. A comparison of the frequency of the district categories in Frankfurt am Main, Amsterdam and Torres Vedras.

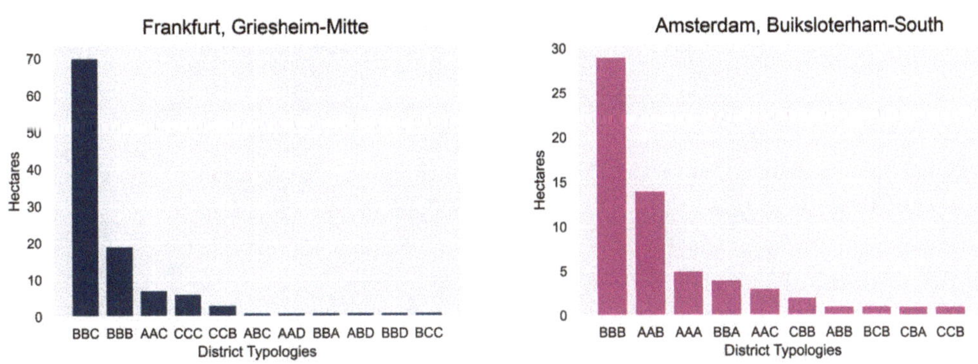

Figure 9. A comparison of the frequency of the district categories in Frankfurt - Griesheim-Mitte am Main and Amsterdam - Buiksloterham-South.

Figures 10 and 11 show aerial views of Griesheim-Mitte and the considered part of Buiksloterham, respectively. The figures support our analysis of the district histograms well and, therefore, also act as validation for the created map. First of all, the size difference between the two districts is evident. Moreover, a large share of the BBC areas in Griesheim-Mitte is seen towards the mid-right of the district in Figure 10, which are predominantly made up of residential buildings. On the other hand, southern Buiksloterham has a small residential area in the lower-left corner with more prominent areas of non-residential buildings, according to the aerial view. Finally, the floor space index appears to be lower at first glance. These observations were consistent with the previously discussed histograms in Figure 9.

Figure 10. An aerial view of Griesheim-Mitte [42].

Figure 11. An aerial view of southern Buiksloterham [43].

6. Conclusions

In light of the energy transition process that is partly facilitated by positive energy districts across Europe, methods that enable comparability of such district projects and thus encourage learning among them is becoming vital. One of the important aspects of creating PEDs is the physical energy infrastructure that is necessary for an existing district to transform into a PED. This article presented a method to facilitate district comparison regarding their energy infrastructure requirements. This work derived four indicators from the literature on energy system modelling: climate, heating demands, floor space index and the share of residential buildings. Furthermore, this study applied the developed methodology using QGIS and data that were openly available from the Hotmaps project, which led to a direct visualisation of each hectare of the cities and districts within Europe, thereby enabling an easy and direct comparison of the different areas. This clustering approach could facilitate the comparison of districts that are very similar according to the developed indicator values. The clustering methodology could also help districts to learn from the successes and challenges that arose in previous energy renovation projects that occurred in similar district categories. The example application using the districts of Griesheim-Mitte in Frankfurt and southern Buiksloterham in Amsterdam indicated the potential of this method for comparing districts.

Beyond this, the map also indicated which zones would be more challenging to convert into PEDs and could therefore work complementarily with the method of [9], which also shows which zones would be the most and least suitable for PEDs using a GIS-based approach. In our map, zones that are coloured in orange or red have high heat demands, high floor space indices and high shares of residential buildings. Those attributes increase the difficulty of achieving PED status as energy loads are less distributed and less space is available for energy generation in relation to the higher energy demands. On the contrary, green zones will likely be easier to convert into PEDs. This information could be vital in district-level renovation projects. It could allow for the differentiation of the areas within a district that require the most attention from a technical standpoint from the areas that could provide the energy generation surplus that is needed to achieve a positive balance overall. Based on the same information, would also be possible to evaluate whether a district needs renewable energy production outside of the district's boundaries to offset its consumption. This analysis could estimate to what extent building stocks need to be improved or how much PV production would be required. Assuming that some energy production would need to be positioned outside of a district, this method could also suggest whether regional electrical infrastructures require further improvement. As of now, the aim of the visualisation map is only to compare districts and potentially transfer knowledge among similar projects. These further applications would require more PED (or PED-like) projects in existing areas and, subsequently, validation.

As stated in Section 1, there are currently only two PEDs operating in Europe, which limits the amount of information that is available regarding the implemented infrastructures. Furthermore, these two PEDs are newly built and are not renovation projects. As the number of implemented PEDs increases in the near future, the addition of more marked areas on the map will become possible. As the map shows data from 2015, it shows district typologies from before any renovations. These future additions would allow local policymakers and administrations to locate implemented PED projects that match their district typology and learn from those already existing projects. The two-layer visualisation map could also be used as an initial step for tracking existing PED projects over time and pinpointing them on the map as a working document. Readers should note that this categorisation comprises only four parameters. District renovations often require a holistic approach that also includes the social fabric of the area in question. Although a solution that was derived from another similar project may be technically feasible in a local context, it may be not socially accepted. This issue, combined with the estimated nature of the data that were used to develop the district typology map, should remind readers that this tool is intended for initial planning and policymaking based on findings from other projects. Thus, the map could be used as the first tool for PED planning to identify projects in similarly structured districts for knowledge transfer.

Further detailed research and calculations are necessary for specific districts as there may be other aspects that need to be considered, e.g., the availability of waste heat. Future research should focus on further verifying whether this method leads to two different districts within the same classification having similar infrastructures to achieve PED status. In addition, it is possible to create other typology and raster maps for different aims and visualisations. Other future work, still revolving around the necessary infrastructures, should focus on whether there is enough energy generation potential in relation to demand at a hectare level. It could be possible to compare the solar PV yearly potential raster to the yearly heat demand and obtain a rough estimate of the energy balance. Overall, this would be an initial step towards creating comparability in terms of buildings and energy infrastructures among European districts.

The limitations of our approach were closely related to the limitations of the Hotmaps data. The gross floor area was calculated using the average gross floor area per dwelling and the average persons per household, based on the available statistical data at the NUTS3 level (e.g., "Landkreise" in Germany). While this approach provided a reasonable estimation for residential building stocks at a hectare level, the non-residential gross floor area calculation was less robust [20]. The heat demand data were calculated at the NUTS0 level (country level) from the statistical data on energy consumption, as well as the national building stock characteristics. The grid cell-specific energy demand per floor area data were derived from the surface to volume ratios of buildings from the OpenStreetMap database, the shares per construction period and the heating and cooling degree days [20]. Each of these indicators were estimations in themselves and thus, were limited in their accuracy. Furthermore, the validation of the model was only based on aerial views of the studied districts. While this provided an initial indication of the model's validity, further studies still need to prove this.

Author Contributions: Conceptualisation, A.B., L.C., A.A., L.K. and K.G.; methodology, A.B., L.C., A.A. and L.K.; validation, A.B., L.C. and A.A.; formal analysis, A.B., L.C. and A.A.; investigation, A.B., L.C. and A.A.; resources, A.B., L.C. and A.A.; data curation, A.B., L.C. and A.A.; writing—original draft preparation, A.B., L.C. and A.A.; writing—review and editing, A.B., L.C., A.A., L.K. and K.G.; visualisation, A.B., L.C. and A.A.; supervision, L.K. and K.G.; project administration, K.G.; funding acquisition, K.G. All authors have read and agreed to the published version of the manuscript.

Funding: This research received funding from the European Union's Horizon 2020 research and innovation programme under the Marie Skłodowska-Curie Actions programme (grant number 812730).

Institutional Review Board Statement: Not applicable.

Informed Consent Statement: Not applicable.

Data Availability Statement: The data used in this study is taken from https://gitlab.com/hotmaps accessed on 17 June 2022 and https://doi.org/10.1038/sdata.2018.214. The derived map is publicly accessible under https://doi.org/10.5281/zenodo.6719747.

Conflicts of Interest: The authors declare no conflict of interest.

Abbreviations

The following abbreviations are used in this manuscript:

DCM	District Categorisation Matrix
DH	District Heating
EV	Electric Vehicle
FSI	Floor Space Index
HVAC	Heating, Ventilation and Air Conditioning
KPI	Key Performance Indicator
PED	Positive Energy District
PV	Photovoltaic Panel

Appendix A

Table A1. A list of the reviewed models.

Model	Aim/Output	Requirement/Input	Method	Time Resolution	Spatial Aspect	Sectors
Calliope [44–46]	Energy portfolio and dispatch optimisation	Demand profiles, technology to consider, available area, meteorological data and costs	Bottom-up; MILP	User defined	User defined	Electricity, heating and mobility (limited)
City-BES [47]	User defined, e.g., energy-, emissions- and cost-related KPIs for each retrofit scenario	The footprint, type, height, year of construction and number of stories of the buildings, shading buildings, shared walls and weather	Bottom-up; physics-based (based on EnergyPlus)	Sub-hourly	Cities	Electricity and heating
City Energy Analyst [48]	Building energy consumption patterns in neighbourhoods and districts	Weather data, urban GIS data building archetypes, distributions database (occupancy schedules; 16 types in this case) and measurements database (for non-standardised energy services in the area, e.g., stadia)	Bottom-up (two methods of load calculation: analytical and statistical)	Hourly	Neighbourhoods	Electricity and heating
CitySim [49]	Heating and cooling demands and urban planning	Building characteristics and climate files	Dynamic building energy simulation; reduced-order RC model	1 min–1 h	Streets to districts	Electricity and heating
DER-CAM [50]	Energy portfolio and dispatch optimisation	Demand profiles, technology to consider, available area, meteorological data and costs	Bottom-up; MILP	User defined (reference: days)	Buildings to microgrids (districts)	Electricity, heating and mobility (limited)

Table A1. *Cont.*

Model	Aim/Output	Requirement/Input	Method	Time Resolution	Spatial Aspect	Sectors
DIMOSIM [51]	Raw outputs, i.e., states of each object (e.g., temperature) and energy fluxes (e.g., consumption per fuel) and KPIs generated from the raw outputs that related to thermal indoor comfort, energy, power and costs	Climatic characteristics, building geometry, U-values and surface ratios of the different components within the envelope, HVAC system characteristics, occupancy rates, insulation types (e.g., indoor or outdoor) and inertia level	Bottom-up; simulation; possible optimisation	User defined (range of minutes to hours)	Small neighbourhoods to cities	Electricity and heating
EnergyPlan [52]	Operation of energy systems and environmental and economic impacts	Installed capacity, available energy and energy demands	Bottom-up; simulation (based on heuristic technique)	Hourly	Cities to countries	Electricity and heating
EnergyPlus [53]	Dynamic building simulations and HVAC	Climate data, U- and g-values, heating and cooling systems, temperature set-point (min; max), air change per hour, internal heat gain, external short-wave absorbance and long-wave emissivity	Bottom-up; physics-based	User defined	Buildings	Electricity and heating
ESP-r [54]	Dynamic building simulations and HVAC	Climate data, U- and g-values, heating and cooling systems, temperature set-point (min; max), air change per hour, internal heat gain, external short-wave absorbance and long-wave emissivity	Bottom-up; physics-based	User defined	Buildings to districts	Electricity and heating

Table A1. *Cont.*

Model	Aim/Output	Requirement/Input	Method	Time Resolution	Spatial Aspect	Sectors
Homer [55]	Energy portfolio and dispatch optimisation	Demand profiles, technology to consider, available area, meteorological data and costs	Bottom-up	User defined	Microgrids (districts)	Electricity, heating and mobility (limited)
oemof [56]	Multiple Python libraries for optimisation and modelling of energy systems	Demand profiles, technology to consider, available area, meteorological data and costs	Bottom-up	User defined (reference: days)	Buildings to microgrids (districts)	Electricity, heating and mobility (limited)
Smart-E [57]	Energy demand simulation, implementation of demand–response strategies in cities	Weather data, household composition, envelope characteristics, heating energy demands, location, time of use (schedules and probabilities (household equipment, set points, etc.)	Bottom-up; simulation	Daily	Cities to larger territories	Electricity and heating
TRNSYS [58]	Thermal and electrical energy systems, dynamic systems, traffic flow and biological processes	User defined components and library components	Simulation; linear and nonlinear programming	0.01 s–1 h	Buildings to districts	Electricity, heating and mobility
urbs [59]	Energy portfolio and dispatch optimisation	Demand profiles, technology to consider, available area, renewable energy supplies as time series and costs	Bottom-up	User defined	User defined	Electricity, heating and mobility (limited)
UMI [60]	Walkability, environmental performance and daylight potential	Parks, streets, shadings, boundaries, ground and the geometry, occupancy and fenestration of buildings	Simulation (based on EnergyPlus, rhinoceros and Daysim)	...	Streets to districts	Electricity, heating and mobility

Appendix B

This section explains how we developed the map that categorises the districts and areas in Europe. The reasons for the selection of each parameter and their related thresholds are not subjected to further analysis here but rather the process itself for replication. The analysis that was conducted in Section 3 led to the selection of the relevant raster files from the Hotmaps repository. These raster files were:

- Total gross floor area;
- Residential gross floor area;
- Final heat demand density.

The first two rasters were relevant for producing the floor space index and the residential gross floor area percentage. To generate the latter, we utilised the raster calculator function in QGIS. We divided the residential gross floor area by the total gross floor area. Using a similar process, the total gross floor area was converted into an approximation of the FSI by simply dividing the raster by 10,000, as each pixel equalled one hectare and the pixel's value was expressed in square meters. Section 3 already explained the reasons behind the selection of each threshold; hence, it is not part of this section. This section explains how the rasters were changed to show the thresholds rather than the values. Again, the raster calculator included in QGIS was the tool that was used to generate the typologies of these rasters. Four bands were generated by setting the following conditions on each raster file:

$$((''R_x'' > 0)AND(''R_x'' \leq t1)) * p_{1.x} + ((''R_x'' > t1)AND(''R_x'' \leq t2)) * p_{2.x} + ((''R_x'' > t2)AND \\ (''R_x'' \leq t3)) * p_{3.x} + (''R_x'' > t3) * p_{4.x} \quad \text{(A1)}$$

where $Raster_x$ is the raster subject to be transformed, t1 to t3 are the set thresholds and $p_{1.x}$ to $p_{4.x}$ indicate the prime numbers that were applied to the first to fourth bands of the xth layer. Each threshold was defined by a prime number rather than a letter as QGIS does not support strings as a data type. The use of prime numbers allowed for the creation of the final visualisation map. The final visualisation map raster condensed the information from each pixel in the three rasters (i.e., total gross floor area, share of residential gross floor area and final heat demand density) into one raster, as shown in Figure A1. QGIS does not allow the multiplication or addition characters; hence, we used prime numbers to track the original values.

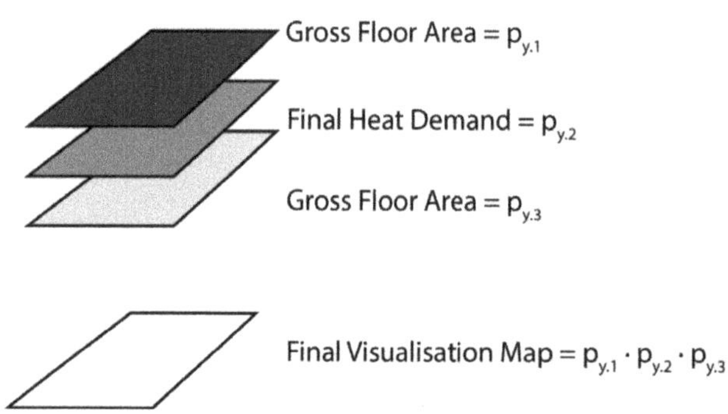

Figure A1. The process for blending the three original layers into the visualisation map. The subscript y indicates the band to which pixel belongs.

The prime numbers allowed us to retain the information from all of the layers that made up the map. The product of three prime numbers could only be obtained by multiplying those exact prime numbers. It is important to note that each threshold had a different prime number related to it, as indicated in Table A2.

Table A2. The code to transform letter indicators into prime numbers and vice versa.

	Thresholds	Letter Indicator	Prime Number Indicator
Final Heat Demand Density	417	A	2
	417–1417	B	3
	1417–2961	C	5
	2961	D	7
Total Gross Floor Area	0.25	A	9
	0.25–1	B	11
	1–2	C	13
	2	D	17
Percentage of Residential GFA	0.25	A	23
	0.25–0.5	B	27
	0.5–0.75	C	29
	0.75	D	31

Let us consider an example in which a pixel on the map (i.e., one hectare) has a final heat demand that is indicated by A, a total gross floor area of B and a percentage of residential GFA of C. The resulting pixel on the visualisation map would have the value of the product of their prime number indicators (in this case, 638). Because this number can only be derived from the multiplication of these three prime numbers, each pixel can be unequivocally identified and categorised.

References

1. United Nations Human Settlements Programme. *Unpacking the Value of Sustainable Urbanization*; World Cities Report 2020; UN: Nairobi, Kenya, 2020; pp. 43–74. [CrossRef]
2. JPI Urban Europe. White Paper on PED Reference Framework for Positive Energy Districts and Neighbourhoods. 2020 . Available online: https://jpi-urbaneurope.eu/wp-content/uploads/2020/04/White-Paper-PED-Framework-Definition-2020323-final.pdf (accessed on 21 June 2022).
3. MAKING CITY. Making City—Energy Efficient Pathway for the City Transformation. Available online : https://makingcity.eu/ (accessed on 2 March 2022).
4. +CityxChange. Available online: https://cityxchange.eu/ (accessed on 2 March 2022).
5. IEA EBC. Annex 83-Positive Energy Districts. Available online: https://annex83.iea-ebc.org/ (accessed on 2 March 2022).
6. POCITYF. Leading the Smart Evolution of Historical Cities. Available online : https://pocityf.eu/ (accessed on 2 March 2022).
7. Bruck, A.; Díaz Ruano, S.; Auer, H. A Critical Perspective on Positive Energy Districts in Climatically Favoured Regions: An Open-Source Modelling Approach Disclosing Implications and Possibilities. *Energies* **2021**, *14*, 4864. [CrossRef]
8. Bossi, S.; Gollner, C.; Theierling, S. Towards 100 positive energy districts in Europe: Preliminary data analysis of 61 European cases. *Energies* **2020**, *13*, 6083. [CrossRef]
9. Alpagut, B.; Lopez Romo, A.; Hernández, P.; Tabanoğlu, O.; Hermoso Martinez, N. A GIS-Based Multicriteria Assessment for Identification of Positive Energy Districts Boundary in Cities. *Energies* **2021**, *14*, 7517. [CrossRef]
10. JPI Urban Europe. *Europe towards Positive Energy Districts*. PED Booklet 2020, 180p. Available online: https://jpi-urbaneurope.eu/wp-content/uploads/2020/06/PED-Booklet-Update-Feb-2020_2.pdf (accessed on 21 June 2022).
11. Zwickl-Bernhard, S.; Auer, H. Open-source modeling of a low-carbon urban neighborhood with high shares of local renewable generation. *Appl. Energy* **2021**, *282*, 116166 . [CrossRef]
12. Fulmer, J.E. What is the World Infrastructure? *PEI Infrastruct. Investor* **2009**, *1*, 30–32.
13. Brozovsky, J.; Gustavsen, A.; Gaitani, N. Zero emission neighbourhoods and positive energy districts—A state-of-the-art review. *Sustain. Cities Soc.* **2021**, *72*, 103013. [CrossRef]
14. Chang, M.; Thellufsen, J.Z.; Zakeri, B.; Pickering, B.; Pfenninger, S.; Lund, H.; Østergaard, P.A. Trends in tools and approaches for modelling the energy transition. *Appl. Energy* **2021**, *290*, 116731. [CrossRef]
15. Siedentop, S.; Schiller, G.; Koziol, M.; Walther, J.; Gutsche, J.M. *Siedlungsentwicklung und Infrastrukturfolgekosten—Bilanzierung und Strategieentwicklung*; Technical Report 3; Bundesamt für Bauwesen und Raumordnung: Bonn, Germany, 2006 .
16. Nzengue, Y.; du Boishamon, A.; Laffont-Eloire, K.; Partenay, V.; Abdelouadoud, Y.; Zambelli, P.; D'Alonzo, V.; Vaccaro, R. Planning city refurbishment: An exploratory study at district scale how to move towards positive energy districts—Approach of the SINFONIA project. In Proceedings of the 2017 International Conference on Engineering, Technology and Innovation (ICE/ITMC), Madeira, Portugal, 27–29 June 2017; pp. 1394–1400. [CrossRef]
17. Hachem-Vermette, C.; Singh, K. Optimization of energy resources in various building cluster archetypes. *Renew. Sustain. Energy Rev.* **2022**, *157*, 112050. [CrossRef]

18. Hachem-Vermette, C.; Singh, K. Energy Systems and Energy Sharing in Traditional and Sustainable Archetypes of Urban Developments. *Sustainability* **2022**, *14*, 1356. [CrossRef]
19. Pezzutto, S.; Croce, S.; Zambotti, S.; Kranzl, L.; Novelli, A.; Zambelli, P. Assessment of the space heating and domestic hot water market in Europe—Open data and results. *Energies* **2019**, *12*, 1760 . [CrossRef]
20. Hotmaps Project. Library-Hotmaps Project. Available online: https://www.hotmaps-project.eu/library/ (accessed on 21 June 2022).
21. Müller, A.; Hummel, M.; Kranzl, L.; Fallahnejad, M.; Büchele, R. Open Source Data for Gross Floor Area and Heat Demand Density on the Hectare Level for EU 28. *Energies* **2019**, *12*, 4789 . [CrossRef]
22. Markovic, D.; Cvetkovic, D.; Masic, B. Survey of software tools for energy efficiency in a community. *Renew. Sustain. Energy Rev.* **2011**, *15*, 4897–4903. [CrossRef]
23. Keirstead, J.; Jennings, M.; Sivakumar, A. A review of urban energy system models: Approaches, challenges and opportunities. *Renew. Sustain. Energy Rev.* **2012**, *16*, 3847–3866. [CrossRef]
24. Allegrini, J.; Orehounig, K.; Mavromatidis, G.; Ruesch, F.; Dorer, V.; Evins, R. A review of modelling approaches and tools for the simulation of district-scale energy systems. *Renew. Sustain. Energy Rev.* **2015**, *52*, 1391–1404. [CrossRef]
25. Huang, Z.; Yu, H.; Peng, Z.; Zhao, M. Methods and tools for community energy planning: A review. *Renew. Sustain. Energy Rev.* **2015**, *42*, 1335–1348. [CrossRef]
26. Ferrari, S.; Zagarella, F.; Caputo, P.; Bonomolo, M. Assessment of tools for urban energy planning. *Energy* **2019**, *176*, 544–551. [CrossRef]
27. Abbasabadi, N.; Mehdi Ashayeri, J.K. Urban energy use modeling methods and tools: A review and an outlook. *Build. Environ.* **2019**, *161*, 106270. [CrossRef]
28. Klemm, C.; Vennemann, P. Modeling and optimization of multi-energy systems in mixed-use districts: A review of existing methods and approaches. *Renew. Sustain. Energy Rev.* **2021**, *135*, 110206. [CrossRef]
29. Yazdanie, M.; Orehounig, K. Advancing urban energy system planning and modeling approaches: Gaps and solutions in perspective. *Renew. Sustain. Energy Rev.* **2021**, *137*, 110607. [CrossRef]
30. Bouw, K.; Noorman, K.J.; Wiekens, C.J.; Faaij, A. Local energy planning in the built environment: An analysis of model characteristics. *Renew. Sustain. Energy Rev.* **2021**, *144*, 111030. [CrossRef]
31. Kottek, M.; Grieser, J.; Beck, C.; Rudolf, B.; Rubel, F. World Map of the Köppen-Geiger climate classification updated. *Meteorol. Z.* **2006**, *15*, 259–263. [CrossRef]
32. Beck, H.E.; Zimmermann, N.E.; McVicar, T.R.; Vergopolan, N.; Berg, A.; Wood, E.F. Present and future Köppen-Geiger climate classification maps at 1-km resolution. *Sci. Data* **2018**, *5*, 1–12. [CrossRef]
33. Ascencio-Vásquez, J.; Brecl, K.; Topič, M. Methodology of Köppen-Geiger-Photovoltaic climate classification and implications to worldwide mapping of PV system performance. *Solar Energy* **2019**, *191*, 672–685. [CrossRef]
34. PVSITES Consortium. D2.2 European Climate Zones and Bio-Climatic Design Requirements. Technical Report. 2016. Available online: https://ec.europa.eu/research/participants/documents/downloadPublic?documentIds=080166e5ac7b5027&appId=PPGMS (accessed on 21 June 2022).
35. Leibold, J.; Schneider, S.; Tabakovic, M.; Zelger, T.; Bell, D.; Schöfmann, P.; Bartlmä, N. 'Zukunftsquartier'—On the Path to Plus Energy Neighbourhoods in Vienna. In *Sustainability in Energy and Buildings*; Springer: Singapore , 2020; pp. 199–209. [CrossRef]
36. Caves, R.W. *Encyclopedia of the City*; Routledge : Abingdon, UK, 2005.
37. Hachem-Vermette, C.; Singh, K. Optimization of the mixture of building types in a neighborhood and their energy and environmental performance. *Energy Build.* **2019**, *204*, 109499. [CrossRef]
38. Yuan, J.; Zhang, G.; Yu, S.S.; Chen, Z.; Li, Z.; Zhang, Y. A multi-timescale smart grid energy management system based on adaptive dynamic programming and Multi-NN Fusion prediction method. *Knowl. Based Syst.* **2022**, *241*, 108284 . [CrossRef]
39. Smart-BEEjS. Innovative Training Network—Smart Value Generation by Building Efficiency and Energy Justice for Sustainable Living. Available online: https://smart-beejs.eu/ (accessed on 21 June 2022).
40. ATELIER. ATELIER-Positive Energy Districts. Available online: https://smartcity-atelier.eu/ (accessed on 21 June 2022).
41. Stadt Frankfurt am Main. *Frankfurt a.M. Griesheim-Mitte—Integriertes Städtebauliches Entwicklungskonzept*; Technical Report; Stadtplanungsamt Frankfurt am Main: Frankfurt, Germany, 2019 .
42. Google Maps. Griesheim-Mitte—Google Maps. Available online: https://www.google.com/maps/place/Griesheim,+Frankfurt,+Germany/@50.0983916,8.5985339,15.3z/data=!4m5!3m4!1s0x47bd0a2fd4be4e49:0x522435029b39a00!8m2!3d50.0965406!4d8.5987565 (accessed on 21 June 2022).
43. Google Maps. Buiksloterham-South—Google Maps . Available online: https://www.google.com/maps/place/Buiksloterham,+Amsterdam,+Netherlands/@52.3901559,4.8951995,15.56z/data=!4m5!3m4!1s0x47c6083a59bcc9b5:0x5b0aa59876a0ead1!8m2!3d52.401543!4d4.8907417 (accessed on 21 June 2022).
44. Pfenninger, S.; Pickering, B. Calliope: A multi-scale energy systems modelling framework. *J. Open Source Softw.* **2018**, *3*, 825. [CrossRef]
45. Calliope: A multi-Scale Energy Systems Modelling Framework—Calliope 0.6.7 Documentation. Available online: https://calliope.readthedocs.io/en/stable/ (accessed on 15 December 2021).
46. Luz, G.P.; E Silva, R.A. Modeling Energy Communities with Collective Photovoltaic Self-Consumption: Synergies between a Small City and a Winery in Portugal. *Energies* **2021**, *14*, 323. [CrossRef]

47. Chen, Y.; Hong, T.; Piette, M.A. Automatic generation and simulation of urban building energy models based on city datasets for city-scale building retrofit analysis. *Appl. Energy* **2017**, *205*, 323–335. [CrossRef]
48. Fonseca, J.A.; Nguyen, T.A.; Schlueter, A.; Marechal, F. City Energy Analyst (CEA): Integrated framework for analysis and optimization of building energy systems in neighborhoods and city districts. *Energy Build.* **2016**, *113*, 202–226. [CrossRef]
49. Walter, E.; Kämpf, J.H. *A Verification of CitySim Results Using the BESTEST and Monitored Consumption Values*; Bozen-Bolzano University Press: Bozen-Bolzano , Italy, 2015; pp. 215–222.
50. DER-CAM | Grid Integration Group. Available online: https://gridintegration.lbl.gov/der-cam (accessed on 15 December 2021).
51. Garreau, E.; Abdelouadoud, Y.; Herrera, E.; Keilholz, W.; Kyriakodis, G.E.; Partenay, V.; Riederer, P. District MOdeller and SIMulator (DIMOSIM)—A dynamic simulation platform based on a bottom-up approach for district and territory energetic assessment. *Energy Build.* **2021**, *251*, 111354. [CrossRef]
52. Lund, H.; Thellufsen, J.Z.; Østergaard, P.A.; Sorknæs, P.; Skov, I.R.; Mathiesen, B.V. EnergyPLAN—Advanced analysis of smart energy systems. *Smart Energy* **2021**, *1*, 100007. [CrossRef]
53. Crawley, D.B.; Pedersen, C.O.; Lawrie, L.K.; Winkelmann, F.C. EnergyPlus: Energy Simulation Program. *ASHRAE J.* **2000**, *42*, 49–56.
54. Energy Systems Research Unit, University of Strathclyde. A Tour of ESP-r. Available online: https://www.esru.strath.ac.uk//Courseware/ESP-r/tour/ (accessed on 21 June 2022).
55. HOMER—Hybrid Renewable and Distributed Generation System Design Software. Available online: https://www.homerenergy.com/ (accessed on 15 December 2021).
56. Hilpert, S.; Kaldemeyer, C.; Krien, U.; Günther, S.; Wingenbach, C.; Plessmann, G. The Open Energy Modelling Framework (oemof)—A new approach to facilitate open science in energy system modelling. *Energy Strategy Rev.* **2018**, *22*, 16–25. [CrossRef]
57. Berthou, T.; Duplessis, B.; Rivière, P.; Stabat, P.; Casetta, D.; Marchio, D. Smart-E: A tool for energy demand simulation and optimization at the city scale. In Proceedings of the 14th Conference of International Building Performance Simulation Association, Hyderabad, India, 7–9 December 2015; pp. 1782–1789.
58. Klein, S.A. *TRNSYS 18: A Transient System Simulation Program*; Solar Energy Laboratory, University of Wisconsin: Madison, WI, USA, 2018 .
59. Dorfner, J. Urbs: A Linear Optimisation Model for Distributed Energy Systems 2022 . Available online: https://urbs.readthedocs.io/en/latest/ (accessed on 21 June 2022).
60. Reinhart, C.F.; Cerezo Davila, C. Urban building energy modeling—A review of a nascent field. *Build. Environ.* **2016**, *97*, 196–202. [CrossRef]

energies

MDPI

Article

Do Smart Cities Restrict the Carbon Emission Intensity of Enterprises? Evidence from a Quasi-Natural Experiment in China

Yituan Liu [1], Qihang Li [2,3] and Zheng Zhang [4,*]

[1] School of Business Administration, Shandong University of Finance and Economics, Jinan 250014, China; 20190711329@mail.sdufe.edu.cn
[2] Longshan Honors School, Shandong University of Finance and Economics, Jinan 250014, China; lqh@sdufe.edu.cn
[3] School of Economics, Shandong University of Finance and Economics, Jinan 250014, China
[4] School of Management Science and Engineering, Shandong University of Finance and Economics, Jinan 250014, China
* Correspondence: john_zhang@sdufe.edu.cn

Abstract: The concept of "smart cities" plays a positive role in the overall green and sustainable development of a nation. However, it is still debated whether smart cities can restrain the carbon emission intensity at the micro-level and promote the green transformation of enterprises. To this end, based on China's smart city policy (SCP) and regional enterprise data from 2008 to 2015, we study the impact of SCP on the carbon emission intensity of local enterprises, using the difference-in-differences method. The results show that SCP significantly reduces the carbon emission intensity of enterprises, and the estimated results remain significant after the propensity score matching. The mechanism analysis finds that digital transformation, innovation by enterprises, and urban green innovation all strengthen the impact of SCP on the carbon emission intensity of enterprises. The conclusions extend the scope of the existing research and provide suggestions for micro-enterprises to take advantage of SCP for better development.

Keywords: smart city policy; carbon emission intensity; digital transformation; green innovation; difference-in-differences

Citation: Liu, Y.; Li, Q.; Zhang, Z. Do Smart Cities Restrict the Carbon Emission Intensity of Enterprises? Evidence from a Quasi-Natural Experiment in China. *Energies* **2022**, *15*, 5527. https://doi.org/10.3390/en15155527

Academic Editors: Benedetto Nastasi, Nuno Carlos Leitão and Andrea Mauri

Received: 29 June 2022
Accepted: 27 July 2022
Published: 29 July 2022

1. Introduction

A smart city is a new, efficient, and technologically advanced city that integrates green and social development [1]. Such cities effectively alleviate the inability to process information related to urban environmental protection and resource utilization efficiency [2]. Through the full application of a new generation of information technology, smart cities can effectively optimize urban services and operations [3], alleviate the information processing capacity contradiction between urban environmental protection and resource utilization efficiency, and achieve the benefits of extensive informatization, industrialization, and urbanization [4]. With the rise of the internet and mobile technology, the smart city is an advanced form of information-based city development.

Owing to its potential advantages brought by digital technologies, the smart city has attracted the attention of many countries, such as the USA, Germany, Japan, and China [5]. However, some studies indicate that environmentally friendly smart cities may exacerbate urban pollution. For example, the large-scale information and communication technology (ICT) and industrial construction brought about by smart city policy (SCP) can lead to an increase in electricity consumption, which, in turn, increases carbon emissions [6]. Although the increases in ICT and internet penetration driven by SCP are not always beneficial to the environment, other studies indicate that, in general, smart cities positively impact the urban environment (e.g., carbon emission reduction) [7].

The above views are quite divergent, and the reason is that many influencing factors contribute to the total carbon emissions in a city. Among these, enterprises, which are the key drivers of an urban economy, generate most of the city's carbon emissions. To some extent, the carbon emission intensity of enterprises directly determines that of the entire city. Therefore, it is necessary to study the impact of SCP on the carbon emission intensity of enterprises at the micro-level. Furthermore, it is worth noting that a smart city is a recently adopted concept that relies mainly on digital transformation to realize city smartness [8]. Digital technologies are closely associated with the concept of a smart city [9]. We posit that the relationship between smart cities and digitalization can promote the adjustment of the city's industrial structure, thereby influencing the city's carbon emissions. This speculation is reasonable from a macro perspective, that is, thinking from the perspective of the whole city. However, does this speculation hold true for micro-enterprises in the city? In other words, does SCP influence the digital transformation of the micro-enterprise and subsequently impact its carbon emission intensity? Most mainstream research uses the construction of information infrastructure in a city to measure its degree of digital transformation. This method, however, does not adequately reflect the degree of digital transformation of urban enterprises.

As mentioned, even if SCP may directly influence some economic and environmental indexes at the city level, their influence at the enterprise level remains ambiguous. There are some specific examples in China showing that SCP has impacted the activity of pollution emission for micro-enterprises. A representative example is the State Grid Corporation of China (SGCC). In 2018, the SGCC completed a total of 38,000 electric energy replacement projects with the support of the "Digital New Infrastructure" policy brought by smart cities, replacing 135.3 billion kWh of electricity, equivalent to reducing coal burning by 75.77 million tons and reducing carbon dioxide emissions by 1.35 billion tons. In November 2020, the SGCC developed China's first globally empowered industrial internet platform in the energy sector. This platform can provide digital intelligent services in energy production and consumption, which promote the digital transformation of the enterprise.

Thus, to investigate whether this impact is common and infer the causal relationship between SCP and the carbon emission intensity of enterprises, we use a difference-in-differences (DID) method to study a quasi-natural SCP experiment based on China's SCP and regional enterprise data from 2008 to 2015. Furthermore, we use propensity score matching (PSM) as part of a robustness test. The DID is a useful and popular method to evaluate policies, as it can significantly reduce problems that arise from endogeneity. The specific process of our DID method is described in Section 3.1. Our research period from 2008 to 2015 was selected because 2015 is the latest year that enterprise-level carbon emission intensity data are available (explained in detail in Section 3.1). During this period, SCP significantly reduced the carbon emission intensity of enterprises. Furthermore, research has found that the digital transformation of enterprises, enterprise innovation, and urban green innovation have strengthened the inhibitory effect of SCP on the carbon emission intensity of enterprises.

Our study's novelty is established in two aspects. First, we directly study the effect of SCP at the micro-level, whereas most of the extant literature focuses only on the effect of SCP at the macro-level, including cities' pollution emissions, energy consumption, and green development. Thus, we extend the impact of SCP to the micro-enterprise level and study how SCP influences enterprises' activities and strategies. We believe this is a key contribution to the literature. Second, we analyze the relationship between SCP, digital transformation, and green innovation, looking at enterprise-level data. In the mechanism analysis, we find that SCP significantly improves green innovation performance and enhances the speed of digital transformation for micro-enterprises. Consequently, our results provide new perspectives for further research studying SCP.

Additional contributions are as follows: (1) Using the empirical method, we analyze the impact of SCP on carbon emission intensity at the enterprise level and expand the conclusions of existing studies at the city level; (2) We use digital word frequency to measure the digital transformation of enterprises, which we find to have strengthened the inhibitory effect of SCP on corporate carbon emissions. This finding also verifies Bhujabal et al.'s [7] conclusion regarding reducing carbon emissions through urban information infrastructure construction at the micro-level. As Xu et al. [10] indicate, the positive impact of innovation affects the carbon emission intensity of cities and enterprises through the adjustment of both the industry structure and enterprise production methods. Accordingly, our results verify the positive impact of SCP on enterprise innovation.

The rest of this paper is presented as follows. Section 2 reviews the relevant literature and introduces the implementation of China's SCP. Section 3 introduces the data sources, variable measurement, and econometric model. Section 4 conducts empirical analyses, robustness tests, a heterogeneity analysis, and a mechanism analysis. Section 5 presents discussions. Section 6 provides conclusions and corresponding suggestions.

2. Literature Review and Hypothesis Development

2.1. Smart Cities and Ecological Environment

The prominent feature of smart cities is the application of ICT and the large-scale construction of related infrastructure. ICT represents the image and expectations of the future [11]. Building a smart city is one significant achievement in the path toward realizing sustainable urban development, which leads to the application of information technology and an increase in the overall competitiveness of cities [12]. Under rapid economic development, the deterioration of the urban environment caused by human activities has become a problem that cannot be ignored [13]. Therefore, smart cities have become a key direction in transforming urban development in many countries [5]. Regions with developed economies and rapid information technology development usually enjoy a high level of social informatization and digitalization. This means they have good basic conditions for smart development and relatively stable planning schemes for smart city construction. Developing a digitally driven smart city is an important way to promote green and sustainable economic growth [5]. Smart cities include the characteristics of urbanization (infrastructure construction) and informatization (digital facility construction and application) because, to a certain extent, they are a product of the combination of these two elements.

Many scholars have researched the impact of urbanization on the environment. Urbanization is seen as an important factor in the soaring global energy consumption and the rapid increase in carbon dioxide emissions [11]. Early studies have found that, since urbanization often accompanies industrialization, there is a close correlation between urbanization and the greenhouse effect [1], which leads to increased urban pollution. Research in developing countries shows that the impact of urbanization on carbon emissions forms an inverted "U"-shaped relationship [6]. However, some scholars posit that environmental pollution will ease with urban expansion, mainly because wealthy cities can transfer part of the environmental cost to other regions [15]. In particular, urbanization measured by different indicators has various impacts on carbon dioxide emissions [14], which explains the differences in research conclusions regarding the impact of urbanization and carbon emissions.

As mentioned, the smart city is the embodiment of in-depth contemporary urban development. Through the combination of the Internet of Things (IoT) and big data, smart cities have greatly promoted the use of professional technology to solve environmental pollution problems [16], which is crucial to urban development and planning [2,16]. Existing research can be divided into two categories. The first explores the factors that promote the construction of smart cities, primarily knowledge [17], IoT systems [5,7,18], and information technology construction [3,9]. The second focuses on the impact of smart cities on urban development. Representative studies have demonstrated the role of smart cities in promoting urban innovation [10,15] and urban competitiveness [12].

Based on the feature of smart cities and their impact on the ecological environment, we propose hypothesis 1.

Hypothesis 1 (H1). *The impact from SCP on the carbon emission intensity of enterprises is significantly negative.*

2.2. *Smart City* Policy *and Digitalization*

We should consider the potential influence that SCP will bring. The digital transformation of an enterprise refers to the application of digital technology and equipment in the process of business improvement [19–21]. Modern society has entered the era of the digital economy [20], and digital technology has significantly changed original production methods, business models, and organizational patterns and has even subverted basic assumptions in many innovation theories [21]. One of the salient features of smart cities is the large-scale application of digital and information technology [22], wherein digital technology and traditional production models are inter-embedded, and production resources are reorganized and optimized with the help of technological innovation [23]. The rapid development of the digital economy has also stimulated research on the digital transformation of enterprises; for example, Yoo et al. [20], Nambisan et al. [21], Li et al. [24], Libert et al. [25], Vial [26], and Nwankpa and Roumani [27]. Thus, we find that the prominent feature for enterprises brought by SCP is to process digital transformation. As mentioned, digital technologies usually lead to less carbon emissions, so we develop hypothesis 2.

Hypothesis 2 (H2). *Digital transformation enhances the negative impact from SCP on the carbon emission intensity of enterprises.*

2.3. *Smart City* Policy *in China*

It should be noted that the characteristics and effects of SCP may vary from country to country. SCP appears to be the "patent" of developed countries, as few developing countries implement it. China is among the few developing countries that implement smart cities, which provides a reference for in-depth research on the application of smart cities in different countries.

China's smart city construction began relatively late but has developed rapidly. In January 2013, the Ministry of Housing and Urban-Rural Development of the People's Republic of China (MOHURD) announced China's smart city pilot list, which included 37 prefecture-level cities, 50 districts (counties), and 3 towns. A total of 90 cities formed the initial wave of smart city construction in China. In March 2021, the 14th Five-Year Plan for the National Economic and Social Development of the People's Republic of China and the Outline of Vision 2035, promulgated by the Chinese government, further emphasized the importance of building many smart cities. Since then, local governments have responded positively and have released development plans with the long-term goal of building smart cities by 2035. With the exception of Shaanxi and Jilin, the other 29 provinces have pointed to the importance of promoting smart city construction and improving the level of intelligence in social governance. At present, China is still promoting the construction of smart cities nationwide.

Qian et al. [5] employed the PSM-DID method to study the impact of China's SCP on economic green growth based on SCP implementation data. They found that smart cities can reduce urban unit gross domestic product (GDP), energy consumption, and waste emissions significantly. Xu et al. [10] and Xin and Qu [19] also used the PSM-DID method to study the impact mechanism of China's smart city construction on the level of urban innovation and green total factor productivity. Yu and Zhang [4] examined the impact of China's SCP on city-level energy emission intensity based on the DID method and found that smart city construction had a significant positive impact on city-level energy efficiency. The above studies analyze the impact of China's smart cities on urban construction and future development from different perspectives and find diverse impacts of China's SCP. The effects are more significant in China's megacities and central cities. Based on these studies, we find that SCP can promote green innovation at the city level. However, it is worth studying whether SCP can promote green-related innovation at the enterprise level. Hence, we develop Hypothesis 3.

Hypothesis 3 (H3). *The impact from SCP on green innovation at the enterprise level is significantly positive.*

Based on our literature review, we found that most relevant studies use samples and data at the prefecture or city level, whereas research on the impact of smart cities on carbon emission intensity at the enterprise level is still scarce. Based on China's SCP and the listed companies' micro-level data, we study the impact of SCP on enterprises' carbon emission intensity. This information can enrich the micro-level analysis based on existing macro-level research.

3. Materials and Methods

According to a previous analysis, we found that SCP was not enacted in all cities in China and that smart cities remain a small proportion of all cities. For convenience, the cities that have become smart city pilot cities based on SCP were the treatment group, and the rest were the control group. As many studies have shown, enterprises' carbon emission intensity remains steady over time; that is, the trends of carbon emission intensity in the treatment and control groups would be parallel, in theory (we test this in Section 4.2). Hence, China's SCP can be an ideal quasi-natural experiment, suitable for the DID method to infer causality between SCP and the carbon emission intensity of enterprises.

The specific estimation model is shown in Section 2.3. First, we conduct a baseline regression to test whether there is a causal relationship between SCP and carbon emission intensity. Second, if the estimated result is significant and passes the parallel trend test, the DID method will verify whether a causal relationship exists. Third, we conduct placebo tests to check whether the relationship is fake, as the existence of a relationship does not equate to robustness. Fourth, we use the PSM-DID method to alleviate the self-selection problem. Fifth, the heterogeneity analysis is needed to test whether the former relationship was influenced by the degree of industrialization, the enterprise scale, and the enterprise age, all of which are important. Finally, through the mechanism analysis, we explain why SCP could affect the carbon emission intensity of enterprises from the perspective of digital transformation and innovation. Our framework and steps are shown in Figure 1.

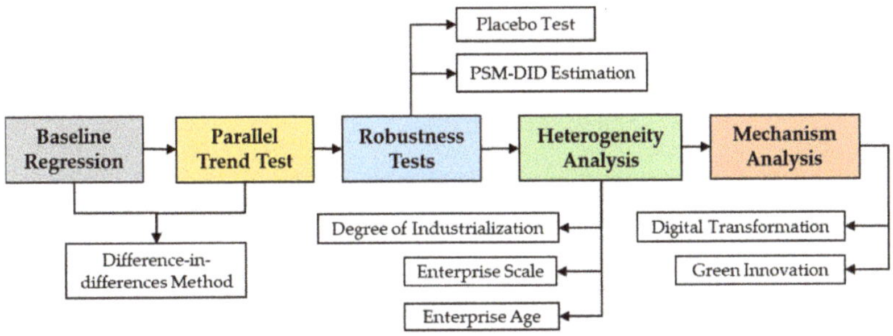

Figure 1. Research framework and processes.

3.1. Data Resource

Our data consist of three parts. The first part is the Chinese National Tax Survey Database (CNTSD) from 2008 to 2015. Each enterprise needs to pay an energy tax if it consumes energy. This database contains the consumption of three types of fossil fuels by energy-consuming enterprises. In addition, it should be noted that 2015 is the latest year in this database. The second part is the official website of the MOHURD; 2008–2015 SCP data for China were obtained through manual sorting. Finally, the other data of the listed companies and other prefecture-level cities come from the China Stock Market and Accounting Research Database (CSMAR) and various statistical yearbooks of the provinces in China. Links to the above data are given in the Data Availability Statement at the end of the article.

3.2. Variables Selection

3.2.1. Enterprise Carbon Emission Intensity

Referring to Cui et al. [28], the dependent variable in this study is enterprise carbon emission intensity (*lnCarbonEff*). The current method for measuring carbon emissions (CO_2) is based on the Guidelines for National Greenhouse Gas Inventories, published by the United Nations Intergovernmental Panel on Climate Change (IPCC) in 2006. The calculation method is proposed as

$$CO_2 = \sum_{i=1}^{n} E_i \times NCV_i \times CEF_i \times COF_i \times \frac{44}{12} \tag{1}$$

where E represents final energy consumption; NCV is the net calorific value of energy (called the average low calorific value in the Chinese national standard GB/T2589-2008); CEF is the carbon emission factor per unit of calorific value equivalent; COF is the carbon oxidation factor (99–100% of the carbon in fossil fuels is oxidized; so, according to IPCC, the default value of COF is set to 1); 44 and 12 are the carbon dioxide and carbon molecular weights, respectively; and i represents the category of various energy sources. According to the formula provided by the IPCC, the carbon dioxide emissions of enterprises can be calculated when their energy consumption is known.

However, owing to the difficulty in obtaining enterprise power consumption data and the large spatial and temporal differences between carbon emissions generated by power production and consumption, we focus only on the direct carbon emissions generated by the direct fossil energy consumption of enterprises. By dividing fossil fuel carbon emissions by the enterprises' gross output value (*GOV*), the carbon emission intensity of industrial enterprises is measured by the logarithm of carbon emissions per unit of output value. The calculation method is presented as

$$\ln CarbonEff = \ln \frac{CO_2}{GOV} \tag{2}$$

3.2.2. Smart City Policy

The independent variable in our study is SCP (*Policy*), and it is a dummy variable. If the city c becomes a smart city in year t, it is assigned a value of 1. Otherwise, it is assigned a value of 0.

3.2.3. Control Variables

To study the precise effect of SCP on the carbon emission intensity of enterprises, we need to control some factors at the enterprise and city level. We now describe the added control variables and selection bases. Large-scale enterprises are subject to more social supervision, and more shareholders pay attention to enterprise environmental information to evaluate their environmental performance [29]. Therefore, we control the size of the enterprise (*Size*), measured by the logarithm of the total assets of the enterprise. Generally, enterprises with better profitability can afford the expenditures required to reduce carbon emissions and optimize enterprise carbon emission intensity [30,31]. Therefore, we control the return on assets (*ROA*) and debt-to-asset ratio (*DAR*). Furthermore, fixed assets, such as plants, machines, and equipment, can reflect capital intensity, which is usually related to polluting activities [32]. Therefore, we control both enterprise fixed assets (*lnFX*) and regional fixed asset investment (*lnFI*). Considering that property rights will significantly affect the enterprise's business performance and strategy formulation [33], which will potentially affect the enterprise's carbon emissions and pollution indicators, the company attribute (*Equity*) is controlled as the last control variable at the enterprise level. Considering that the level of economic development will significantly affect the business activities of local enterprises, and a large number of studies have shown that the level of economic development is also the main factor affecting urban carbon emissions [6,34–37], we add the logarithm of prefecture-level city GDP (*lnGDP*) to control differences in the levels of regional economic development. Table 1 presents the variables used in our study, and Table 2 presents the descriptive statistics of these variables.

Table 1. Primary variables and explanations.

Variable Type	Symbol	Variable Name	Explanations
Dependent variable	*lnCarbonEff*	Enterprise carbon emission intensity	See Section 3.2.1.
Independent variable	*Policy*	Smart city policy	Dummy variable. See Section 3.2.2.
Enterprise-level variables	*Size*	Enterprise scale	Total assets of the enterprise.Logarithmic value
	ROA	Return on assets	Net profit/total assets of the enterprise
	DAR	Debt-to-asset ratio	Total liabilities/total assets of the enterprise
	Equity	Enterprise property	Stated-owned, non-stated-owned, overseas-funded enterprise, and so on.
	lnFX	Fixed assets	Fixed asset logarithm of the enterprise.Logarithmic value.
	DTR	Digital transformation ratio	See Section 4.6.1.
	RDSSR	R&D sum spend ratio	See Section 4.6.2.
	PQS	Patent quote sum	See Section 4.6.2.
City-level variables	*lnGDP*	Gross regional product	Logarithmic value. See Section 4.4.
	lnGDP1	Gross regional product of primary industry	Logarithmic value. See Section 4.4.
	lnGDP2	Gross regional product of secondary industry	Logarithmic value. See Section 4.4.
	lnGDP3	Gross regional product of tertiary industry	Logarithmic value. See Section 4.4.
	GDPR1	Proportion of GDP of primary industry	Gross regional product of primary industry/GDP
	GDPR2	Proportion of GDP of secondary industry	Gross regional product of secondary industry/GDP
	GDPR3	Proportion of GDP of tertiary industry	Gross regional product of tertiary industry/GDP
	GreenPatent	Green patent applications	See Section 4.6.2.
	GreenUtility	Green utility model applications	See Section 4.6.2.
	lnFI	Fixed asset investment at the city level	Logarithmic value.

Table 2. Descriptive statistics.

Variable Type	Symbol	Sample *Size*	*Mean*	Standard Deviation	Min.	Max.
Dependent variable	*lnCarbonEff*	8362	0.10	0.42	0.00	8.69
Independent variable	*Policy*	8362	0.30	0.46	0.00	1.00
Enterprise-level variables	*Size*	8362	14.45	1.72	0.00	23.55
	ROA	8362	0.04	0.09	−3.20	0.21
	DAR	8362	0.44	0.24	0.00	1.46
	Equity	8362	4.63	2.50	1.00	8.00
	lnFX	8362	11.80	2.57	0.00	20.25
	DTR	8362	0.39	1.13	0.00	32.92
	RDSSR	4897	4.76	5.25	0.00	58.85
	PQS	5980	110.30	300.30	0.00	2274
City-level variables	*lnGDP*	8362	8.49	1.12	3.85	10.20
	lnGDP1	8362	14.18	0.77	9.73	16.18
	lnGDP2	8362	16.82	1.06	11.91	18.27
	lnGDP3	8362	16.97	1.41	12.39	19.12
	GDPR1	8362	4.95	5.36	0.03	48.64
	GDPR2	8362	45.03	11.45	17.02	91.00
	GDPR3	8362	50.05	13.80	8.50	86.60
	GreenPatent	11,682	1829.62	2759.47	0.00	13,971.00
	GreenUtility	11,682	1252.67	1541.71	0.00	7268.00
	lnFI	8362	17.00	0.89	13.48	18.85

Note: This table includes all the variables used in this study, some of which are added as control variables in the regression. Variables *lnGDP1* to *GDPR3* are used as covariates in propensity score matching. Explanations of the rest of the important abbreviations: SCP: smart city policy; ICT: information and communication technology; DID: difference-in-differences method; PSM: propensity score matching method; IPCC: Intergovernmental Panel on Climate Change.

As shown in Table 2, the sample size for our main variables is 8376, which is sufficient to ensure the validity of our results. For the main variables, the standard deviations of *lnCarbonEff* and *Policy* are comparatively low, which means they are fewer extreme values. Other variables have similar features. Therefore, the data we use are comparatively suitable, and our later research results are credible.

3.3. Econometric Model

Referring to Yu and Zhang [4], we use the DID method to examine the impact of China's SCP on the carbon emission intensity of enterprises. By adding time, individual, regional, and industry fixed effects and clustering to cities, the designed estimation model is shown as

$$\ln CarbonEff_{i,t,j,c} = \alpha_0 + \alpha_1 Policy_{t,c} + \sum_k \alpha_k X_k + \gamma_i + \lambda_t + \mu_c + \nu_j + \varepsilon_{i,t,j,c} \tag{3}$$

where subscripts i, t, j, and c represent the enterprise, year, industry, and (prefecture-level) city, respectively. *lnCarbonEff* is the carbon emission intensity of the enterprise; *Policy* is SCP; X_k is a series of control variables; γ_i, λ_t, μ_c, ν_j represent the individual (enterprise), time (year), region (city), and industry fixed effects; $\varepsilon_{i,t,j,c}$ is a random disturbance term; and α_0 is a constant. The core coefficient we focus on is α_1, whose economic implication is the impact rate of SCP on the carbon emission intensity of enterprises.

4. Results and Analysis

4.1. Baseline Regression

According to model (3), the estimation results of SCP on the carbon emission intensity of enterprises are shown in Table 3. Without any control variables, the estimated coefficient of α_1 is −0.052, as shown in column (1). This indicates the originally negative and significant relationship between SCP and the carbon emission intensity of enterprises. The control variables are gradually added to columns (1) to (8). We can see in column (8) that, after adding control variables, fixed effects, and clustering to cities, the estimated coefficient becomes −0.078 and is significant at the 5% level. That is, SCP has effectively reduced the carbon emission intensity of enterprises. Hence, this verifies the relationship between SCP and the carbon emission intensity of enterprises, and the hypothesis H1 is accepted.

Table 3. Regression results of smart city policies and carbon emission intensity.

Variable	*lnCarbonEff*							
	(1)	(2)	(3)	(4)	(5)	(6)	(7)	(8)
Policy	−0.052 **	−0.052 **	−0.052 **	−0.052 **	−0.057 **	−0.058 **	−0.070 **	−0.078 **
	(−2.02)	(−2.02)	(−2.02)	(−2.04)	(−2.19)	(−2.24)	(−2.46)	(−2.02)
Size		−0.005	−0.005	−0.006	−0.006	−0.011	−0.010	−0.010
		(−0.43)	(−0.47)	(−0.54)	(−0.52)	(−0.94)	(−0.86)	(−0.79)
ROA			−0.089	−0.069	−0.076	−0.082	−0.101	−0.100
			(−1.09)	(−0.83)	(−0.83)	(−0.88)	(−1.01)	(−0.79)
DAR				0.071	0.092	0.084	0.102	0.100
				(1.19)	(1.52)	(1.37)	(1.56)	(1.33)
Equity					0.023 **	0.024 **	0.019	0.019
					(1.99)	(2.06)	(1.50)	(0.99)
lnFX						0.008	0.006	0.006
						(1.47)	(1.04)	(0.89)
lnGDP							0.131	0.166
							(0.73)	(0.82)
lnFI								−0.067
								(−0.52)
Year fixed effect	Yes	Yes	Yes	Yes	Yes	Yes	Yes	Yes
City fixed effect	Yes	Yes	Yes	Yes	Yes	Yes	Yes	Yes
Enterprise fixed effect	Yes	Yes	Yes	Yes	Yes	Yes	Yes	Yes
Industry fixed effect	Yes	Yes	Yes	Yes	Yes	Yes	Yes	Yes
Constant	0.115 ***	0.180	0.189	0.171	0.048	0.024	−1.023	−0.169
	(15.39)	(1.18)	(1.24)	(1.12)	(0.29)	(0.15)	(−0.69)	(−0.09)
Observations	7871	7871	7869	7869	7797	7783	7281	7247
Adj. R-squared	0.132	0.132	0.132	0.132	0.138	0.140	0.143	0.106
F Statistics	4.076	2.129	1.816	1.717	2.353	2.351	1.914	0.958

Note: *t*-values are reported in parentheses. The robust standard errors are clustered at the city level. *** and ** represent significance at the levels of 1% and 5%, respectively. *Policy*: SCP; *lnCarbonEff*: carbon emission intensity of enterprises; *Size*: enterprise scale; *ROA*: return on assets; *DAR*: debt-to-asset ratio; *Equity*: enterprise property; *lnFX*: fixed assets; *lnGDP*: gross regional product; *lnFI*: fixed assets investment at the city level; Adj.: adjusted.

4.2. Parallel Trend Test

The premise of the validity of the DID method is to pass the parallel trend test. Referring to Beck et al. [28], we conduct a parallel trend test on the treatment group (*Policy* = 1) and the control group (*Policy* = 0). The design estimation model is shown in formula (4), and the results are shown in Figure 2. The horizontal axis of Figure 2 is the periods before and after the establishment of SCP, and the vertical axis is the percentage change of the estimated coefficient of the test statistic a_m. We can see that, before the SCP took place, the 95% confidence intervals of the estimated coefficients of the treatment group and the control group were estimated to include the 0 in the vertical axis; that is, the difference between the treatment and the control group was not significant. Thus, the parallel trend test passed.

$$\ln CarbonEff_{i,j,t,c} = a_0 + \sum_{m=1}^{4} a_m Policy_{t,c}^{-m} + \sum_{n=1}^{3} a_n Policy_{t,c}^{+n} + \sum_k a_k X_k + \gamma_i + \lambda_t + \mu_c + \nu_j + \varepsilon_{i,t,j,c} \quad (4)$$

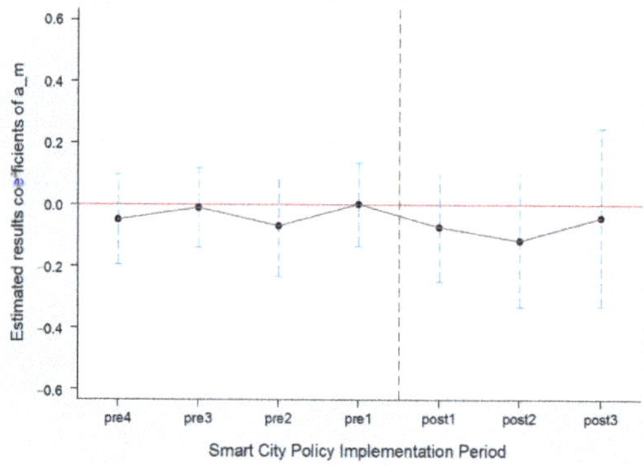

Figure 2. Result of the parallel trend test. a_m is the difference in carbon emission intensity between the treatment and control groups before the policy (i.e., SCP) happens. The blue dashed line is the 95% confidence interval of the estimated a_m. The red line marked the zero value of the estimated a_m. The black dashed line connects the estimates in different periods.

4.3. Placebo Test

The placebo test is a robustness test that changes the timing of policy shocks artificially [38]. To further test the robustness of the results, we refer to the method of Lu et al. [39] and use the placebo test. The specific method is to select the known cities without SCP as the treatment group for regression within the research interval. Repeating this random sampling process 1000 times, the normal distribution kernel density of the 1000 "pseudo-policy" estimated coefficients is shown in Figure 3, and the estimated coefficient −0.078 of the baseline regression in this study is marked with a dashed line in the figure. We can see in Figure 3 that the placebo estimation results are significantly different from the baseline regression results, indicating that there is no false association between SCP and the carbon emission intensity of enterprises.

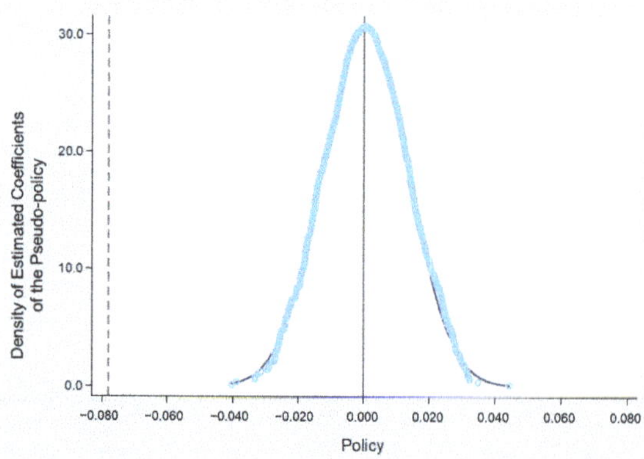

Figure 3. Result of the placebo test. The vertical axis is the density of estimated coefficients of the corresponding pseudo-policy. The blue holly circle represents the distribution of the pseudo-policy coefficients. The grey dashed line marks the estimated coefficient in the baseline regression. The grey line is the zero value of the pseudo-policy coefficients.

4.4. PSM-DID Estimation

As mentioned above, China's SCP reflects obvious regional heterogeneity, mainly determined by the level of local economic development. Therefore, we refer to the ideas of Xin and Qu [1], Yu and Zhang [4], Qian et al. [5], and Xu et al. [10] and re-estimate the model (4) by PSM to reduce the estimation error caused by potential sample selection bias. Then, the matched samples undergo the DID estimation; the results can better reflect the real impact of SCP on the carbon emission intensity of enterprises.

As covariates, we select the gross regional product (*GDP*), the gross regional product of each industry sample (*GDP1, GDP2, GDP3*), and their ratio to the *GDP* (*GDPR1, GDPR2, GDPR3*) in the study interval. We use the 1:1 nearest neighbor method to match the samples. PSM is conducted, and economic development indicators after matching are shown in Figure 4.

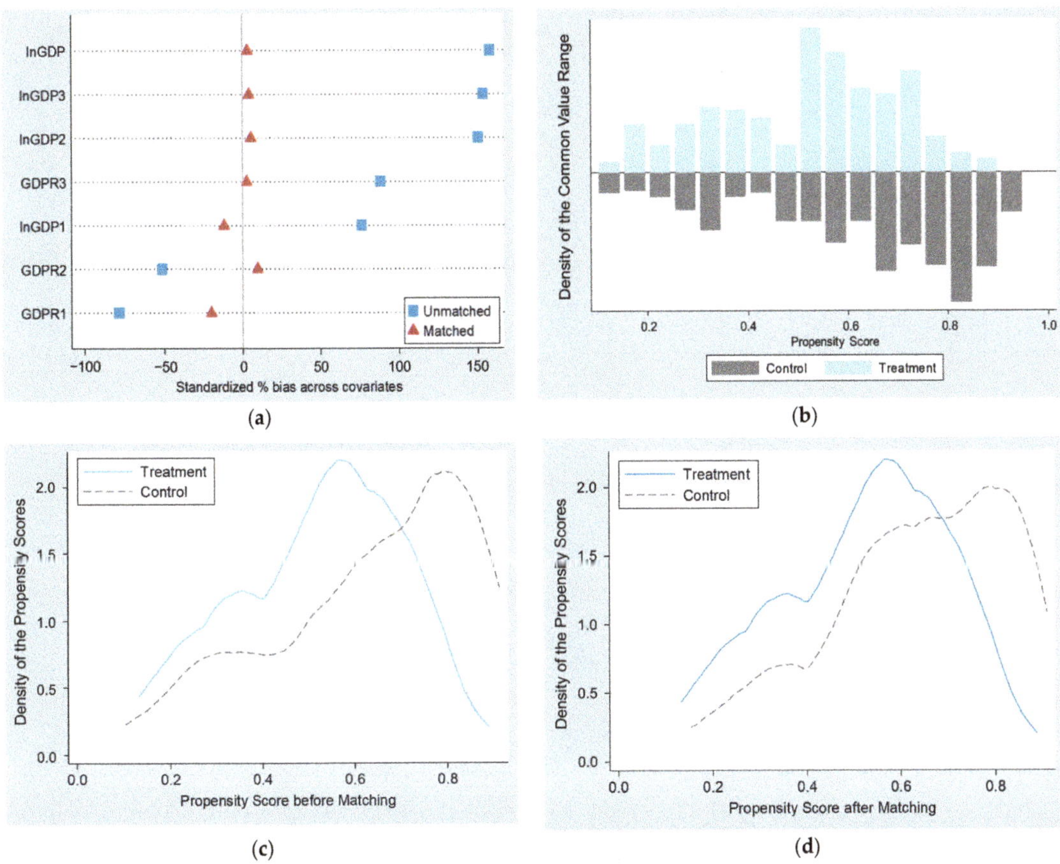

Figure 4. Results of the Propensity Score Matching (PSM). (**a**) Standardized bias across covariates; (**b**) common value range density of treatment and control groups; (**c**,**d**) density for treatment and control groups before and after the PSM, respectively. The grey line in (**a**) is the zero value of the standardized bias across covariates.

Part (a) of Figure 4 shows that the standard deviation of each indicator is reduced significantly after matching. Specifically, all covariate errors are reduced to within 20%, which indicates that the PSM method is suitable for our research. Part (b) of Figure 4 represents the propensity score of the treatment and control groups. In each score interval, the density of both the treatment and control groups is moderate. Parts (c) and (d) of Figure 4 show that, after the PSM, the treatment and control groups are approximately gathered at the intervals [0.5, 0.7].

The matched samples are re-estimated using DID, and the results are presented in Table 4. The results show that, in the samples matched by the economic development level of prefecture-level cities, the influence of SCP on the carbon emission intensity of enterprises remains significantly negative. After adding control variables, the coefficient estimated by PSM-DID is −0.133 (shown in column (8) of Table 4), which is higher than the baseline regression value of −0.078. Hence, the robustness of the results is strong regarding the influence of SCP on the carbon emission intensity of enterprises.

Table 4. Regression results of the PSM-DID estimation.

Variable	*lnCarbonEff*							
	(1)	(2)	(3)	(4)	(5)	(6)	(7)	(8)
Policy	−0.080 *	−0.082 *	−0.082 *	−0.081 *	−0.091 *	−0.093 *	−0.124 **	−0.133 *
	(−1.69)	(−1.72)	(−1.72)	(−1.71)	(−1.91)	(−1.95)	(−2.35)	(−1.89)
Size		−0.026	−0.027	−0.028	−0.027	−0.028	−0.013	−0.013
		(−1.27)	(−1.31)	(−1.36)	(−1.32)	(−1.26)	(−0.55)	(−0.41)
ROA			−0.103	−0.084	−0.100	−0.194	−0.323	−0.320
			(−0.85)	(−0.67)	(−0.67)	(−0.97)	(−1.20)	(−0.96)
DAR				0.068	0.086	0.058	0.096	0.102
				(0.66)	(0.81)	(0.53)	(0.73)	(0.70)
Equity					0.009	0.010	0.003	0.003
					(0.52)	(0.55)	(0.15)	(0.17)
lnFX						0.000	−0.011	−0.010
						(0.01)	(−1.07)	(−1.30)
lnGDP							0.590 *	0.633 *
							(1.73)	(1.72)
lnFI								−0.134
								(−0.94)
Year fixed effect	Yes	Yes	Yes	Yes	Yes	Yes	Yes	Yes
City fixed effect	Yes	Yes	Yes	Yes	Yes	Yes	Yes	Yes
Enterprise fixed effect	Yes	Yes	Yes	Yes	Yes	Yes	Yes	Yes
Industry fixed effect	Yes	Yes	Yes	Yes	Yes	Yes	Yes	Yes
Constant	0.136 ***	0.508 *	0.524 *	0.511 *	0.450	0.474	−4.244	−2.403
	(12.44)	(1.73)	(1.78)	(1.73)	(1.44)	(1.51)	(−1.56)	(−0.80)
Observations	3557	3557	3557	3557	3520	3509	2999	2976
Adj. R-squared	0.131	0.131	0.131	0.131	0.141	0.140	0.123	0.033
F Statistics	2.871	2.238	1.732	1.407	1.315	1.180	1.798	0.842

Note: *t*-values are reported in parentheses. The robust standard errors are clustered at the city level. ***, **, and * represent significance at the levels of 1%, 5%, and 10%, respectively. *Policy*: SCP; *lnCarbonEff*: carbon emission intensity of enterprises; *Size*: enterprise scale; *ROA*: return on assets; *DAR*: debt-to-asset ratio; *Equity*: enterprise property; *lnFX*: fixed assets; *lnGDP*: gross regional product; *lnFI*: fixed assets investment at the city level; Adj.: adjusted.

4.5. Heterogeneity Analysis

4.5.1. Degree of Industrialization

Directly correlated with carbon intensity, the production structure of enterprises is related to the industry in which the enterprise operates. Generally, the carbon emission intensity of heavy industrial enterprises should be greater than that of light industrial enterprises. To verify whether the impact of SCP on the carbon emissions of enterprises is heterogeneous due to the degree of industrialization, we divide all the enterprise samples into two groups, heavy-industry and light-industry, according to the degree of industrialization and in accordance with the Chinese industry classification standards. The regression results show that SCP has a more significant effect on the carbon emission intensity of heavy-industry enterprises. This indicates that the high requirements of SCP for environmental protection have forced heavy-industry enterprises to adjust their production structure to promote carbon emission reduction. The regression results are shown in columns (1) and (2) in Table 5.

Table 5. Regression results of the heterogeneity analysis.

Variable	*lnCarbonEff*					
	Light-Industry	**Heavy-Industry**	**Large-Scale**	**Small-Scale**	**Young Enterprise**	**Old Enterprise**
	(1)	**(2)**	**(3)**	**(4)**	**(5)**	**(6)**
Policy	−0.157	−0.102 *	−0.152 **	0.018	−0.025	−0.062 *
	(−1.61)	(−1.93)	(−2.49)	(1.14)	(−0.55)	(−1.69)
Control	Yes	Yes	Yes	Yes	Yes	Yes
Year fixed effect	Yes	Yes	Yes	Yes	Yes	Yes
City fixed effect	Yes	Yes	Yes	Yes	Yes	Yes
Enterprise fixed effect	Yes	Yes	Yes	Yes	Yes	Yes
Industry fixed effect	Yes	Yes	Yes	Yes	Yes	Yes
Constant	9.241	5.014 **	−3.150	0.454	3.698	−0.658
	(1.26)	(1.96)	(−0.73)	(0.27)	(0.91)	(−0.24)
Observations	887	2933	3961	2972	2436	4414
Adj. R-squared	0.150	0.191	0.121	0.207	0.245	0.095
F Statistics	1.015	1.517	1.425	0.744	0.361	1.045

Note: *t*-values are reported in parentheses. The robust standard errors are clustered at the city level. ** and * represent significance at the levels of 5% and 10%, respectively. *Policy*: SCP; *lnCarbonEff*: carbon emission intensity of enterprises; Control: control variables, including *Size, ROA, DAR, Equity, lnFX, lnGDP*, and *lnFI*; Adj.: adjusted.

4.5.2. Enterprise Scale and Age

There is also a direct relationship between the carbon emission intensity of an enterprise and its scale and age, so we group the sample according to enterprise size and age. The enterprises with a larger-than-average size, namely, *Size* > mean (*Size*), are large-scale enterprises, and those with a smaller-than-average size are called small-scale enterprises. Columns (3) and (4) in Table 5 present the regression results and show that small-scale enterprises have a more significant effect on carbon emission intensity than large-scale enterprises. Moreover, as the inherent production capacity of large-scale enterprises often leads to greater carbon emissions, SCP has generated more digital and information transformation opportunities for such enterprises. This stimulates them to reduce production costs and improve production efficiency.

Similarly, an enterprise with a higher-than-average age, namely, *Age > mean (Age)*, is considered an old enterprise, and that with a lower-than-average age is considered a young enterprise. The regression results are shown in columns (5) and (6) in Table 5. The results indicate that SCP has a more significant effect on the carbon intensity of old enterprises, indicating that enterprises with higher age (or a longer history of development) are more inclined to change their development methods and use more digital production methods to maintain their core competitiveness, under the impact of SCP.

We also categorized a sample of Chinese cities by location but did not find significant geographic heterogeneity in the SCP impact on the carbon emission intensity of enterprises.

4.6. Mechanism Analysis

To analyze the potential mechanism behind SCP's impact on the carbon emission efficiency of enterprises, we try to explain it from the perspective of digital transformations and innovations. The research concept is shown in Figure 5.

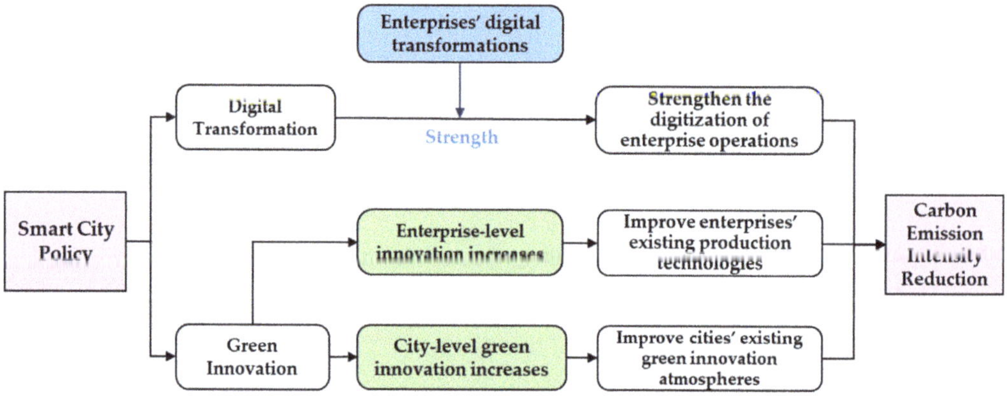

Figure 5. The mechanism analysis of the effect of SCP on carbon emission intensity.

4.6.1. Digital Transformation

The construction of digital infrastructure mandated by SCP provides enterprises with the conditions for digital transformation. Using digital technology, enterprises can combine their own advantages with smart cities to effectively transform production methods and strategies [24]. Enterprises can use this information to serve their production decisions, market-oriented tracking, and production process optimization [40]. Therefore, digital transformation and platform-based enterprise development are inevitable requirements based on strategic iterative upgrades and represent the general direction of the future development of the environmental industry. With the help of digital transformation, enterprises can fully tap the value of data and facilitate product and service innovation and upgrading, thereby generating new digital businesses. Digitalization can improve operational intensity internally and improve customer satisfaction externally; moreover, digital transformation can bring more core competitiveness to enterprises under the derivation of SCP.

More importantly, in an era of a digital economy, enterprises adhering to the traditional operation model will face pressure from competitors' digital transformation. The traditional production and operation mode of enterprises may significantly limit the development of enterprises [21]. The revolutionary impact of digital transformation leads to the elimination of outdated production models. Consequently, high-pollution and high-emission production models will no longer be the main force for urban economic development driven by SCP.

Based on the above analysis, we find that SCP may further promote the digital transformation of enterprises in the cities they operate in; that is, digital transformation enhances the impact of SCP on the carbon emission intensity of enterprises. To test that impact, we refer to Xue et al. [41] to select the digital-related word frequency to measure the level of digital transformation of enterprises. Formula (5) represents the model used to estimate the impact of digital transformation on enterprises.

$$\ln CarbonEff_{i,t,j,c} = \beta_0 + \beta_1 Policy_{t,c} * DTR + \sum_k \beta_k X_k + \gamma_i + \lambda_t + \mu_c + \nu_j + \varepsilon_{i,t,j,c} \quad (5)$$

$$DTR = \frac{\text{Digital related words in enterprise annual report}}{\text{Total words in enterprise annual report}} * 1000 \quad (6)$$

where *DTR* is the degree of the digital transformation of enterprises, and the calculation formula is shown in Equation (6). Because digital words are rare in the annual report, which leads to an extremely low coefficient, we multiply 1000 by the digital-related words in the enterprise annual report in Equation (6). We measured the degree of enterprises' digital transformation by using Python to mine and count the digital vocabulary (cloud computing, big data, Internet of Things, etc.) in the annual report of the enterprise, and we obtain the digitized word frequency for each enterprise during the year. The digital words we selected are shown in Table 6. We added the interactive items (i.e., *DTR**Policy*) and their original items (i.e., *Policy* and *DTR*) to the model. If the interactive item was significant, the moderating effect of digital transformation was established. The estimated results of this model are shown in column (5) of Table 7. The coefficient of the interactive item is −22.45, with the same direction and significance of coefficient −0.078 in column (8) of Table 3. Therefore, the digital transformation at the enterprise level enhanced the reducing effect of SCP on the carbon emission intensity of enterprises. Hence, the hypothesis H2 is accepted.

Table 6. Types of digitization categories and specific vocabularies.

Type	Digital-Related Words
Artificial intelligence	Business intelligence, image understanding, investment decision assistance systems, intelligent data analysis, intelligent robots, machine learning, deep learning, semantic search, biometrics, face recognition, speech recognition, identity verification, autonomous driving, natural language processing.
Blockchain technology	Digital currency, distributed computing, differential privacy technology, smart financial contracts.
Cloud computing technology	Stream computing, graph computing, memory computing, multi-party secure computing, brain-like computing, green computing, cognitive computing, fusion architecture, billion-level concurrency, EB-level storage, Internet of Things, cyber-physical systems.
Big data technology	Data mining, text mining, data visualization, heterogeneous data, credit investigation, augmented reality, mixed reality, virtual reality.
Digital technology applications	Mobile internet, industrial internet, internet medical, e-commerce, mobile payment, third party payment, NFC payment, smart energy, B2B, B2C, C2B, C2C, O2O, network connection, smart wear, smart agriculture, smart transportation, smart medical care, smart customer service, smart home, smart investment, smart tourism, smart environmental protection, smart grid, smart marketing, digital marketing, unmanned retail, internet finance, digital finance, fintech, financial technology, quantitative finance, open banking.

Note: the original digital-related words are in Chinese, and this table lists their English translation. We should point out that most of the existing research on digital transformation is at the qualitative level, and few studies adopt quantitative methods. Although the indicator used in this study (i.e., *DTR*) does not accurately reflect the degree of the digital transformation of enterprises, it can still reflect enterprises' perceptions of digital transformation to a certain extent [42]. Therefore, it has certain rationality and reference values.

4.6.2. Green Innovation

Another distinctive feature of the digital economy is innovation, which includes technological innovation at the enterprise level and green innovation at the city level. At the enterprise level, the digital transformation generated by the SCP can significantly enhance the enterprise's operational intensity. This enables the enterprise to achieve greater output performance under the resource boundary of the original research and development (R&D) innovation [43]. Generally, the R&D spend sum reflects the importance that enterprises place on innovation. In addition, the number of patent citations, compared with the number of patents, can reflect the enterprise patent quality and better reflect the level of enterprise innovation.

At the city level, SCP has produced many innovations, especially green innovation patents, which play a significant role in urban environmental improvement and ecological governance [10]. For example, "the digital intelligent operation system for rural sewage treatment terminal facilities" developed by the China State Construction Engineering Group Co., Ltd. is based on the basic network built by the IoT. It uses the cloud platform as a data center and is supplemented by inspections of intelligent equipment, the automatic transformation of on-site sewage treatment terminals, and on-site safety protection. This system represents a case of digital innovation in rural revitalization, the ecological environment, and the intelligent operation of rural pollution control, and it reflects the main characteristics of smart cities.

Based on the above information, referring to the analyses of Hall and Jaffe [44], Zahra and Bogner [45], and Meuleman and De Maeseneire [46], we select the R&D spend sum ratio (*RDSSR*) and the number of patent citations (*PQS*) as proxy variables for innovation at the enterprise level. Referring to Xu et al. [10], we select the number of green patent applications (*GreenPatent*) and the number of green utility model applications at the city level (*GreenUtility*) as proxy variables for green innovation at the city level. We select green patents because they can better reflect the innovation level of urban environmental protection and ecological governance and are more relevant to SCP. The estimated results are shown in columns (1) to (4) of Table 7.

In Table 7, we can see that the estimated coefficients of SCP at the enterprise level and city level are all significant. At the enterprise level, implementing SCP will bring about a 25.5% incremental increase in the R&D spend sum ratio of enterprises and an incremental increase of 39.66 in the sum of patent quotes. At the city level, implementing SCP will increase the number of cities' green patents by about 223.91 and the number of green utilities by about 123.79. These all reflect the significant and enormous relationship between SCP and green innovation. Based on this, we can find an influencing path from SCP to green innovation and finally reduce the carbon emission intensity of enterprises. That is, the hypothesis H3 is verified.

$$RDSSR_{i,t,j,c} = \phi_0^R + \phi_1^R Policy_{t,c} + \sum_k \phi_k^R X_k + \gamma_i + \lambda_t + \mu_c + v_j + \varepsilon_{i,t,j,c} \quad (7)$$

$$PQS_{i,t,j,c} = \phi_0^P + \phi_1^P Policy_{t,c} + \sum_k \phi_k^P X_k + \gamma_i + \lambda_t + \mu_c + v_j + \varepsilon_{i,t,j,c} \quad (8)$$

$$GreenPatent_{t,c} = \varphi_0^P + \varphi_1^P Policy_{t,c} + \sum_k \varphi_k^P X_k + \gamma_i + \lambda_t + \mu_c + v_j + \varepsilon_{i,t,j,c} \quad (9)$$

$$GreenUtility_{t,c} = \varphi_0^U + \varphi_1^U Policy_{t,c} + \sum_k \varphi_k^U X_k + \gamma_i + \lambda_t + \mu_c + v_j + \varepsilon_{i,t,j,c} \quad (10)$$

Table 7. Regression results of SCP on green innovation and digital transformation.

Variable	Enterprise Level		City Level		Digital Transformation
	RDSSR	*PQS*	*GreenPatent*	*GreenUtility*	*lnCarbonEff*
	(1)	(2)	(3)	(4)	(5)
Policy	0.225 ** (2.22)	39.657 *** (4.76)	223.90 8 *** (14.44)	127.787 *** (11.13)	−0.198 (−0.67)
*Policy*DTR*					−22.449 * (1.74)
Control	Yes	Yes	Yes	Yes	Yes
Year fixed effect	Yes	Yes	Yes	Yes	Yes
City fixed effect	Yes	Yes	Yes	Yes	Yes
Enterprise fixed effect	Yes	Yes	Yes	Yes	Yes
Industry fixed effect	Yes	Yes	Yes	Yes	Yes
Constant	12.417 (0.82)	1701.156 ** (2.10)	6725.162 *** (5.30)	5773.045 *** (4.81)	0.472 (−0.08)
Observations	4897	5980	11682	11682	4328
Adj. R-squared	0.873	0.740	0.995	0.990	0.003
F Statistics	4.229	4.375	49.387	35.201	1.18

Note: *t*-values are reported in parentheses. The robust standard errors are clustered at the city level. ***, **, and * represent significance at the levels of 1%, 5%, and 10%, respectively. The regression results of SCP to green innovations at the enterprise level and city level are shown in column (1) to (2) and column (3) to column (4), respectively. The regression result of the moderating effect of digital transformation between SCP on the carbon emission intensity of enterprises is shown in column (5). *Policy*: SCP; *lnCarbonEff*: carbon emission intensity of enterprises; *DTR*: the degree of the digital transformation of enterprises; *RDSSR*: R&D spend sum ratio; *PQS*: the number of patent citations; *GreenPatent*: the number of green patent applications at the city level; *GreenUtility*: the number of green utility model applications at the city level; Control: control variables, including *Size*, *ROA*, *DAR*, *Equity*, *lnFX*, *lnGDP*, and *lnFI*; Adj.: adjusted.

5. Discussion

5.1. Implications for Research

We examine China's SCP and data on the carbon emission intensity of regional enterprises from 2008 to 2015 and adopt the DID method to empirically analyze the impact of China's SCP on these emissions. The results show that SCP has a significant negative impact; that is, SCP inhibits the carbon emissions of enterprises. Specifically, SCP can reduce the carbon emission intensity of enterprises by about 7.8% in our research samples and periods. However, we should emphasize that the absolute number is only for reference, because the value may change differently based on different calculation methods, samples, and periods. However, we have confidence in this number because the value of 7.8% is not much different from the estimated results obtained by Yu and Zhang [4] and Cui et al. [28].

Our research results are consistent with the conclusion of Qian et al. [5] and extend the conclusion of Yu and Zhang [4] on a micro level. Simultaneously, we find that SCP affects the carbon emission intensity of enterprises by increasing R&D investment, promoting innovation, and enhancing green innovation at the city level. These results provide evidence that SCP can stimulate green total factor productivity [10] and green innovation at the city level [19] and further indicate that SCP can stimulate innovation at the enterprise level. In addition, our heterogeneity analysis finds that the impact of SCP on the carbon emission intensity of enterprises is heterogeneous in terms of industrialization, enterprise size, and enterprise age, whereas the heterogeneity of the urban geographic location is not obvious. Our mechanism analysis finds that digital transformation strengthens the inhibitory effect of SCP on the carbon emission intensity of enterprises. That is, digital transformation stimulated by SCP is an important method to adjust production and operation processes for enterprises. This result provides evidence that smart cities are the large-scale application of digital and information technology, which corresponds to the conclusions of van den Buuse [22].

However, we should note that SCP may have a different impact on carbon or pollution emissions due to the differences in regions and countries. For instance, Yigitcanlar and Kamruzzaman [47], based on the data in the United Kingdom from 2005 to 2013, found that the impact of SCP on the carbon emission levels in a city is not a simply linear relation. Considering the characteristics of the UK, Cavada et al. [48] propose the interdependent triptych philosophy for "smart city" resilience to offer tailored solutions for cities, communities, and individuals. There are many similar cases, which indicates that it is important to focus on the differences in regions or countries and provide suitable SCP to better implement in order to reduce carbon emissions.

5.2. Marginal Implications and Limitations

We believe that our research can inform government SCP policies by providing a new perspective and reference on achieving success. Environmental sustainability is vital for humans' wellbeing. SCP is one of the pathways to achieving that objective. SCP can significantly reduce the carbon emission intensity of enterprises, increase energy efficiency, make our cities more livable, less polluted, and energy efficient, and, with innovative planning, make cities physically and psychologically healthier for inhabitants.

This study has a few limitations that should be mentioned. First, the research period in this paper is 2008–2015, which is limited by the CNTSD and may not reflect recent trends. However, our results show a significant negative relationship between SCP and the carbon emission intensity of enterprises. We believe that this effect is still present. Second, we use digital word frequency in the annual reports of listed companies to measure the digital transformation of enterprises. In our case, this serves to verify the moderating role of digital transformation between SCP and enterprise carbon efficiency. However, although useful, there are limitations to such an analysis of digital transformation indicators. We hope to analyze the internal impact mechanism of enterprises' digital transformation using better indicators. Third, we use green patent-related indicators to measure the level of innovation at the city level. Compared with other patents, green patents can reflect smart city characteristics to a certain extent, but not fully. Fourth, we only focus on the direct fossil fuel consumption and exclude the electricity in the carbon emissions. In our research period, electricity is not a huge part of enterprises' carbon emissions. In recent years, however, it has become a major proportion of enterprises' carbon emissions, which means that future research that focuses on this field should note the electricity consumption in carbon emissions. Fifth, the most estimated coefficients obtained by PSM-DID only represent a significance level of 0.10. To test the extreme situation, we select the strictest method for the propensity score matching (i.e., we use the 1:1 nearest neighbor method to match the samples), which ensures that the matched samples are more representative. However, the process of matching may eliminate many samples. In this case, the significance of estimated coefficients may increase (from 0.05 to 0.1), but it is only to show that the robustness of our estimated results in the baseline regression is strengthened.

6. Conclusions and Recommendations

The main objective of this study is to verify the causal relationship between SCP and the carbon emission intensity of enterprises by using the DID method. Based on a quasi-natural experiment in China's SCP from 2008 to 2015, we find that SCP significantly reduces the carbon emission intensity of enterprises, which indicates that hypothesis H1 is verified positively. Furthermore, we find that green innovation and digital transformation are important mechanism pathways, indicating that hypotheses H2 and H3 are verified positively.

Based on the above conclusions and discussions, we conclude with the following suggestions. First, at the city level, we believe that SCP is beneficial to improving the urban environment. Among businesses, however, the benefits are only more pronounced for heavy-industry, large-scale, and older firms. Such enterprises are among the main driving forces of urban economic development. Therefore, when formulating SCP rules, local governments should focus on analyzing the type and structure of their local enterprises, supporting their digital transformation in the creation of solid digital transformation. Second, as innovation is an important factor affecting carbon emission intensity, local governments should emphasize auxiliary policies related to their SCP to promote innovation among local enterprises. Reasonable financial subsidies should also be provided to encourage innovation. Third, the selected empirical proxy variables for digital transformation and innovation enrich the existing literature and provide pointers for research. We conclude that, when governments formulate green policies, they should consider the impacts on smart city construction, digital transformation, and innovation equally, as they are interconnected.

Author Contributions: Conceptualization, Y.L. and Q.L.; Data curation, Y.L. and Q.L.; Formal analysis, Q.L. and Z.Z.; Investigation, Y.L., Q.L. and Z.Z.; Methodology, Y.L. and Q.L.; Project administration, Q.L. and Z.Z.; Resources, Q.L. and Z.Z.; Software, Y.L.; Validation, Z.Z.; Visualization, Y.L. and Q.L.; Writing—original draft, Y.L. and Q.L.; Writing—review & editing, Y.L., Q.L. and Z.Z. All authors have read and agreed to the published version of the manuscript.

Funding: The research project is supported by the Natural Science Foundation of Shandong Province, China (ZR2019MG040), and the Ministry of education of Humanities and Social Science project, China (19YJAZH063).

Institutional Review Board Statement: Not applicable.

Informed Consent Statement: Not applicable.

Data Availability Statement: The official website of the MOHURD is https://www.mohurd.gov.cn/ (accessed on 10 May 2022), which includes the latest information about SCP in China and the list of pilot cities. The Chinese National Tax Survey Database (CNTSD) can be made available upon request to the State Administration of Taxation of China or the Ministry of Finance of China. The official website of the China Stock Market and Accounting Research Database (CSMAR) is https://www.gtarsc.com/ (accessed on 10 May 2022). The relevant information about the IPCC Guidelines for National Greenhouse Gas Inventories can be found at https://www.scirp.org/reference/ReferencesPapers.aspx?ReferenceID=221977 (accessed on 10 May 2022).

Acknowledgments: The authors wish to acknowledge Jindong Liu of Shandong University of Finance and Economics for his data support and suggestions.

Conflicts of Interest: The authors declare no conflict of interest.

References

1. Vanolo, A. Smartmentality: The smart city as disciplinary strategy. *Urban Stud.* **2013**, *51*, 883–898. [CrossRef]
2. Li, X.; Fong, P.S.W.; Dai, S.; Li, Y. Towards sustainable smart cities: An empirical comparative assessment and development pattern optimization in China. *J. Clean. Prod.* **2019**, *215*, 730–743. [CrossRef]
3. Harrison, C.; Eckman, B.; Hamilton, R.; Hartswick, P.; Kalagnanam, J.; Paraszczak, J.; Williams, P. Foundations for smarter cities. *IBM J. Res. Dev.* **2010**, *54*, 1–16. [CrossRef]
4. Yu, Y.; Zhang, N. Does smart city policy improve energy efficiency? Evidence from a quasi-natural experiment in China. *J. Clean. Prod.* **2019**, *229*, 501–512. [CrossRef]
5. Qian, Y.; Liu, J.; Cheng, Z.; Forrest, J.Y.L. Does the smart city policy promote the green growth of the urban economy? Evidence from China. *Environ. Sci. Pollut. Res.* **2021**, *28*, 66709–66723. [CrossRef]
6. Hamdi, H.; Sbia, R.; Shahbaz, M. The nexus between electricity consumption and economic growth in Bahrain. *Econ. Model.* **2014**, *38*, 227–237. [CrossRef]
7. Bhujabal, P.; Sethi, N.; Padhan, P.C. ICT, foreign direct investment and environmental pollution in major Asia Pacific countries. *Environ. Sci. Pollut. Res.* **2021**, *28*, 42649–42669. [CrossRef]
8. Osman, A.M.S.; Elragal, A.A.; Ståhlbröst, A. Data-driven decisions in smart cities: A digital transformation case study. *Appl. Sci.* **2022**, *12*, 1732. [CrossRef]
9. Hämäläinen, M.; Ratten, V. (Eds.) A framework for a smart city design: Digital transformation in the Helsinki Smart City. In *Entrepreneurship and the Community*; Contributions to Management Science; Springer: Cham, Switzerland, 2020; pp. 63–86.
10. Xu, N.; Ding, Y.X.; Guo, J.H. Do smart city policies make cities more innovative: Evidence from China. *J. Asian. Public. Polic.* **2022**, *15*, 1–17. [CrossRef]
11. Caruso, L. Digital innovation and the fourth industrial revolution: Epochal social changes? *AI Soc.* **2018**, *33*, 379–392. [CrossRef]
12. Caragliu, A.; del Bo, C.; Nijkamp, P. Smart cities in Europe. *J. Urban Technol.* **2011**, *18*, 65–82. [CrossRef]
13. Chen, B.; Cheng, Y.S. The impacts of environmental regulation on industrial activities: Evidence from a quasi-natural experiment in Chinese prefectures. *Sustainability* **2017**, *9*, 571. [CrossRef]
14. Wang, S.J.; Zeng, J.Y.; Huang, Y.Y.; Shi, C.Y.; Zhan, P.Y. The effects of urbanization on CO_2 emissions in the Pearl River Delta: A comprehensive assessment and panel data analysis. *Appl. Energy* **2018**, *228*, 1693–1706. [CrossRef]
15. Satterthwaite, D. Environmental transformations in cities as they get larger, wealthier and better managed. *Geogr. J.* **1997**, *163*, 216–224. [CrossRef]
16. Allam, Z.; Dhunny, Z.A. On big data, artificial intelligence and smart cities. *Cities* **2019**, *89*, 80–91. [CrossRef]
17. Shobande, O.A. Decomposing the persistent and transitory effect of information and communication technology on environmental impacts assessment in Africa: Evidence from Mundlak specification. *Sustainability* **2021**, *13*, 4683. [CrossRef]
18. Jin, J.; Gubbi, J.; Marusic, S.; Palaniswami, M. An information framework for creating a smart city through Internet of Things. *IEEE Internet Things J.* **2014**, *1*, 112–121. [CrossRef]
19. Xin, B.; Qu, Y. Effects of smart city policies on green total factor productivity: Evidence from a quasi-natural experiment in China. *Int. J. Environ. Res. Public Health.* **2019**, *16*, 2396. [CrossRef]
20. Yoo, Y.; Henfridsson, O.; Lyytinen, K. Research commentary-the new organizing logic of digital innovation: An agenda for information systems research. *Inf. Syst. Res.* **2010**, *21*, 724–735. [CrossRef]
21. Nambisan, S.; Lyytinen, K.; Majchrzak, A.; Song, M. Digital innovation management: Reinventing innovation management research in a digital world. *MIS Q.* **2017**, *41*, 223–238. [CrossRef]
22. van den Buuse, D.; Kolk, A. An exploration of smart city approaches by international ICT firms. *Technol. Forecast. Soc. Chang.* **2019**, *142*, 220–234. [CrossRef]
23. Acemoglu, D. Labor- and capital-augmenting technical change. *J. Eur. Econ. Assoc.* **2003**, *1*, 1–37. [CrossRef]
24. Li, L.; Su, F.; Zhang, W.; Mao, J.-Y. Digital transformation by SME entrepreneurs: A capability perspective. *Inf. Syst. J.* **2018**, *28*, 1129–1157. [CrossRef]
25. Libert, B.; Beck, M.; Wind, Y. Questions to ask before your next digital transformation. *Harv. Bus. Rev.* **2016**, *60*, 11–13.
26. Vial, G. Understanding digital transformation: A review and a research agenda. *J. Strateg. Inf. Syst.* **2019**, *28*, 118–144. [CrossRef]
27. Nwankpa, J.K.; Roumani, Y. *IT Capability and Digital Transformation: A Firm Performance Perspective*; ICIS: London, UK, 2016.
28. Cui, J.B.; Wang, C.H.; Zhang, J.J.; Zheng, Y. The effectiveness of China's regional carbon market pilots in reducing firm emissions. *Proc. Natl. Acad. Sci. USA* **2022**, *118*, e2109912118. [CrossRef]
29. Cowen, S.S.; Ferreri, L.B.; Parker, L.D. The impact of corporate characteristics on social-responsibility disclosure—A typology and frequency-based analysis. *Account. Org. Soc.* **1987**, *12*, 111–122. [CrossRef]

30. Stanny, E.; Ely, K. Corporate environmental disclosures about the effects of climate change. *Corp. Soc. Responsib. Environ. Manag.* **2008**, *5*, 338–348. [CrossRef]
31. Bewley, K.; Li, Y. Disclosure of environmental information by Canadian manufacturing companies: A voluntary disclosure perspective. *Adv. Environ. Account. Manag.* **2000**, *1*, 201–226.
32. Aerts, W.; Cormier, D.; Magnan, M. Corporate environmental disclosure, financial markets and the media: An international perspective. *Ecol. Econ.* **2008**, *64*, 643–659. [CrossRef]
33. Bhattacharya, P.S.; Graham, M.A. On institutional ownership and firm performance: A disaggregated view. *J. Multinatl. Financ. Manag.* **2009**, *19*, 370–394. [CrossRef]
34. Poumanyvong, P.; Kaneko, S. Does urbanization lead to less energy use and lower CO_2 emissions? A cross-country analysis. *Ecol Econ.* **2010**, *70*, 434–444. [CrossRef]
35. Pao, H.T.; Tsai, C.M. CO_2 emissions, energy consumption and economic growth in BRIC countries. *Energy Policy* **2010**, *38*, 7850–7860. [CrossRef]
36. Martinez-Zarzoso, I.; Maruotti, A. The impact of urbanization on CO_2 emissions: Evidence from developing countries. *Ecol. Econ.* **2011**, *70*, 1344–1353. [CrossRef]
37. Beck, T.; Levine, R.; Levkov, A. Big bad banks? The winners and losers from bank deregulation in the United States. *J. Financ.* **2010**, *65*, 1637–1667. [CrossRef]
38. Fan, Z.; Tian, B. Tax competition, tax enforcement and tax avoidance. *Econ. Res. J.* **2013**, *9*, 99–111.
39. Lu, Y.; Tao, Z.; Zhu, L. Identifying FDI spillovers. *J. Int. Econ.* **2017**, *107*, 75–90. [CrossRef]
40. Liu, D.; Chen, S.; Chou, T. Resource fit in digital transformation. *Manag. Decis.* **2011**, *49*, 1728–1742. [CrossRef]
41. Xue, L.; Zhang, Q.; Zhang, X.; Li, C. Can digital transformation promote green technology innovation? *Sustainability* **2022**, *14*, 7497. [CrossRef]
42. Karaboga, T.; Gurol, Y.D.; Binici, C.M.; Sarp, P. Sustainable digital talent ecosystem in the New Era: Impacts on businesses, governments and universities. *Istanb. Bus. Res.* **2021**, *49*, 365–384. [CrossRef]
43. Loebbecke, C.; Picot, A. Reflections on societal and business model transformation arising from digitization and big data analytics. *J. Strateg. Inf. Syst.* **2015**, *24*, 149–157. [CrossRef]
44. Hall, B.H.; Jaffe, A.; Trajtenberg, M. Market value and patent citations. *RAND J. Econ.* **2005**, *36*, 12–38.
45. Zahra, S.A.; Bogner, W.C. Technology strategy and software new ventures' performance: Exploring the moderating effect of the competitive environment. *J. Bus. Ventur.* **2000**, *15*, 135–173. [CrossRef]
46. Meuleman, M.; De Maeseneire, W. Do R&D subsidies affect SMEs' access to external financing? *Res. Policy* **2012**, *41*, 580–591.
47. Yigitcanlar, T.; Kamruzzaman, M. Does smart city policy lead to sustainability of cities? *Land Use* Policy **2018**, *73*, 49–58. [CrossRef]
48. Cavada, M.; Hunt, D.V.L.; Rogers, C.D.F. Do smart cities realise their potential for lower carbon dioxide emissions? *Proc. Inst. Civ. Eng. Eng. Sustain.* **2016**, *169*, 243–252. [CrossRef]

MDPI
St. Alban-Anlage 66
4052 Basel
Switzerland
Tel. +41 61 683 77 34
Fax +41 61 302 89 18
www.mdpi.com

Energies Editorial Office
E-mail: energies@mdpi.com
www.mdpi.com/journal/energies

www.ingramcontent.com/pod-product-compliance
Lightning Source LLC
Chambersburg PA
CBHW061837210926
51850CB00006B/911

* 9 7 8 3 0 3 6 5 5 9 6 4 3 *